Symposium on Palynology of the Late Cretaceous and Early Tertiary

Sponsored by the Coal Division of The Geological Society of America, San Francisco, California, November 14, 1966

Edited by

Robert M. Kosanke

U.S. Geological Survey, Denver, Colorado

and

Aureal T. Cross

Michigan State University, East Lansing, Michigan

THE GEOLOGICAL SOCIETY OF AMERICA

SPECIAL PAPER 127

Library of Congress Catalog Card Number 79-111435
S.B.N. 8137-2127-X

Published by
THE GEOLOGICAL SOCIETY OF AMERICA, INC.
Colorado Building, P.O. Box 1719
Boulder, Colorado 80302

Printed in The United States of America

The printing of this volume has been made possible
through the bequest of
Richard Alexander Fullerton Penrose, Jr.,
and is partially supported by a grant from
The National Science Foundation.

Introduction*

At the 1963 business meeting of the Coal Geology Division of The Geological Society of America, a palynological committee was appointed to investigate the desirability of sponsoring a symposium. The committee consisted of R. M. Kosanke, chairman, J. M. Schopf, and W. R. Evitt. In August 1965, this committee proposed that the Coal Geology Division sponsor a symposium to be held in San Francisco in 1966 on Cretaceous-Tertiary palynology. To quote from this proposal: "More than four and one-half years will have elapsed since the last palynological symposium was held in this country. Much new data has become available from various parts of the world and a forum for its presentation and synthesis is considered desirable at this time." The proposal was approved by The Geological Society of America.

Aureal T. Cross, chairman of the Coal Geology Division for the San Francisco meetings, joined the committee as an active member and participated in making arrangements for the meetings and obtaining papers for the symposium.

The theme of the symposium, Late Cretaceous-early Tertiary palynology, represents a narrow span of geologic time which is in part parallel to the Laramide orogeny. Although this time span is short compared to the geologic history of the earth, a number of significant biological events are attributed to this interval. Among these events are the extinction of dinosaurs, pterodactyls, toothed birds, and ammonites, and the continued rise and diversification of flowering plants. The literature describing these milestones of historical geology is well known. The use of mollusks, particularly ammonites of the Western Interior and gastropods from the Mississippi Embayment area and the eastern United States to help delineate the Cretaceous-Tertiary boundary is well known through the works of W. A. Cobban, N. F. Sohl, and others. The published

---

*Publication authorized by the Director, U. S. Geological Survey.

record concerning the value of palynology in helping to differentiate Cretaceous from Tertiary strata is not well known.

Within the past decade there have been 10 or more meetings at which paleontological differentiation of the Cretaceous from the Tertiary has been either the primary theme or an important aspect of the meetings. These meetings have been held in India, Sweden, Hawaii, The Netherlands, West Africa, and western Europe. We would like to call attention to a recent publication concerned in a large measure with the differentiation of the Cretaceous from the Tertiary. This is Part III, Proceedings of Section 3 of the 1964 International Geological Congress which was recently distributed (November 1968) in this country. The editors are M. V. A. Sastry and S. B. Bhatia with R. K. Sandaram serving as general editor. Thirteen papers are contained in this volume and are devoted to a discussion of the Cretaceous-Tertiary. These 13 papers can be grouped as follows: seven are largely concerned with paleontological evidence but not palynological, four with sedimentation or mineralogy, and two are concerned with geochemical data.

Eight papers were presented at the San Francisco symposium and five of these are herewith published. Two additional papers are included in this publication and are by (1) B. D. Tschudy and E. B. Leopold, and (2) J. B. Zaitzeff and A. T. Cross. In a sense the Tschudy and Leopold paper was presented at the symposium using a large stratigraphic chart and numerous palynological photomicrographs. The paper by Zaitzeff and Cross was not a part of the symposium because of a lack of space on the program. One of the papers presented at the symposium and not published in this volume has been published in another journal as a result of a prior commitment. This is the paper by Warren S. Drugg on "Palynology of the Upper Moreno Formation (Late Cretaceous-Paleocene) Escarpado Canyon, California" (Palaeontographica, Bd. 120, Abt. B, Liefg. 1-4, p. 1-71, 1967).

At the business meeting of the Coal Geology Division after the symposium, it was decided to inquire into the possibility of publication of the symposium papers by the Society. Approval was obtained for publication as a Special Paper of The Geological Society of America.

The papers included in this publication record information on palynomorphs from the Mississippi Embayment westward, including Canada. The paper by Helen Tappan and A. R. Loeblich, Jr., is an exception in that it has a much broader scope both geographically and stratigraphically than the rest of the papers in this symposium. In a sense this paper represents a change of pace from the narrow geographic and stratigraphic circumscription of most of the other papers.

The first two papers in this volume are similar in approach in that palynomorph assemblages both below and above the Cretaceous-Tertiary boundary have been studied in detail and comparisons made with assemblages from other areas. The first of these two papers is by H. A. Leffingwell who has recognized three palynological zones within the stratigraphic interval of the upper Lance (Late Cretaceous) and Fort Union (Paleocene) Formations. Similar assemblages exist between the Powder River Basin of Wyoming and the Cannonball area of North Dakota. R. H. Tschudy delineates the palynological assemblages of the Hell Creek-Tullock interval of Montana and compares these assemblages with those of a similar stratigraphic interval from the Mississippi Embayment area. Tschudy can distinguish the Cretaceous from the Tertiary in both areas, but the assemblages from the two areas reveal significant differences.

The paper by B. D. Tschudy and E. B. Leopold on *Aquilapollenites* from the Rocky Mountain region is unlike the other papers published herein because a single genus has been studied in detail rather than an assemblage. This is an ideal genus for such consideration in connection with the delineation of the Cretaceous-Tertiary boundary in the Rocky Mountain region. This is because *Aquilapollenites* is abundantly represented by several species with restricted range zones in Cretaceous rocks in the Rocky Mountain region. This paper is largely taxonomic, but the stratigraphic distribution of all species is clearly set forth.

The paper by P. H. Griggs on the palynomorphs of the lower 5,000 feet of the Chuckanut Formation in Washington indicates an age assignment of Late Cretaceous to early Eocene. G. E. Rouse, W. S. Hopkins, and K. M. Piel report palynological evidence from British Columbia and Alberta suggesting major differences in floral composition between coast and interior regions. These differences are attributed to different floral provinces. The age of the samples examined range from Late Cretaceous to middle Eocene.

The paper by Tappan and Loeblich on fossil phytoplankton has already been mentioned. Fossil phytoplankton evolution is discussed on the basis of distribution in time and space together with similar information about modern representatives.

The last paper, that of Zaitzeff and Cross, demonstrates the usefulness of dinoflagellates and acritarchs for zonation and correlation of the Navarro Group (Maestrichtian) of Texas. A total of 81 species of the dinoflagellates and acritarchs are identified from two sections about 135 miles apart. Three assemblage zones are distinguished spanning the interval of Neylandville Marl, Nacatoch Sand, Corsicana Marl, and the Kemp Clay.

On behalf of the Coal Geology Division the editors wish to express their appreciation to the Society for publication of this symposium, to the contributors to the symposium, and to the various individuals who served as reviewers of the papers. Palynology has a contribution to make in the delineation of the Cretaceous-Tertiary boundary. Palynology can be utilized to help solve some of the problems not resolved by other facets of paleontology.

ROBERT M. KOSANKE
U. S. Geological Survey
Denver, Colorado
AUREAL T. CROSS
Michigan State University
East Lansing, Michigan

Contents

Palynology of the Lance (Late Cretaceous) and Fort Union (Paleocene) Formations of the Type Lance Area, Wyoming

HARRY A. LEFFINGWELL

Union Oil Company of California Research Center, Brea, California

ABSTRACT

Three palynological assemblage zones, based on two major floral changes, are established within the upper Lance (Late Cretaceous) and Fort Union (Paleocene) Formations in and adjacent to the type Lance area in southeastern Wyoming. The lower palynological change essentially coincides with the Cretaceous-Paleocene (Maestrichtian-Danian) boundary, which is based on the vertebrate, fossil leaf, and foraminiferal evidence of previous workers. The upper palynological change occurs in the lower part of the Lebo Shale Member of the Fort Union Formation.

The oldest palynological assemblage (Assemblage A) is of Maestrichtian age and is characterized by diverse and abundant angiosperm pollen, particularly species of the genera *Gunnera, Proteacidites, Aquilapollenites,* and *Cranwellia.* The middle assemblage (Assemblage B) is, in part, of Danian age and is composed mostly of fern and moss spores and small tricolpate and tricolporate angiosperm pollen. The youngest assemblage (Assemblage C) is of Paleocene age and consists primarily of taxodiaceous-cupressaceous pollen with secondary but significant quantities of betulaceous and juglandaceous pollen.

These defined floral assemblages occur regionally and are considered to be synchronous. The Assemblage A-Assemblage B boundary coincides with the Hell Creek (Lance)-Fort Union contact in the Cave Hills area of South Dakota (Stanley, 1965; this study), in the type area of the Hell Creek Formation in Montana (Norton, 1963; Tschudy, 1968, oral commun.), and in the type area of the Cannonball Member of the Fort Union Formation in North Dakota. The Assemblage B-Assemblage C boundary occurs in the lower part of the Tongue River Member of the Fort Union Formation in the type Cannonball area and in the lower Fort Union Formation in central and western Wyoming.

Occurrences of Assemblage B in the lower Tongue River Member in the Cave Hills and type Cannonball areas of northwestern South Dakota and central North Dakota indicate that the lower Tongue River in these areas is older than the lower Tongue River in the Powder River Basin of Wyoming.

Two new genera, *Maceopolipollenites* and *Insulapollenites*, and sixteen new species have been described, and the generic description of *Kurtzipites* has been emended.

CONTENTS

INTRODUCTION

The rocks of latest Cretaceous and early Tertiary age in the Rocky Mountain region are predominately continental. Previous paleontological criteria used in determining the position of the Cretaceous-Tertiary boundary have been leaf floras, vertebrates, and freshwater invertebrates. Paleontologists (for example, Stanton, *in* Stanton and Knowlton, 1897, p. 135; Lull, 1915, p. 327; Russell, 1930, p. 134, 141) placed the Cretaceous-Tertiary boundary at the top of the dinosaur-bearing Lance Formation, while Knowlton (1909, p. 180-181; 1919, p. 10; 1930, p. 10), a paleobotanist responsible for many of the early studies of the Fort Union and Lance leaf floras, placed the boundary at the base of the Lance. This inconsistency was resolved by Dorf (1940, p. 219) when he demonstrated that at least 10 of Knowlton's 58 Lance leaf localities were from the Fort Union Formation. Dorf's study of many new leaf assemblages proved that the Lance floras could be readily distinguished from those of the overlying Fort Union. In many areas, however, the precise relation between the Cretaceous-Tertiary boundary and the Lance-Fort Union contact is obscure because of the absence of vertebrate and leaf fossils at the systematic boundary and because of gradational lithologic changes at the Lance-Fort Union contact. Because pollen was known to occur abundantly in Lance and Fort Union rocks, it was believed that the boundary position in many areas of this region might be more accurately determined by comprehensive palynologic studies.

The type Lance area was selected for this initial study because it is the type area of one of the important stratigraphic units of Late Cretaceous age in the Rocky Mountains and also because both the vertebrates and the fossil leaves have been investigated in sufficient detail to provide basic paleontological control.

The objective of this study was to determine whether or not palynological changes occur within the Upper Cretaceous and Paleocene rocks in the Rocky Mountain region and to relate any such changes to the stratigraphic units involved and to the Cretaceous-Tertiary boundary as determined by paleobotanical and vertebrate fossil evidence.

STRATIGRAPHY

The age relations of the Fox Hills-Fort Union interval in and adjacent to the type Lance area of the Powder River Basin, Wyoming, are summarized on Figure 1. This entire sequence is apparently conformable and is composed of continental beds except for the Fox Hills Sandstone, which contains an abundant marine fauna.

Fox Hills Sandstone

The Fox Hills Sandstone is a buff, brown to light-gray sandstone. The presence of the ammonite *Sphenodiscus* cf. *S. lenticularis* in these rocks a few miles northeast of the type Lance area (Cobban, 1958, p. 114) indicates an early late Maestrichtian age. The Fox Hills Sandstone has been estimated to be from 300 to 750 feet thick, depending upon whether the Lance-Fox Hills contact is placed above or below several massive gray sandstone beds (Dorf, 1942, p. 89).

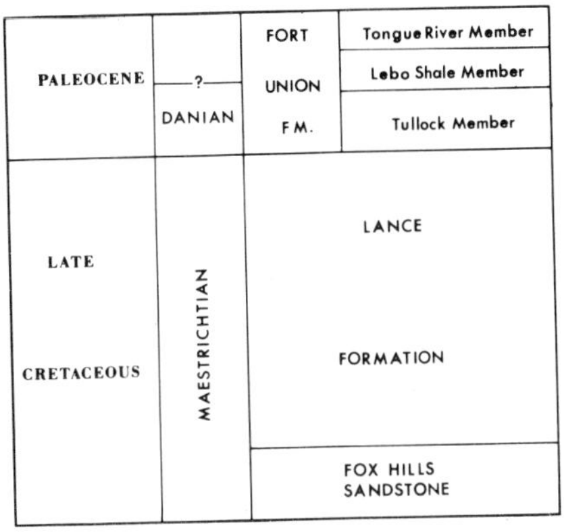

Figure 1. Generalized stratigraphic section of the Fox Hills Sandstone-Fort Union interval, Powder River Basin, Wyoming.

Lance Formation

The Lance in the type area is approximately 2500 feet thick and consists of dark-gray to brown sandy shales and fine-grained, cross-bedded, massive, yellow-brown sandstones. These strata were referred to by earlier workers as the "somber beds" because of the over-all drab color of the exposures.

Marsh (1889, p. 501-502) applied the name "*Ceratops* Beds" to deposits of Late Cretaceous age in Wyoming, Colorado, and Montana that contain remains of horned dinosaurs. Hatcher (1893, p. 135-136; 1896, p. 120) described and sketched an area located in the vicinity of Lance Creek in eastern Wyoming, shown on Figure 2, and stated that "fully ninety percent of all vertebrate fossils described by Professor Marsh from the *Ceratops* Beds were found in this region." These deposits were originally referred to by Hatcher as the "*Ceratops* Beds" of Converse (now Niobrara) County, Wyoming, and later as the "Lance Creek Beds." The U.S. Geological Survey shortened the latter term to "Lance Formation" and officially adopted this name in 1910. The type area of the Lance is considered to be essentially that area enclosed by the *Ceratops* border illustrated on Hatcher's 1896 sketch map, except for the southeastern part of Hatcher's *Ceratops* boundary, which requires modification (Clemens, 1964, p. 7). Hatcher did not designate a type section. The type Lance area is located adjacent to Lance Creek, approximately 25 miles northeast of Lusk, Wyoming (Fig. 2).

Fort Union Formation

The Fort Union Formation in the southeastern part of the Powder River Basin is divided into the Tullock Member and an upper unnamed member on the geologic map of Wyoming (Love and others, 1955). The Tullock consists of 400 to 600 feet of yellow, fine-grained, scarp-forming sandstones, interbedded light-colored

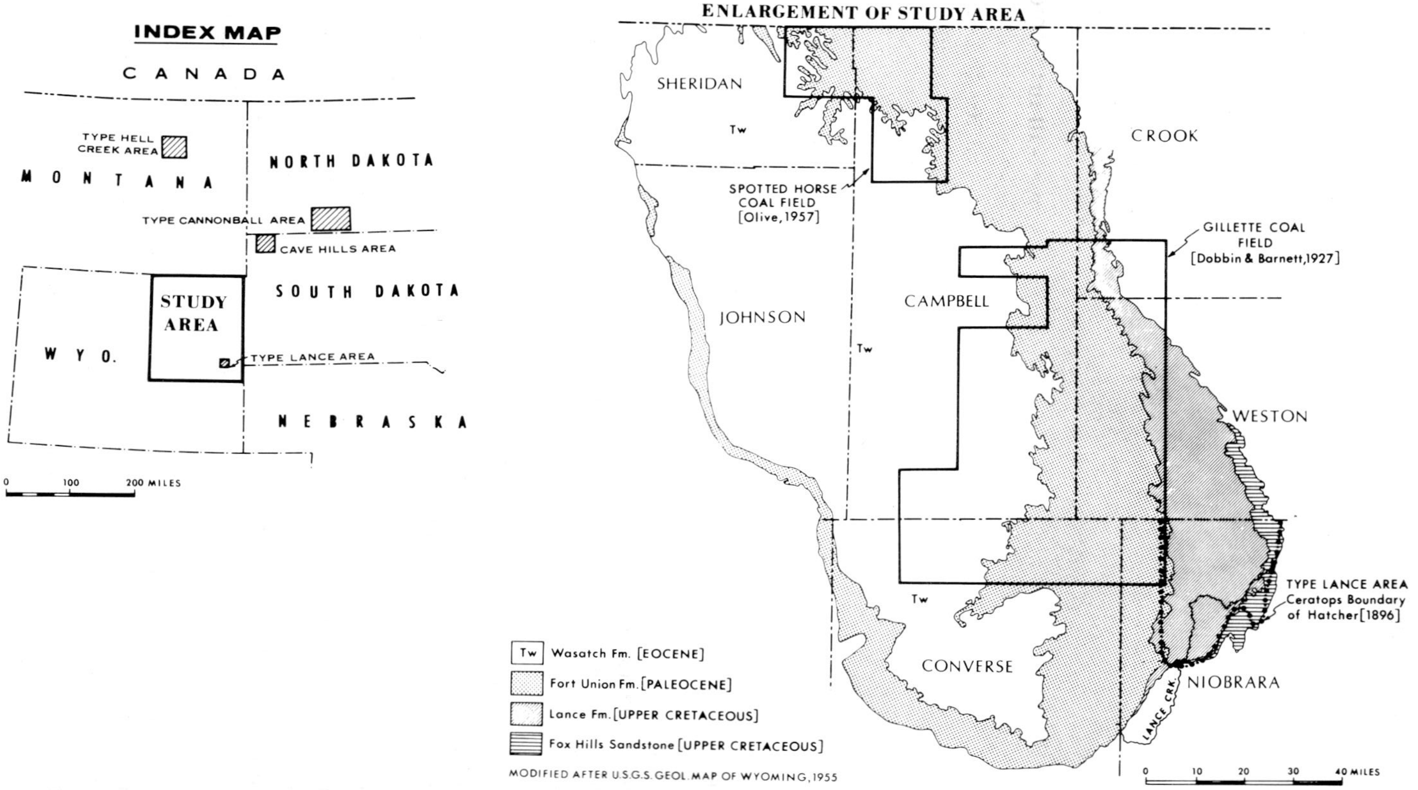

Figure 2. Map showing the location of the Cave Hills area and the type areas of the Lance and Hell Creek Formations and the Cannonball Member of the Fort Union Formation, with enlargement of study area in and adjacent to the type Lance area.

sandy shales, and numerous thin lignite beds. In the Gillette coal field, directly northwest of the type Lance area (Fig. 2), Dobbin and Barnett (1927, p. 9) divided the upper unnamed member of the Fort Union into the Lebo Shale and Tongue River Members. The lower unit, the Lebo Shale, consists of 250 to 400 feet of dark-gray shales and numerous thin lignites, while the upper unit, the Tongue River, consists of 250 to 400 feet of ferruginous sandstone, gray shales, and thick lignites. The Roland Coal, which attains a thickness of 90 feet in the northern part of the Gillette coal field, is considered to be the top of both the Tongue River Member and the Fort Union Formation. Olive (1957, p. 13) suggests that the Roland Coal correlates with the Anderson Bed in the Spotted Horse coal field, just north of the Gillette coal field, and that the Eocene Wasatch Formation truncates part of the upper Fort Union in the northern part of the Gillette coal field (Fig. 3). The Tongue River is a thicker, more complete section in the Spotted Horse coal field and contains more individual coal beds there than it does in the Gillette coal field.

Lance-Fort Union Contact

Various criteria have been used, singly or in combination, to establish the Lance-Fort Union contact. For example, it has been placed at the highest occurrence of dinosaur remains, at particular changes of rock color and lithology, and at the base of the lignitic beds of the Fort Union. Hatcher (1893, p. 140) stated that in contrast to those of the Lance the Fort Union sandstones are more massive, the shales are sandier, and the lignites are more abundant, thicker, and of better quality and that vertebrate and invertebrate fossils are absent. He failed, however, to designate a specific upper limit for the Lance in the type area. Although Hatcher's generalizations are valid, Clemens (1960, p. 21; 1964, p. 12) pointed out that the lithologies near the contact are transitional in the type area and that the highest occurrence of dinosaur remains is generally set by the highest occurrence of sandstones, in which lithology most remains have been discovered. Clemens placed the Lance-Fort Union contact at the lowest persistent lignite bed overlying the highest occurrence of dinosaur remains, a criterion originally used for mapping the contact by Calvert (1912, p. 197) and advocated in recent years by Brown (1952, p. 92; 1958, p. 112).

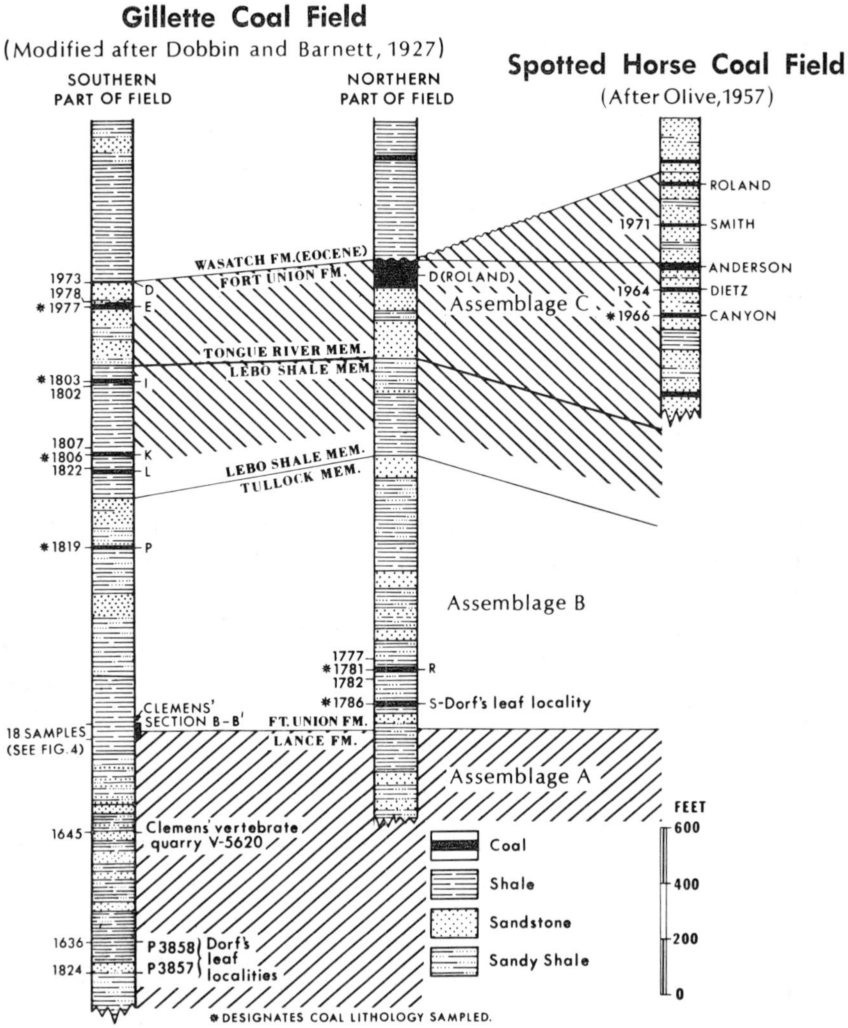

Figure 3. Composite stratigraphic section of the upper Lance-Fort Union Formations showing the stratigraphic position of the palynological samples, Dorf's (1942) leaf localities, and Clemens' (1964) vertebrate locality.

PALYNOLOGICAL CONTROL SECTION

The control section for this study is a composite stratigraphic section (Fig. 3) embodying samples from Dorf's (1942) fossil leaf localities, Clemens' (1964) vertebrate localities and measured sections from the type Lance area, Dobbin and Barnett's (1927) coal localities in the Gillette coal field, and Olive's (1957) coal localities in the Spotted Horse coal field. The geographic location of each of these areas is shown in Figure 2, and the sample localities are described in Appendix 1.

A composite stratigraphic section was necessary because the pertinent fossil localities are widely scattered and complete, well-exposed sections are virtually impossible to find because of grassy, rolling topography and low dips.

Samples from three of Dorf's nine fossil leaf localities were used in the control section; two are from the upper half of the Lance, approximately 1700 feet and 1800 feet from the base of the formation, and the third is from the Tullock, approximately 100 feet above the Lance-Fort Union contact. Another upper Lance locality did not yield palynomorphs. His other localities are in the lower half of the Lance and are not pertinent to this study.

Samples from one of Clemens' three vertebrate quarries in the Lance yielded well-preserved palynological assemblages and therefore was used for control. This quarry (V5620) is 2139 feet above the base of the Lance. Samples from only one (V5611) of the remaining two quarries yielded palynomorphs, but this assemblage was not used because of its somewhat questionable stratigraphic position and poor preservation. Its pollen assemblage, however, is very similar to the one from quarry V5620. In addition to these vertebrate localities, Clemens' measured section B-B′ was sampled for palynomorphs. This section traverses the Lance-Fort Union contact and contains dinosaur remains in a sandstone bed 39 feet below the Lance-Fort Union lithologic contact (Fig. 4).

Neither Dorf nor Clemens reported fossils from the upper part of the Fort Union. Consequently, palynological control from that part of the section was obtained from coals and associated lithologies whose stratigraphic positions within the Fort Union Formation are well-established. Historically, coal beds are a principal means of correlation within this interval because they are persistent, are easily mapped, and were of great economic

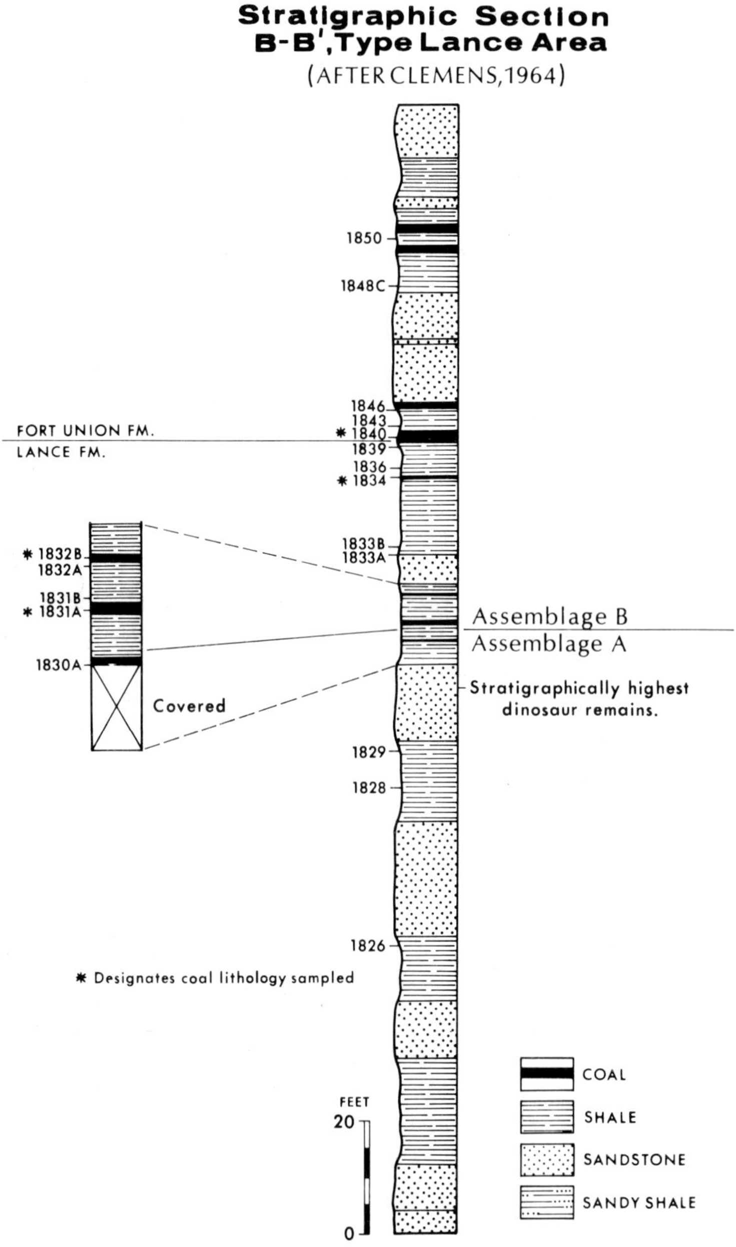

Figure 4. Clemens' (1964) stratigraphic section B-B′, showing detailed stratigraphic position of palynological samples collected across the Lance-Fort Union contact.

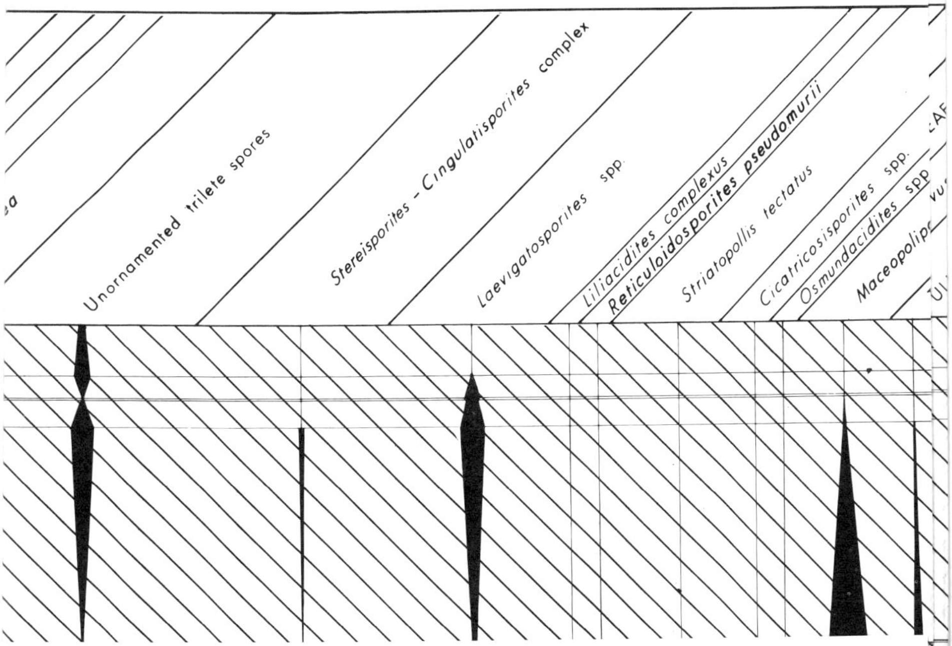

G THE RELATIVE ABUNDANCES OF THE
FOR QUANTITATIVE ANALYSES WITHIN
UNION INTERVAL, EXCLUSIVE OF THE
EAE-CUPRESSACEAE COMPLEX

LANCE — FORT UNION CONTACT

The quantitative data were also arranged to show only the dominant taxonomic groups represented in the analysis of the 51 taxa. This arrangement is illustrated as histograms on Plate 3 and tabulated in Appendix 2, Table 3.

To reduce the depositional environmental effects on the quantitative analyses, clastics were selected in preference to coals whenever possible. A clastic sample is more likely to contain a regional assemblage than is a coal because the streams carrying the clastic load to the depositional site will introduce pollen from a much larger area. For some sample localities, both coal and clastic samples were analyzed in an effort to compare the influence of lithologic type on the pollen assemblage.

Sample Preparation and Photography

The general sample preparation procedures described by Funkhouser and Evitt (1959) were used, with the exception that zinc bromide replaced zinc chloride as a heavy liquid. Many samples required little or no oxidative treatment. Those requiring oxidation were treated for short periods (1 to 3 minutes) in Schulze's solution (2 parts nitric acid, 1 part potassium chlorate). When additional oxidation was required, a short treatment of 5 percent sodium hypochlorite was used. All residues were first mounted in Clearcol on coverslips that in turn were cemented to glass slides with Permount.

Photographs were taken with a Zeiss Photomicroscope, in bright field, phase contrast, and interference contrast. Bright-field photographs were taken with a 6-volt light source, and a 12-volt system was used for phase and interference contrast. Interference contrast was achieved with a Zeiss interference-contrast system (*after* Nomarski). The resulting image gives the appearance of being sidelighted, which effectively emphasizes the surface sculpturing of the pollen (for example, Pl. 7, fig. 6).

DISCUSSION

Two important floral changes occur in the upper Lance-Fort Union interval, dividing it into three zones (Pls. 1-3). The assemblages characteristic of each zone are designated A, B, and C, from oldest to youngest.

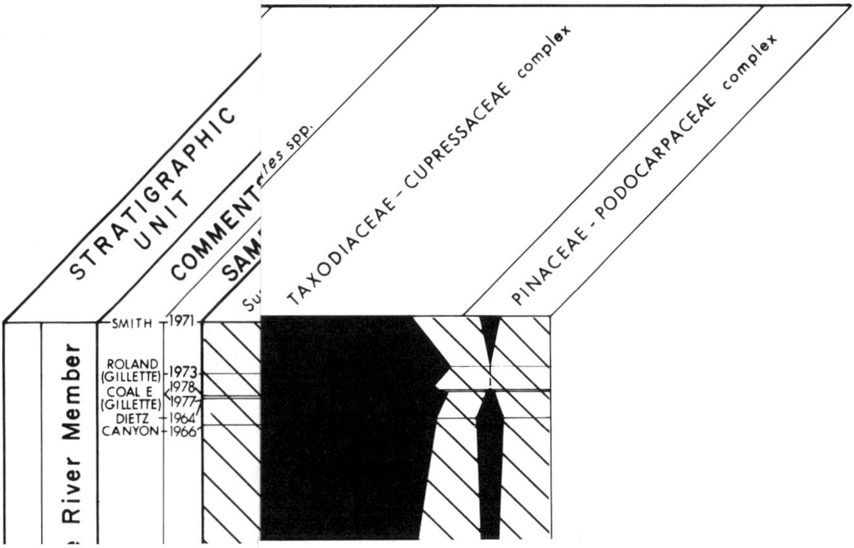
STRATIGRAPHIC UNIT
COMMENT
TAXODIACEAE - CUPRESSACEAE complex
PINACEAE - PODOCARPACEAE complex
River Member
SMITH
1971
ROLAND (GILLETTE)
1973
COAL E (GILLETTE)
1978
1977
DIETZ
1964
CANYON
1966

Assemblage A

The top of the Assemblage A zone coincides with the top of the Lance and occurs in samples RC 1824 to RC 1830A in the composite stratigraphic control section (Fig. 3). The base of the zone has not been precisely determined, but a cursory examination of assemblages from the lower part of the Lance indicates that Assemblage A is present there. Assemblage A is characterized by its diversity of angiosperm pollen. Some of the most common and characteristic elements of the assemblage are:

**Gunnera microreticulata*
**Liliacidites altimurus*
**Foraminisporis undulatus*
**Cranwellia striata*
**Liliacidites complexus*
**Proteacidites* spp.
†*Aquilapollenites* spp.
†*Wodehouseia spinata*
†*Tricolpites parvistriatus*
†*Zlivisporis novamexicanum*
†*Ghoshispora* spp.
†*Arecipites columellus*
Kurtzipites trispissatus
Kurtzipites sp.
Triporopollenites plektosus
Ulmipollenites (<25 u)

The assemblages most typical of this flora were recovered from samples taken at Dorf's plant localities P3857 and P3858 and from Clemens' vertebrate quarry V5620. The sample from plant locality P3857 (RC 1824) was taken from a horizon containing abundant freshwater invertebrates, which indicate these beds were lake deposits. Similar deposits are probably represented at plant locality P3858 and at the vertebrate quarry V5620, as spores and massulae of the floating fern *Azolla* are present in all three assemblages. In the uppermost Lance samples (RC 1826 to RC 1830A), the frequency of fern and sphagnum spores increases sharply, which suggests that ecologic conditions changed in latest Lance time. In spite of the increased number of spores in these samples, angiosperm pollen characteristic of Assemblage A occur in relatively high frequencies.

Assemblage B

Assemblage B occurs in samples RC 1831A to RC 1822, which represent all of the Tullock Member and the lower part of the Lebo Shale Member of the lower Fort Union Formation in the composite control section (Fig. 3). This assemblage is also present in the

*Restricted to Lance.
†Common in Lance, rare in Fort Union.

upper 30 feet of the Lance as defined by Clemens (1964), but the results of this study indicate that the Lance-Fort Union boundary should be placed at the base of the Assemblage B zone (*see* pgs. 15-16).

Assemblage B is characterized by a predominance of fern and moss spores, a marked increase in species and numbers of reticulate, granulate, and striate tricolpate and tricolporate angiosperm pollen, an increase in the number of *Kurtzipites* pollen (primarily of the *K. simplex* type), and infrequent occurrences of *Cicatricosisporites* (that is, *Anemia* and/or *Mohria*) spores.

Quantitatively, Assemblage B differs markedly from Assemblage A in essentially lacking pollen of *Gunnera, Proteacidites, Aquilapollenites,* and *Cranwellia,* and it differs from Assemblage C in lacking diverse and abundant juglandaceous pollen. The boundaries of the Assemblage B zone are thus readily identified. For example, the boundary between Assemblages A and B falls within a 3-foot interval in Clemens' section B-B′, and the magnitude of the floral change within this 3-foot interval suggests the presence of a hiatus at this horizon, although none could be detected in the field. The qualitative composition of Assemblage B is intermediate between those of Assemblages A and C, because none of the 51 taxa is restricted to Assemblage B. Many taxa, such as *Cicatricosisporites* and *Kurtzipites* pollen, occur in both Assemblage A and Assemblage B but are not present in Assemblage C. Conversely, other pollen taxa occur in the upper two assemblages but not in the lower one. Some of the common elements of Assemblage B are:

**Kurtzipites* (all species, but particularly *K. simplex*)
**Cicatricosisporites* spp.
**Wodehouseia spinata*
**Arecipites tenuiexinous*
**Zlivisporis novamexicanum*
†*Tricolpites angululuminosus-T. bathyreticulatus* complex
†*Ulmipollenites* (>25 μ)
†*Triatriopollenites granulatus*
Triporopollenites plektosus
Unornamented trilete spores
Stereisporites-Cingulatisporites complex
Reticuloidosporites pseudomurii
Laevigatosporites spp.
Striatopollis tectatus
Maceopolipollenites tenuipolus
Tricolpites mutabilis
Fraxinoipollenites variabilis
Syncolporites minimus
Ericaceae-Typhaceae complex

*Taxa not observed in the upper Fort Union Formation.
†Taxa not observed in the Lance Formation.

Assemblage C

Assemblage C is restricted to the upper Fort Union Formation. This assemblage occurs in samples RC 1802 to RC 1806 and from RC 1964 to RC 1978, which represent all of the Tongue River Member and all but the lower 100± feet of the Lebo Shale Member in the composite control section. Although the stratigraphic position of the Assemblage C top has not been precisely determined, the occurrence of a distinctively different flora in the Eocene Felix Coal near Gillette, Wyoming, indicates the top may be essentially coincident with the Fort Union-Wasatch contact. Assemblage C is characterized by a predominance of taxodiaceous-cupressaceous pollen with secondary but significant quantities of juglandaceous and betulaceous pollen. Several of the juglandaceous species have been observed only in the Assemblage C zone. Some of the most common and characteristic elements of this assemblage are:

**Maceopolipollenites amplus*
**M. rotundus*
**M. leboensis*
**M. triorbicularis*
**Carya veripites*
**C. viridifluminipites*
Insulapollenites rugulatus
Alnus spp.
Juglans spp.
Pterocarya spp.
Fraxinoipollenites pachyexinous
Betulaceae-Myricaceae-Casuarinaceae complex
Taxodiaceae-Cupressaceae complex

Paleontological and Stratigraphic Boundaries

The precise location of the Lance-Fort Union contact in the type Lance area is difficult to find because the lithologies of the two formations are transitional. Clemens followed the procedure advocated by Brown (1952, p. 92) and placed the contact at the base of the first persistant coal overlying the highest occurrence of dinosaur remains. The coal Clemens selected for the basal Fort Union bed is a 2-foot lignite that occurs 39 feet above the top of the highest dinosaur-bearing sandstone in his B-B′ section.

A marked change in the quantitative composition of the pollen floras occurs between samples RC 1830A and RC 1831A, which suggests the presence of a depositional hiatus within a 3-foot

*Taxa not observed below Assemblage C zone.

interval. Unfortunately, this could not be confirmed in the field. Sample RC 1830A was taken immediately below a 1-inch lignite and sample RC 1831A was taken from a 7-inch lignite that occurs approximately 8 feet above the top of Clemens' highest dinosaur-bearing sandstone bed. In order for it to coincide with this major palynological break, I believe the Fort Union contact in this area should be relocated at the base of the 7-inch lignite, which is 31 feet below the contact designated by Clemens.

The Assemblage B-Assemblage C boundary marks the second palynological change within the upper Lance-Fort Union interval in the Powder River Basin. It occurs 60 to 130 feet above the base of the Lebo Shale Member, between Coals L and K in the Gillette coal field. Consequently, the floral and stratigraphic boundaries in this part of the section are not coincident.

Regional Correlations

The Lance-Fort Union floral succession described in this paper has also been observed in other sedimentary basins in Wyoming, Montana, and North and South Dakota. These major floral changes are therefore considered synchronous and can clarify some of the time-stratigraphic relations of the Lance-Fort Union strata in the Rocky Mountain region (Fig. 5). For example, the Assemblage

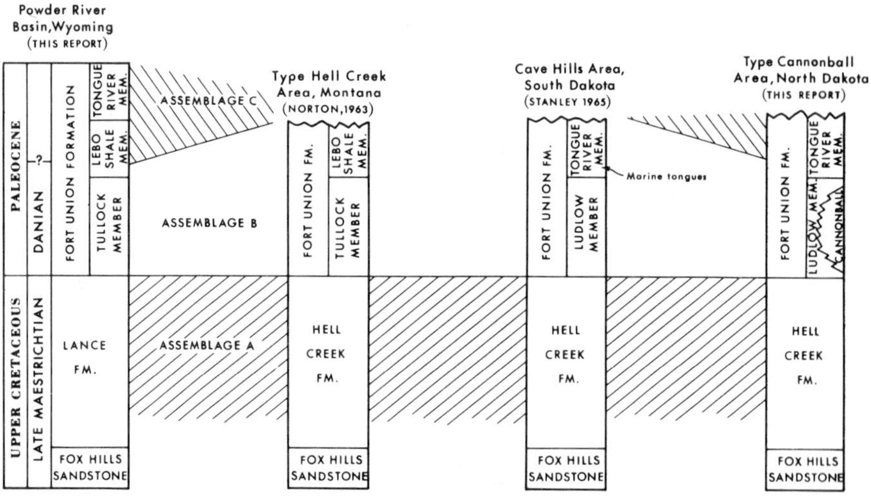

Figure 5. Cross section showing the correlation of the Upper Cretaceous and Paleocene rocks in the Powder River Basin to adjacent areas and the stratigraphic distribution of the palynological assemblages.

A-Assemblage B boundary is essentially coincident with the Lance (Hell Creek)-Fort Union contact, not only in the type Lance area of Wyoming but also in areas of Montana and North and South Dakota (Fig. 5). In the type Hell Creek area of Montana, a floral change equivalent to the one described in the type Lance area has been reported by Norton (1963, p. 48) and verified in a number of additional sections studied by Donald Oltz (1966, oral commun.) and R. H. Tschudy (1968, oral commun.).

Stanley (1965, Table 4) observed a similar floral change in the Cave Hills area of South Dakota between his highest Hell Creek and his lowest Fort Union samples. Unfortunately, the stratigraphic interval containing the lithologic boundary was not exposed in his sections. However, I have studied a section in this area described by Pipiringos and others (1965, Pl. 2, Sec. 19) in which the Hell Creek-Fort Union contact is exposed, and here it coincides with a major change in palynological assemblages. I have also observed a similar relation in the type Cannonball area of North Dakota.

In contrast to the regional synchroneity of the Lance-Fort Union contact, palynological correlations also document instances in which stratigraphic boundaries are not synchronous. For example, a lignite in the basal 50 feet of the Tongue River Member in the New Salem coal field, which is located immediately north of the type Canonball area, contains a flora equivalent to that of the Tullock-basal Lebo Shale section (that is, Assemblage B) in the Powder River Basin. The lignite was named and mapped as "Lignite A" by Hancock (1921, p. 28), and the sample was collected between Hancock's localities 26 and 27. Similarly, a thin shale bed between two massive sandstone beds in the basal Tongue River Member in the vicinity of the Riley Pass (Pipiringos and others, 1965, p. A22), in the Cave Hills area of South Dakota, also yielded a flora equivalent to Assemblage B. Consequently, the basal Tongue River Member in these areas of North and South Dakota are older than the basal Tongue River Member in the Powder River Basin (Fig. 5).

In the New Salem coal field, the boundary between Assemblage B and Assemblage C occurs in the Tongue River Member, within a 180-foot interval between Lignites A and D of Hancock (1921, p. 13). Lignite A, sampled between Hancock's localities 26 and 27, contains a diagnostic Assemblage B flora, and a shale sampled 3 feet above Lignite D in the Kaelberer strip mine (SW¼, sec. 9,

T. 138 N., R. 85 W.) contains an Assemblage C flora. Assemblage C also occurs in the Haynes lignite bed of the Cannonball coal field, which was mapped by Lloyd (1914). The Haynes bed may be correlative with Hancock's Lignite C in the New Salem field. If so, the interval in which this major floral boundary occurs in the type Cannonball area is reduced to approximately 100 feet. The Haynes lignite samples were collected from the Tipke Mines locality of Lloyd (SW¼, sec. 7, T. 135 N., R. 88 W.) and at the Coffin Butte strip mine (NE¼, sec. 3, T. 131 N., R. 90 W.), 2.5 feet above the main 10-foot lignite seam.

The Assemblage B-Assemblage C boundary also occurs within the lower part of the Fort Union Formation exposed in the Waltman area on the eastern edge of the Wind River Basin (Newman, in prep.), and I have observed this major floral boundary within the lower Fort Union Formation in unpublished studies of central and western Wyoming.

The composition of pollen assemblages reported by Norton and Hall (in press) indicates that the Assemblage A-Assemblage B boundary does not occur within the lower 90 feet of the Lebo Shale in the type Hell Creek area.

Similarly, this major floral boundary is not present in the Cave Hills area, because only Assemblage B floras were recovered from the youngest fossiliferous Fort Union rocks (Coals E and F and associated lithologies at Riley Pass; *see* Pipiringos and others, 1965, Pl. 2). These beds are approximately 140 feet above the base of the Tongue River Sandstone and within 40 feet of the youngest Tongue River exposures in this area. The uppermost 40 feet are unfossiliferous sandstones.

Ages of the Palynological Assemblages

No comparison of latest Cretaceous mammalian faunas of the Rocky Mountain area with those of the European Maestrichtian-Landenian sequence can be made because no mammal remains have been described from the European sections (Clemens, 1960, p. 490). The absence of marine invertebrate fossils throughout the uppermost Cretaceous-lower Tertiary interval in the Rocky Mountain region also precludes a direct correlation of these beds with marine faunas of the late Maestrichtian, Danian, and Landenian both here and in Europe, except for the marine Cannonball Member of the Fort Union Formation. Evidence presented by Jeletzky (1960, 1962), however, indicates that the Lance and Hell Creek Formations are of Maestrichtian age and that the Cannonball Member of the Fort Union Formation is of Danian age. Jeletzky's Maestrichtian age for the Hell Creek and Lance is based primarily on dinosaur remains and an age determination of ammonites in the Fox Hills Sandstone, which laterally replaces part of the Hell Creek and Lance. His supporting arguments include:

(1) The Hell Creek and Lance conformably overlie the late lower Maestrichtian beds of the type Fox Hills Sandstone and underlie the Danian beds of the Cannonball.

(2) An ammonite shell of *Scaphites* sp., associated with the *Triceratops* fauna in the upper part of the Lance (Clemens, 1960, p. 489), was interpreted as contemporaneous with the enclosing sediments (Jeletsky and Clemens, 1965, p. 954).

Jeletsky (1962, p. 1007) assigned a Danian age to the upper part of the Cannonball, based on the planktonic foraminiferal evidence provided by Dr. Steven Fox and Noel K. Brown. Additional planktonic foraminiferal studies by Fox and Olsson (Steven K. Fox, 1968 oral commun.), indicate that the entire Cannonball is of lower Danian age.

The upper part of the Fort Union Formation is of middle and late Paleocene age, based on the vertebrate fossil studies of a number of workers (Wood and others, 1941; Jepsen, 1940; Brown, 1962, p. 7-8).

Assemblage A is of Maestrichtian age, since this pollen flora occurs in association with leaf fossils and vertebrates of Late Cretaceous age and in beds that are laterally equivalent to beds containing Maestrichtian ammonites. The top of the Assemblage A zone is essentially equivalent to the highest occurrence of Late Cretaceous vertebrates.

The occurrence of Assemblage B floras above and below the Cannonball Member at its type locality demonstrates that at least part of the Assemblage B zone in the Powder River Basin is of lower Danian age. The Assemblage B flora has been observed in the basal 50 feet of the Tongue River in the New Salem coal field, which overlies the Cannonball throughout this entire area. The same flora has also been observed in numerous samples of the Ludlow Shale Member of the Fort Union Formation, which overlies the Hell Creek Formation and underlies the Cannonball Member in this area. Since no significant floral change occurs within the Ludlow, the Maestrichtian-Danian boundary is placed at the major palynological change at the Hell Creek-Ludlow contact. This boundary is approximately 62.7 m.y., based on the radiometric dating of a bentonite bed immediately above the basal Fort Union coal in the type Hell Creek area (Shafiqullah and others, 1964, p. 13, 19).

Assemblage C is of Paleocene age, but no studies have been undertaken to correlate it with the succession of vertebrate faunas in the Rocky Mountain area. The top of the Assemblage C zone may be essentially coincident with the base of the Eocene, although the precise relation has not been determined. The Roland Coal, which contains an Assemblage C flora, marks the top of the Fort Union in the Gillette, Wyoming area, and the Eocene Felix Coal from the Vines Mine locality of Dobbin and Barnett (1927, Pl. 12: NE¼, sec. 29, T. 50 N., R. 72 W.) just west of Gillette contains a flora of significantly different composition than that of the Roland Coal. The Felix Coal pollen flora includes *Platacarya* and *Tilia crassipites* Wodehouse, which have not been reported in rocks older than Eocene in Wyoming (Hail and Leopold, 1960).

SUMMARY

Two major floral changes within the late Cretaceous-Paleocene interval in the Rocky Mountain area divide this sequence into three palynological zones. These changes are regional in nature and are considered synchronous. A high degree of precision in locating the floral boundaries is possible because the rocks contain abundant palynomorphs.

The boundary between the lower and middle palynological assemblages (Assemblage A-Assemblage B) essentially coincides with the Lance (Hell Creek)-Fort Union contact in the type Lance area of Wyoming, the type Hell Creek area of Montana, the type Cannonball area of North Dakota, and the Cave Hills area of South Dakota. This floral boundary is essentially equivalent to the Cretaceous-Tertiary (Maestrichtian-Danian) boundary, a finding consistent with the vertebrate, foraminiferal, and leaf evidence. The radiometric age of a bed near this contact in the type Hell Creek area is approximately 62.7 m.y.

The boundary between the middle and upper palynological assemblages (Assemblage B-Assemblage C) occurs in the lower part of the Lebo Shale Member of the Fort Union Formation in the Powder River Basin, in the lower part of the Tongue River Member of the Fort Union Formation in the type Cannonball area, and in the lower Fort Union Formation in central and western Wyoming. The regional occurrence of these two major, synchronous floral changes can clarify some of the time-stratigraphic relations of the Lance (Hell Creek)-Fort Union strata in the Rocky Mountain region.

TAXONOMY[2]

Previously reported stratigraphic occurrences are included for each species, and, if higher taxa are restricted within the Lance-Fort Union interval, other reported occurrences of comparable age in the Rocky Mountain area are given. Only taxa identified to the specific level have been illustrated.

Pteridophyta
Family: Salviniaceae
Genus: *Azolla* L.
Azolla cretacea Stanley, 1965
(Pl. 4, figs. 1, 2)

Azolla sp. Tschudy, Wyoming Geol. Assoc. Guidebook, p. 54, Pl. 1, figs. 7, 8, 1961.

Azolla cretacea Stanley, Am. Paleontology Bull., v. 49, no. 222, p. 256-257, Pl. 33, figs. 1-5, 1965.

Stratigraphic Occurrence. Lance and Fort Union Formations.

Tschudy (1961): Fort Union Formation (Paleocene), Colorado and Wyoming.

Stanley (1965): Hell Creek Formation (Late Cretaceous) and Ludlow Member (early Paleocene) of the Fort Union Formation, South Dakota.

Srivastava (1966, 1967): Edmonton Formation (Late Cretaceous), Alberta, Canada.

Norton and Hall (1967): Hell Creek Formation (Late Cretaceous) and Tullock Member (early Paleocene) of the Fort Union Formation, Montana.

Family: Osmundaceae
Organ Genus: *Osmundacidites* Couper, 1953
Osmundacidites spp.

Stratigraphic Occurrence. Lance and Fort Union Formations.

Family: Schizaeaceae
Organ Genus: *Cicatricosisporites* Potonié and Gelletich, 1933
Cicatricosisporites spp.

---

[2]See Appendix to References, p. 60.

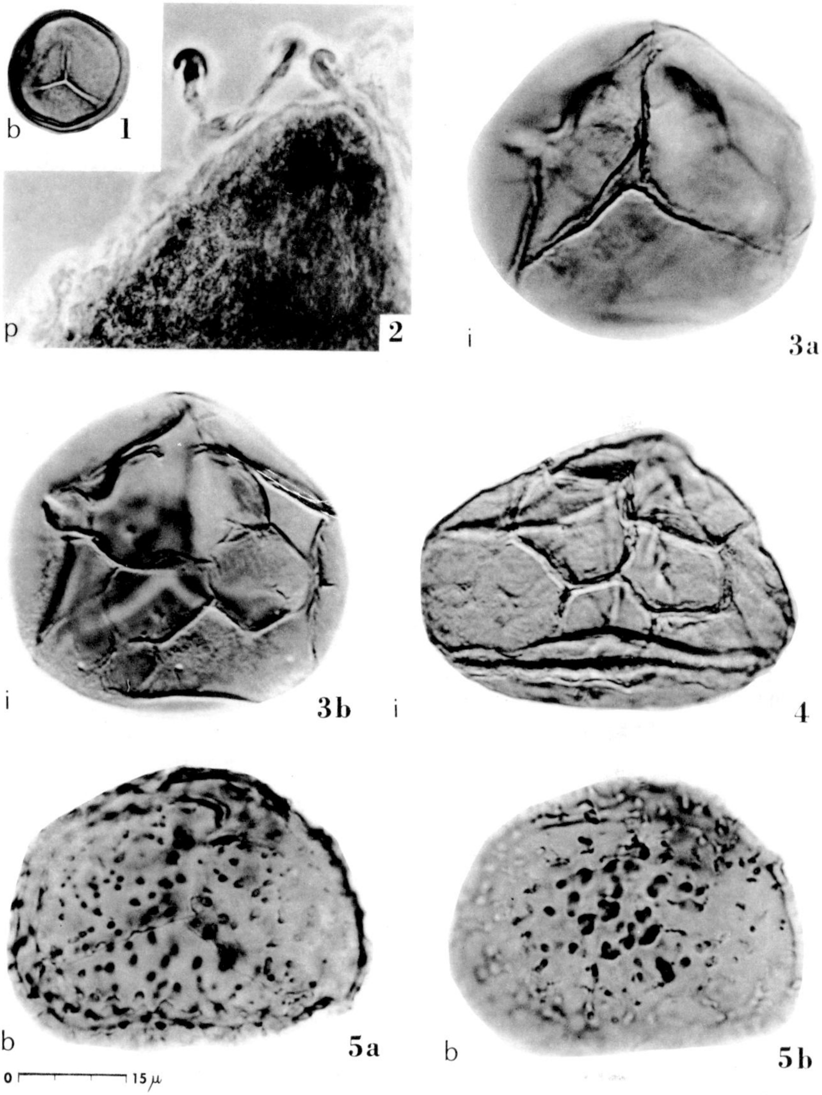

ILLUSTRATIONS OF PALYNOMORPHS

PLATE 4. ILLUSTRATIONS OF PALYNOMORPHS

All illustrations ×1000. Notations b, p, and i indicate photographs taken in bright field, phase contrast, and interference contrast, respectively.

Figures 1, 2. *Azolla cretacea* Stanley. Figure 1, microspore. Figure 2, massula illustrating shape of glochidial stalks, Page 21.

Figures 3, 4. *Zlivisporis novamexicanum* (Anderson). Figure 3a proximal; figure 3b, distal. Figure 4, distal. Page 25.

Figure 5. *Foraminisporis undulosus* sp. nov. Figure 5, holotype; 5a, proximal; 5b, distal. Page 25.

Stratigraphic Occurrence. Lance Formation and Tullock Member of the Fort Union Formation.

Rouse (1957, 1962): Burrard Formation (Late Cretaceous), British Columbia, Canada; Brazeau and Oldman Formations (Late Cretaceous), Canada; Princeton Formation (middle Eocene), Canada.

Sarmiento (1957, as *Anemia*): Mancos Shale (Late Cretaceous), Utah.

Stanley (1965, as *Anemia*): Hell Creek Formation (Late Cretaceous) and Ludlow Member (early Paleocene) of the Fort Union Formation, South Dakota.

Family: Gleicheniaceae

Organ Genus: *Gleicheniidites* Ross ex Delcourt and Sprumont emend. Dettman, 1963

Gleicheniidites spp.

Stratigraphic Occurrence. Lance and Fort Union Formations.

Spores of Uncertain Familial Affinity

Form Genus: *Reticuloidosporites* Pflug in Thomson and Pflug, 1953

Reticuloidosporites pseudomurii Elsik, 1968

(Pl. 5, figs. 1a, 1b)

Reticuloidosporites pseudomurii Elsik, Pollen et Spores, v. 10, no. 2, p. 290, Pl. 7, fig. 2, 1968.

The specimens of *R. pseudomurii* encountered in this study are ornamented with low verrucae and rugulae that, in some areas of the spore, are joined by narrow exinous thickenings to form a partial pseudoreticulum. The verrucae and rugulae are of variable size, characteristically with a few large (approximately 4- to 8-μ) elements scattered among smaller (approximately 1- to 3-μ) elements. The ornamentation is reduced and of equal size adjacent to the monolete scar. Size (10 specimens) 34 to 37 μ x 22 to 36 μ.

Stratigraphic Occurrence. Lance and Fort Union Formations.

Form Genus: *Laevigatosporites* Ibrahim, 1933

Stratigraphic Occurrence. Lance and Fort Union Formations.

Form Genus: *Ghoshispora* Srivastava, 1966

Ghoshispora spp.

Styx Norton (*in* Norton and Hall, 1967) is very similar to *Ghoshispora*, and the two genera are probably synonymous. Three species have been described thus far, two by Norton and one by Srivastava. No specific determinations were attempted in the present study.

Stratigraphic Occurrence. Lance Formation and basal Tullock Member of the Fort Union Formation.

Srivastava (1966, 1967): Edmonton Formation (Late Cretaceous), Alberta, Canada.

Norton and Hall (1967, as *Styx*): Hell Creek Formation (Late Cretaceous), Montana.

Form Genus: *Zlivisporis* Pacltova, 1961
Zlivisporis novamexicanum (Anderson) n. comb.
(Pl. 4, figs. 3a-4)

Lycopodium novamexicanum Anderson, New Mexico Bur. Mines and Mineral Resources Mem. 6, p. 14-15, Pl. 1, fig. 2; Pl. 8, fig. 1, 1960.
Lycopodiumsporites novamexicanum (Anderson) Drugg, Palaeontographica, bd. 120, abt. B, p. 40, Pl. 6, fig. 27, 1967.

In polar view, a thin membrane extends beyond the equatorial spore margin in places. This is interpreted as part of the reticulum viewed in optical section. Muri sometimes can be observed in optical section and traced back onto the distal surface (Pl. 4, fig. 4).

Stratigraphic Occurrence. Lance and Fort Union Formations.
Anderson (1960): Kirkland Shale (Late Cretaceous) and Nacimiento Shale (early Paleocene), New Mexico.
Drugg (1967): Upper Moreno Formation (Maestrichtian-Danian), California.

Form Genus: *Foraminisporis* Krutzsch, 1959b
Foraminisporis undulatus sp. nov.
(Pl. 4, figs. 5a, 5b)

Description. Trilete, cingulate spores, laesurae undulose to straight, extending to equator, cingulum 1.5 to 2 μ wide; equatorial outline circular to rounded triangular; proximal surface spinate, spines 1.5 μ high and of approximately equal size; larger sculptural elements of varying size present distally, consisting of spines, verrucae, and baculae, average size 4 μ long $\times$1.5 to 2.5 μ high; elongate sculpture elements are often formed from single elements fused together; unfused spines straight or curved; size (10 specimens) 35 to 60 μ.

Holotype. Plate 4, figures 5a, 5b; Slide 1825-4/106.4 $\times$ 6.1; Lance Formation (Maestrichtian).

Stratigraphic Occurrence. Lance Formation.

Differential Diagnosis. This species is distinguished from *Foraminisporis wonthaggiensis* Dettman, 1963, in having less densely spaced spines on its proximal surface and larger sculptured elements on its distal surface.

Remarks. The spines in some samples are foveolate, but the absence of foveolae in other samples indicates this is probably a preservational feature.

Stereisporites-Cingulatisporites Complex

Several morphologically similar spores have been grouped for this study, including the following species: *Stereisporites antiquasporites* (Wilson and Webster) Krutzsch, *Stereisporites* (*Sphagnum*) *australe,* and *S. regium* as figured by Stanley, 1965; *Cingulatisporites dakotaensis* Stanley, 1965; and *C. radiatus* Stanley, 1965.

Stratigraphic Occurrence. Lance and Fort Union Formations.

PLATE 5. ILLUSTRATIONS OF PALYNOMORPHS

All illustrations ×1000 unless otherwise noted. Notations b, p, and i indicate photographs taken in bright field, phase contrast, and interference contrast, respectively.

Figure 1. *Reticuloidosporites pseudomurii* Elsik. Page 24.
Figure 2. *Cycadopites follicularis* Wilson and Webster. Page 28.
Figures 3-5. *Pandaniidites radicus* sp. nov. In figure 3a, note pore on left margin; figure 3b illustrates longitudinal fold. Figure 4, holotype; figure 4b, enlargement of 4a, showing spines projecting below exine. Page 42.
Figures 6, 7. *Liliacidites altimurus* sp. nov. Figure 6, holotype; 6a, distal; 6b, equatorial; 6c, proximal. Figure 7a, distal; 7b, equatorial; 7c, proximal. Page 41.
Figures 8, 9. *Arecipites tenuiexinous* sp. nov. Figure 8, holotype. Page 41.

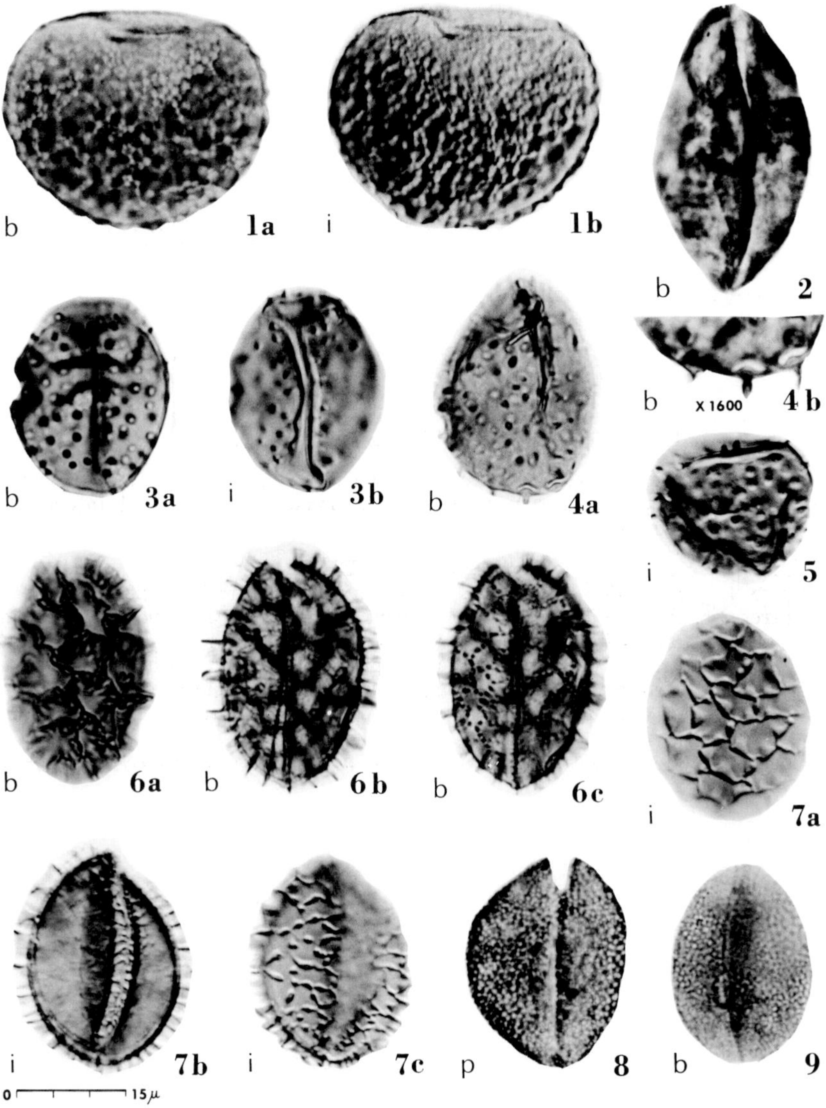

ILLUSTRATIONS OF PALYNOMORPHS

LEFFINGWELL, PLATE 5
Geological Society of America Special Paper 127

Gymnospermae

Family: Cycadaceae

Organ Genus: *Cycadopites* Wodehouse ex Wilson and Webster, 1946

Cycadopites follicularis Wilson and Webster, 1946

(Pl. 5, fig. 2)

Cycadopites folicularis Wilson and Webster, Am. Jour. Botany, v. 33, p. 274-275, fig. 7, 1946.

The specimens observed in this study lower the minimum length of this species to 34 μ.

Stratigraphic Occurrence. Lance and Fort Union Formations.

Radforth and Rouse (1956); Kootenay and Tantalus Formations (Early Cretaceous), Oldman and Brazeau Formations (Late Cretaceous), and Onakawana Formation (Tertiary), western Canada.

Rouse (1957): Oldman Formation (Late Cretaceous), Alberta, Canada.

Srivastava (1966, 1967): Edmonton Formation (Late Cretaceous), Alberta, Canada.

Taxodiaceae-Cupressaceae Complex

Pollen referable to the families Taxodiaceae and Cupressaceae have been tabulated as a single entity in this study. The fossil pollen included in this group are related, at least in part, to *Sequoia, Metasequoia, Cryptomeria, Taxodium, Glyptostrobus,* and *Thuja,* all of which have been reported in the leaf floral studies of this region (Dorf, 1942; Bell, 1949; Brown, 1962).

Stratigraphic Occurrence. Lance and Fort Union Formations.

Podocarpaceae-Pinaceae Complex

Stanley (1965) and Hall and Norton (1967) observed that bi-bladdered conifers occur more frequently in their early Paleocene assemblages than in their Late Cretaceous assemblages. This numerical increase of conifer pollen was much less pronounced in the early Paleocene assemblages in the type Lance area (Pl. 3).

Stratigraphic Occurrence. Lance and Fort Union Formations.

Angiospermae

Family: Juglandaceae

Genus: *Carya* Nutt.

Carya veripites Wilson and Webster, 1946

(Pl. 7, fig. 3)

Carya veripites Wilson and Webster, Am. Jour. Botany, v. 33, p. 276, Fig. 14, 1946.

Carya veripites occurs much more frequently in my Fort Union assemblages than does *C. viridifluminipites* (Wodehouse) Wilson and Webster. Conversely, the latter species occurred more frequently in both the Kolarich Coal flora described by Wilson and Webster and the Green River Shale flora described by Wodehouse.

I have not observed any typical *Carya* pollen (that is, with definate subequatorial pores) below the uppermost Lebo Shale of middle-to-late Paleocene age. Stanley (1965, p. 299, Pl. 45, figs. 3-7) assigned one species to *Carya* from the lower Fort Union Formation (Ludlow Member) of early Paleocene age, but its sculpture is more typical of the Ulmaceae.

Stratigraphic Occurrence. Uppermost Lebo Shale and Tongue River Members of the Fort Union Formation.

Wilson and Webster (1946): Kolarich Coal, upper Fort Union Formation (late Paleocene), Red Lodge, Montana.

Carya viridifluminipites (Wodehouse) Wilson and Webster, 1946
(Pl. 7, fig. 4)

Hicoria viridifluminipites (Wodehouse) Torrey Bot. Club Bull., v. 60, p. 503, Fig. 29, 1933.

Carya viridifluminipites (Wodehouse) Wilson and Webster, Am. Jour. Botany, v. 33, p. 276, Fig. 13, 1946.

Stratigraphic Occurrence. Tongue River Member of the Fort Union Formation.

Wodehouse (1933): Parachute Creek Member of the Green River Formation (middle Eocene), Colorado and Utah.

Wilson and Webster (1946): Kolarich Coal, upper Fort Union Formation (late Paleocene), Red Lodge, Montana.

Organ Genus: *Maceopolipollenites* gen. nov.

Type Species. *Maceopolipollenites triorbicularis* sp. nov.

Generic Diagnosis. Triporate, pores equatorial, with large atriae, circular to meridionally elongated; grain shape oblate, outline in polar view concavely to convexly triangular; exine in polar area of one hemisphere thinned in various patterns (circles, triangles, y shapes), very finely scabrate, observed clearly as small granules only under oil in bright field or phase contrast, sexine thicker than nexine; sexine thickens very slightly or not at all at the pores.

This genus is distinguished from *Engelhardtioidites*, *Engelhardtioipollenites*, and *Momipites* by the polarly thinned areas on one hemisphere.

Discussion. This genus is established for a group of fossil pollen species that are well represented in the Paleocene floral assemblages of the Rocky Mountain area and are characterized by polarly thinned areas of various shapes on one hemisphere, a feature not found in the morphologically similar fossil pollen genera of *Engelhardtioidites*, *Engelhardtioipollenites*, and *Momipites*. Potonié did not mention the presence of a polarly thinned area in his original descriptions of *Engelhardtioidites* and *Engelhardtioipollenites* or in his subsequent Synopsis (1960, p. 117, 118), and Stanlcy (1965, p. 301) stated that pollen of *Momipites* *coryloides* Wodehouse, which he examined from samples of the Green River Shale, lacked any exinous thinning in the polar area.

Maceopolipollenites is considered to be of juglandaceous affinity because the fossil pollen possess several morphologic characters that are found in several extant juglandaceous genera. Morphologically, the fossil pollen are intermediate between the modern pollen of *Engelhardtia* and *Alfaroa* and that of *Carya*. The similarity with the former two genera is primarily in the location and structure of the pores, and the similarity with *Carya* is the thinned polar exinous area present on one hemisphere. In addition, *Maceopolipollenites* possess morphologic characters in common with *Alfaroa, Engelhardtia,* and *Carya,* including a very finely granular sculpture that is clearly discernible only under oil, wide pore atriae, and the absence of any significant protrusion or exine thickening at the pores. Whitehead's (1960, unpublished report) study of juglandaceous pollen has been a particularly useful reference in studying the affinities of *Maceopolipollenites*. The juglandaceous affinity of *Maceopolipollenites* also appears to be supported by its stratigraphic occurrence. This pollen first occurs in significant numbers—both species and individuals—in the Lebo Shale Member of the Fort Union Formation, which is of middle Paleocene age. This first quantitatively significant occurrence immediately precedes the first quantitatively significant occurrence of *Carya* pollen of modern aspect. The name *Maceopolipollenites* is derived from maceo = meager, thin, and polus = pole, which refer to the thinned exinous polar areas that characterize the genus.

Maceopolipollenites triorbicularis sp. nov.
(Pl. 6, figs. 5a, 5b)

Description. Triporate, pores equatorial, atriate; shape oblate, triangular to concavely triangular in polar view; exine 1.2 to 1.5 μ thick, thickening slightly or not at all at the pores, nexine less than 0.5 μ thick and stopping about 5 μ from the pores; exine on one hemisphere thinned in three small circular areas approximately 4 μ in diameter and located about the pole; scabrate, formed of very fine granules about 0.2 μ high; size (maximum diameter, 28 specimens) 20.2 to 26.4 μ.

Holotype. Plate 6, figures 5a, 5b; Slide RC 1803-7/97.7 × 16.3; Fort Union Formation (Paleocene).

Differential Diagnosis. This species is distinguished from *M. rotundus* sp. nov. by its smaller size and its triangular to concavely triangular equatorial outline.

Stratigraphic Occurrence. Lebo Shale and Tongue River Members of the Fort Union Formation.

Maceopolipollenites amplus sp. nov.
(Pl. 7, figs. 1-2b, 5)

Description. Triporate, pores equatorial, atriate; shape oblate, rounded triangular in polar view; exine 1 to 1.5 μ thick, thickening slightly or not at all at pores, nexine less than 0.5 μ thick and stopping about 8 μ from the pores; exine on one hemisphere thinned to form a circular to triangular circumpolar area about 12 μ in diameter; scabrate, formed of very fine granules about 0.2 μ high; size (maximum diameter, 25 specimens) 28.7 to 37.3 μ.

Holytype. Plate 7, figures 2a, 2b; Slide RC 1803-7/99.5 x 8.4, Fort Union Formation (Paleocene).

Differential Diagnosis. This species is distinguished from *M. tenuipolus* by its larger size.

Stratigraphic Occurrence. Fort Union Formation.

Maceopolipollenites tenuipolus (Anderson, 1960) n. comb.
(Pl. 6, fig. 4)

Momipites tenuipolus Anderson, New Mexico Bur. Mines and Mineral Resources Mem. 6, p. 25, Pl. 7, fig. 14; Pl. 8, figs. 14, 15, 1960.

The size range of *Maceopolipollenites tenuipolus* in our Wyoming samples is 15 to 23 μ, while Anderson reported a size range of 14 to 20 μ. The average exine thickness is approximately 1 μ, except at the pores, where the thickness is approximately 1.2 μ.

Stratigraphic Occurrence. Fort Union Formation.

Anderson (1960): Nacimiento 1 and 2 florules (lowermost to lower Paleocene), New Mexico.

Drugg (1967): Upper Moreno (Maestrichtian-Danian) Formation, California.

Maceopolipollenites leboensis sp. nov.
(Pl. 6, figs. 6a, 6b)

Description. Triporate, pores equatorial, atriate; shape oblate, triangular to rounded triangular in polar view; exine about 1.2 μ thick, thickening slightly or not at all at the pores, nexine 0.5 μ thick and stopping about 6 μ from the pores; exine on one hemisphere thinned to form a triradiate scar (~10 μ radius) in one polar area, and the opposite hemisphere sometimes thinned to form a triangular-shaped area or triradiate scar at the other pole; scabrate, formed of very fine granules about 0.2 μ high; size (10 specimens) 27 to 31 μ.

Holotype. Plate 6, figures 6a, 6b; Slide RC 1803-7/97.2 × 15.3; Fort Union Formation (Paleocene).

Differential Diagnosis. This species is distinguished from *Engelhardtia triletipollenites* Rouse, 1957, by its larger size.

Stratigraphic Occurrence. Lebo Shale Member of the Fort Union Formation.

Maceopolipollenites rotundus sp. nov.
(Pl. 6, fig. 7)

Description. Triporate, pores equatorial, atriate; shape oblate, rounded triangular in polar view; exine 1 to 1.5 μ thick, thickening

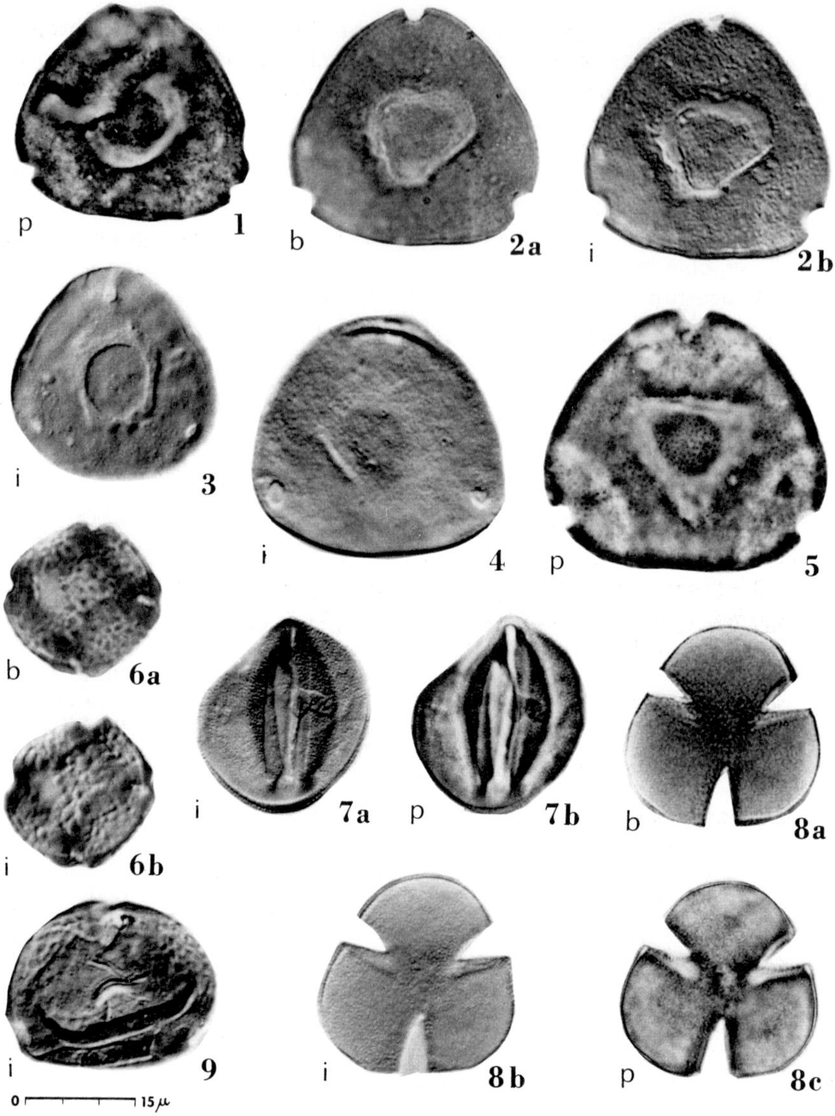

ILLUSTRATIONS OF PALYNOMORPHS

PLATE 6. ILLUSTRATIONS OF PALYNOMORPHS

All illustrations ×1000. Notations b, p, and i indicate photographs taken in bright field, phase contrast, and interference contrast, respectively.

Figures 1, 2, 5. *Maceopolipollenites amplus* sp. nov. Figure 2, holotype. Page 30.
Figure 3. *Carya veripites* Wilson and Webster Page 28.
Figure 4. *Carya viridifluminipites* (Wodehouse) Wilson and Webster. Page 28.
Figure 6. *Ulmipollenites* sp., illustrating group whose size is less than 25 μ. Page 36.
Figures 7, 8. *Gunnera microreticulata* (Belsky, Boltenhagen, and Potonié). Figure 7b, equatorial view; note thickening of exine at pole. Page 37.
Figure 9. *Ulmipollenites* sp., illustrating group whose size is greater than 25 μ. Page 36.

PLATE 7. ILLUSTRATIONS OF PALYNOMORPHS

All illustrations ×1000. Notations b, p, and i indicate photographs taken in bright field, phase contrast, and interference contrast, respectively.

Figures 1, 2. *Arecipites columellus* sp. nov. Figure 2, holotype. Page 40.
Figure 3. *Liliacidites complexus* (Stanley). Figure 3a, proximal; 3b, distal. Page 41.
Figure 4. *Maceopolipollenites tenuipolus* (Anderson), Page 31.
Figure 5. *Maceopolipollenites triorbicularis* sp. nov. Holotype. Page 30.
Figure 6. *Maceopolipollenites leboensis* sp. nov. Holotype. Page 31.
Figure 7. *Maceopolipollenites rotundus* sp. nov. Holotype. Page 31.

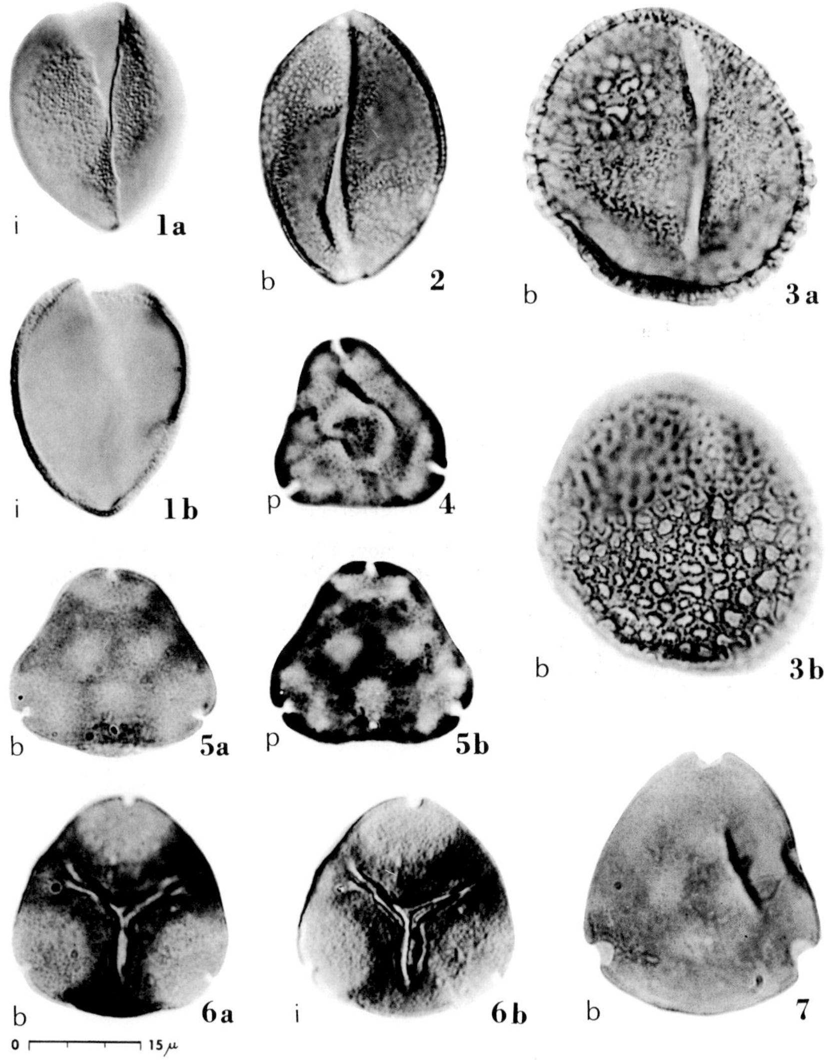

ILLUSTRATIONS OF PALYNOMORPHS

slightly or not at all at the pores, nexine about 0.5 μ thick and stopping about 8 μ from the pores; exine on one hemisphere thinned in three small circular areas about 3 μ in diameter, located about the pole; scabrate, formed of very fine granules about 0.2 μ high; size (maximum diameter, 20 specimens) 24 to 34 μ.

Holotype. Plate 6, figure 7; Slide RC 1803-7/88.1 x 10.7; Fort Union Formation (Paleocene).

Differential Diagnosis. This species is distinguished from *M. triorbicularis* by its slightly larger size and its rounded-triangular equatorial outline.

Stratigraphic Occurrence. Lebo Shale and Tongue River Members of the Fort Union Formation.

Family: Betulaceae
Genus: *Alnus* Mill.
Alnus spp.

Pollen of *Alnus* has not been observed to occur frequently in any of the samples in the present study and consequently no effort has been made to tabulate the quantitative variations of species within this genus.

A species referred to as *Alnipollenites quadrapollenites* (Rouse) Srivastava, 1966, from the Edmonton Formation of Alberta is probably an ulmaceous pollen, based on the exine ornamentation illustrated in the photographs. *Alnus trina* Stanley 1965, a triporate grain from the Ludlow Formation (Paleocene), has not been included in my pollen counts because its morphology is somewhat atypical of *Alnus*. *Alnus* pollen occurs only rarely in rocks of Late Cretaceous age in this area.

Stratigraphic Occurrence. Lebo Shale and Tongue River Members, Fort Union Formation.

Wilson and Webster (1946): Upper Fort Union Formation (late Paleocene), Red Lodge, Montana.

Rouse (1962, as *A. quinquepollenites*): Burrard Formation (Late Cretaceous), British Columbia, Canada.

Stanley (1965, as *A. quaternaria*): Hell Creek Formation (Late Cretaceous), South Dakota.

Family: Ulmaceae
Organ Genus: *Ulmipollenites* Wolff, 1934
Ulmipollenites spp.
(Pl. 7, figs. 6a, 6b, 9)

Species of *Ulmipollenites* observed in this study have been divided into two morphologic groups based on size. The smaller specimens (16 to 25 μ; Pl. 7, figs. 6a, 6b) are most abundant in the Lance and lower Fort Union Formations, while the larger specimens (25 to 35 μ, Pl. 7, fig. 9) occur only in the Fort Union Formation, most commonly in the upper part.

Family: Loranthaceae
Organ Genus: *Cranwellia* Srivastava, 1966
Cranwellia striata (Couper) Srivastava, 1966
(Pl. 8, figs. 13a, 13b)

Elytranthe striatus Couper, New Zealand Geol. Survey Paleontology Bull., v. 22, p. 51-52, Pl. 6, fig. 85, 1953.

Cranwellia striata (Couper) Srivastava, Pollen et Spores, v. 8, p. 537-538, Pl. 11, figs. 1, 4, 1966.

Cranwellia striata has been described from New Zealand, Russia, and western North America. In western North America and Russia, this species has been observed only in rocks of Late Cretaceous age.

Stratigraphic Occurrence. Lance Formation.

Radforth and Rouse (1956): Oldman and Brazeau Formations (Late Cretaceous), western Canada.

Srivastava (1966): Edmonton Formation (Late Cretaceous), Alberta, Canada.

Couper (1953, 1960): Runangan (late Eocene) to Waitotaran (late Pliocene), New Zealand.

Bratzeva (1967): Tsagaiansk deposits (Maestrichtian), Zeya-Bureya depression, U.S.S.R.

Samoilovich (1967): Yenisey-Amur Floral Province (Late Cretaceous), U.S.S.R.

Family: Gunneraceae
Genus: *Gunnera* L.

Gunnera microreticulata (Belsky, Boltenhagen and Potonié, 1965) n. comb.
(Pl. 7, figs. 7a-8b)

Tricolpites microreticulata Belsky, Boltenhagen and Potonié, Paläon. Zeitschr., v. 39, pt. 1/2, p. 75, Pl. 12, figs. 8, 9, 1965.

Tricolpites reticulatus Cookson ex Couper, in Norton and Hall, Rev. Palaeobotany and Palynology, v. 2, p. 105, 106, Pl. 1, fig. E, 1967.

This morphologically distinct fossil pollen species is assigned to the extant genus *Gunnera* because of its marked similarity to *Gunnera*, particularly to *Gunnera petaloidea* Gaus, as illustrated by Selling (1947). The combination of morphologic similarities between the fossil species and *G. petaloidea* includes the interapical position of the colpi, the prolate spheroidal to oblate-spheriodal shape, the sculpture, and the slight exinous thickening at the poles that can be seen in equatorial view (Pl. 7, fig. 7b) and that is indicated in polarly oriented grains by a darker polar area.

Gunnera microreticulata is smaller (equatorial dimension, 22 to 32 μ) than *G. petaloidea* (32 to 41 x 30 to 40 μ).

Our specimens of *Gunnera microreticulata* are more finely reticulate than *Tricolpites interangulas* Newman, 1965. The reticulations of *G.*

ILLUSTRATIONS OF PALYNOMORPHS

LEFFINGWELL, PLATE 8
Geological Society of America Special Paper 127

PLATE 8. ILLUSTRATIONS OF PALYNOMORPHS

All illustrations ×1000. Notations b, p, and i indicate photographs taken in bright field, phase contrast, and interference contrast, respectively.

Figures 1-3. *Tricolpites mutabilis* sp. nov. Figure 3, holotype. Page 44.
Figure 4. *Tricolpites parvus* Stanley. Page 44.
Figure 5. *Tricolpites parvistriatus* Norton. Page 44.
Figures 6, 7. *Quercoidites spissus* sp. nov. Figure 6, holotype. Page 45.
Figures 8-10. *Fraxinoipollenites variabilis* Stanley. Page 45.
Figure 11. *Tricolpites anguloluminosus* Anderson. Page 43.
Figure 12. *Tricolpites bathyreticulatus* Stanley. Page 44.
Figure 13. *Cranwellia striata* (Couper) Srivastrava. Page 37.

microreticulata are difficult, if not impossible, to resolve without oil immersion, while the reticulation of *T. interangulus* can be resolved with high dry objectives. The exine also appears to be thicker at the poles in *G. microreticulata* than in *T. interangulus*.

Stratigraphic Occurrence. Lance Formation.

Belsky, Boltenhagen and Potonié (1965): Senonian-Maestrichtian, Gabon, Equatorial Africa.

Srivastava (1966): Edmonton Formation (Late Cretaceous), Alberta, Canada.

Norton and Hall (1967): Hell Creek Formation (Late Cretaceous), Montana.

Betulaceae-Myricaceae-Casuarinaceae Complex

Triporate, aspidate pollen referable to the Betulaceae, Myricaceae, and/or Casuarinaceae have been grouped as a single entity.

In the Lance assemblages, the majority of the pollen of this type is from a single species. In the middle and late Paleocene assemblages, however, there is a marked increase in both the number of species and the number of specimens of this complex.

Ericaceae-Typhaceae Complex

Ericaceous and typhaceous pollen tetrads were tabulated together for the purposes of this study; they occur in both Lance and Fort Union assemblages.

Angiosperm Pollen of Uncertain Familial Affinity

Form Genus: *Arecipites* (Wodehouse) Anderson, 1960

Anderson emended Wodehouse's description of *Arecipites* by restricting the genus to those monosulcate, reticulate pollen grains whose lumen width does not exceed 0.5 μ. Reticulate monosulcate pollen whose lumen width exceeds 0.5 μ were referred by Anderson to *Liliacidites* Couper, 1953. This procedure is also followed in this paper.

Arecipites columellus sp. nov.
(Pl. 6, figs. 1a, 2)

Description. Monosulcate, sulcus extending the entire length of the grain, never expanding at the ends; shape elongate ellipsoidal; exine 1.2 to 1.5 μ thick, and thinnest adjacent to the sulcus; exine partially tectate, forming distinct ectosexinous and columella layers that are best seen in optical section, sexine thicker than nexine; reticulate, lumina about 0.5 μ wide and of uniform size; grain size (15 specimens) 30 to 40 μ x 20 to 25 μ.

Holotype. Plate 6, figure 2; Slide RC 1825-4/116.1 x 9.8; Lance Formation (Maestrichtian).

Differential Diagnosis. *Arecipites columellus* is distinguished from other species in this genus by its distinct ectosexinous and columella layers, as seen in optical section.

Stratigraphic Occurrence. Lance Formation and Tullock Member of the Fort Union Formation.

Arecipites tenuiexinous sp. nov.
(Pl. 5, figs. 8, 9)

Description. Monosulcate, sulcus extending the entire length of grain, never expanding at the ends; shape ellipsoidal; exine less than 1 μ thick, wall layering obscure; reticulate, lumina about 0.5 μ wide; grain size (10 specimens) 25 to 30 μ × 18 to 23 μ.

Holotype. Plate 5, figure 9; Slide RC 1819-15/107.3 × 14.7; Fort Union Formation (Paleocene).

Differential Diagnosis. *Arecipites tenuiexinous* is distinguished from *A. punctatus* Wodehouse, 1933, and *A. reticulatus* Anderson, 1960, by its larger size.

Stratigraphic Occurrence. Lance Formation and Tullock Member of the Fort Union Formation.

Form Genus: *Liliacidites* Couper, 1953
Liliacidites complexus (Stanley, 1965) n. comb.
(Pl. 6, figs. 3a, 3b)

Schizosporis complexus Stanley, Am. Paleontology Bull., v. 49, no. 222, p. 267, 268, Pl. 36, figs. 7-17, 1965.

This species is transferred from *Schizosporis* to *Liliacidites* because the germinal feature is interpreted to be a sulcus restricted to a single hemisphere rather than a fissure or a tear.

Stratigraphic Occurrence. Lance Formation.
Stanley (1965): Hell Creek Formation (Late Cretaceous), South Dakota.

Liliacidites altimurus sp. nov.
(Pl. 5, figs. 6a, 7c)

Description. Monosulcate, sulcus extending entire length of grain; shape ellipsoidal; exine 3 to 4 μ thick; nexine 0.5 μ thick; reticulate; on the proximal surface, the reticulum incomplete and composed of baculae 3 to 4 μ high × 1 μ wide; on distal surface, reticulum complete and composed of muri 3 to 4 μ high × 2 μ wide, which often expand at the tips and are often branched and fused into a continuous reticulum; lumina of the distal reticulum about 5 μ wide, on proximal surface 2.5 to 3 μ wide; grain size (10 specimens) 25 to 35 μ × 22 to 27 μ.

Holotype. Plate 5, figures 6a-6c; Slide RC 1825-4/124.0 × 6.4; Lance Formation (Maestrichtian).

Differential Diagnosis. This species is distinguished from all other previously described in this genus by its high baculae and its branching distal muri.

Stratigraphic Occurrence. Lance Formation.

Form Genus: *Pandaniidites* Elsik, 1968
Pandaniidites radicus sp. nov.
(Pl. 5, figs. 3a-5)

Description. Spherical to elliptical monoporate pollen, pore circular, 2 to 3.2 μ (rarely 4 to 5 μ) in diameter, annulate, pore rim (annulus) 1.5 to 2 μ thick, remainder of exine thin, less than 0.5 μ, and ornamented predominately with spines projecting 0.5 to 3.2 μ above the exine, spaced 1.5 to 5 μ apart, and sometimes curved or bent; the bases of some spines project below the exine (Pl. 5, fig. 4b); granules about 0.5 μ high and baculae 1 to 3.2 μ high often present as secondary sculptural elements; grains usually folded, sometimes randomly, but predominately parallel to the long axis, or concentrically; size (20 specimens) 15 to 24.9 μ × 21.8 to 31.1 μ.

Holotype. Plate 5, figures 4a, 4b; Slide RC 1825-3/115.6 × 17.5; Lance Formation (Maestrichtian).

Differential Diagnosis. The considerably thinner exine of *P. radicus* distinguishes it from *P. texus* Elsik, 1968, and *P. shiabensis* (Simpson) Elsik, 1968.

Stratigraphic Occurrence. Lance and Fort Union Formations, Wyoming.

Discussion. *P. radicus* is assigned to *Pandaniidites* because of its single pore and its spinate sculpture.

Although Elsik described the exine of *Pandaniidites* as tectate, with two closely appressed wall layers, the exine stratification as exhibited by the holotype and paratype of *P. texus* is difficult to discern. Exine stratification of the thinner walled Rocky Mountain species, *P. radicus,* is impossible to resolve with light microscopy. The lack of differential staining of the wall layers of *P. texus* and *P. radicus* suggests that both are predominately, if not entirely, of sexinous composition. Leffingwell and others (in prep.) have shown that when differential staining occurs in fossil residues comparable staining can be obtained with electron microscope stains. In fossil residues, the nexine accepts stain more readily than does the sexine, which permits differentiation of the foot layer and nexine with both light and transmission electron microscopy. Although differentially stained pollen of various types occurs in Elsik's residue containing the holotype of *P. texus,* the holotype is unstained. If *P. texus* is two-layered, the lack of staining indicates the lower layer is a foot layer rather than a nexine. The nexine, if present, is extremely thin and not clearly discernible with light microscopy. The lack of staining could also be explained by the presence of a single, solid, sexinous layer. Similarly, *P. radicus* pollen did not stain in a Lance residue that contained differentially stained pollen.

Pandaniidites radicus resembles pollen of the extant species *Pandanus odoratissimus* Linn., *P. tahitensis* Martelli, and *P. kafu* Martelli; however, the fossil grains have an annulate, circular pore in contrast to pollen of these three modern species of *Pandanus* that have nonannulate apertures and that in some cases have a circular pore but more commonly have an irregular ill-defined aperture. The grains of *P. odoratissimus, P. tahitensis,* and *P. kafu* are slightly thicker walled than *P. radicus* (except for the apertural area), tend to be more oval, and often have the pore at one of the narrow ends. Erdtman's (1952, Fig. 179A) illustration of *Pandanus odoratissimus* var. *suvaensis* shows that the exine below the spines is depressed inwardly. In *Pandaniidites radicus,* the spine bases project below the exine. Pollen grains of *Freycinetia multiflora* Merr. (Pandanaceae) are smooth and have a large, poorly defined pore. The pore is annulate, but otherwise the grains do not resemble those of *P. radicus.* The fossil species also resemble *Lemna gibba* Weber ex Mert. and Koch. and *L. trisulca* Linn. (Lemnaceae) as illustrated by Erdtman (1952, Fig. 137); however, those species appear to be thicker walled and nonannulate and apparently lack the inward depression of the exine below the spines.

Form Genus: *Aquilapollenites* (Rouse) Funkhouser, 1961

Most species of *Aquilapollenites* reported from the Rocky Mountain area have been observed in rocks of Late Cretaceous age, including all species described by Rouse (1957), Stanley (1965), Norton (1965), and Srivastava (1966) and all but one species described by Funkhouser (1961). The single reported Rocky Mountain Paleocene occurrence is *A. spinulosus* Funkhouser. No specimens of *Aquilapollenites* were observed in rocks younger than Late Cretaceous age in the present study.

Form Genus: *Tricolpites* (Cookson ex Couper) Belsky, Boltenhagen and Potonié, 1965

Tricolpites anguloluminosus Anderson, 1960
(Pl. 8, fig. 11)

Tricolpites anguloluminosus Anderson, New Mexico Bur. Mines and Mineral Resources Mem. 6, p. 26, Pl. 6, figs. 15-17; Pl. 8, figs. 17, 18, 1960.

T. anguloluminosus Anderson, 1960, and *T. bathyreticulatus* Stanley, 1965, although treated as separate species, were combined for the purposes of the quantitative analyses.

Stratigraphic Occurrence. Fort Union Formation.

Anderson (1960): Ojo Alamo 2 florule; Nacimiento 2 florule (lowermost to lower Paleocene), New Mexico.

Norton and Hall (1967): Tullock and Lebo Shale Members (early to middle Paleocene) of the Fort Union Formation, Montana.
Drugg (1967): Upper Moreno Formation (Danian), California.

Tricolpites bathyreticulatus Stanley, 1965
(Pl. 8, fig. 12)

Tricolpites bathyreticulatus Stanley, Am. Paleontology Bull., v. 49, no. 222, p. 320-321, Pl. 47, figs. 18-23, 1965.

Stratigraphic Occurrence. Fort Union Formation.
Stanley (1965): Cannonball and Ludlow Members (early Paleocene) of the Fort Union Formation, South Dakota.
Hall and Norton (1967): Tullock and Lebo Members (early to middle Paleocene) of the Fort Union Formation, Montana.

Tricolpites parvistriatus Norton, 1967
(Pl. 8, figs. 5a, 5b)

Tricolpites parvistriatus Norton, *in* Norton and Hall, Rev. Palaeobotany and Palynology, v. 2, p. 105, Pl. 1, fig. D, 1967.

The size of the specimens from the type Lance area is 16 to 26.5 μ, whereas Norton reported a size range of 18 to 23 μ in the type Hell Creek area.

Stratigraphic Occurrence. Lance and Fort Union Formations, abundant only in the Lance.
Norton and Hall (1967): Hell Creek Formation (Late Cretaceous), Montana.

Tricolpites parvus Stanley, 1965
(Pl. 8, figs. 4a, 4b)

Tricolpites parvus Stanley, Am. Paleontology Bull., v. 49, no. 222, p. 322, Pl. 47, figs. 28-31, 1965.

The thick nexine distinguishes this species from others in this genus. The maximum nexine thickness of our specimens is approximately 1.2 μ measured between the colpi. The nexine thins abruptly to a feather edge at the colpi.

Stratigraphic Occurrence. Lance and Fort Union Formations, common only in the Lance Formation.
Stanley (1965): Ludlow Member (early Paleocene) of the Fort Union Formation, South Dakota.

Tricolpites mutabilis sp. nov.
(Pl. 6, figs.1-3)

Description. Tricolpate, colpi extending nearly to poles; shape oblate, outline triangular to rounded triangular in polar view; exine 1 μ thick between colpi and less than 0.5 μ thick adjacent to colpi, thinning abruptly about 2.5 μ from colpi margins; scabrate in areas between colpi and smooth adjacent to colpi; size (equatorial diameter, 15 specimens) 17 to 21 μ.

Holotype. Plate 8, figure 3; Slide RC 1831B-5/101.5 × 12.5; Fort Union Formation (Paleocene).

Differential Diagnosis. *Tricolpites mutabilis* is distinguished from other species within this genus by its smaller size and the abrupt thinning of the exine adjacent to the colpi.

Stratigraphic Occurrence. Lance and Fort Union Formations.

Form Genus: *Quercoidites* Potonié, Thomson and Thiergart, 1950

Quercoidites spissus sp. nov.

(Pl. 8, figs. 6, 7)

Description. Tricolpate, colpi extending nearly to poles; shape prolate spheroidal; exine 1.2 μ thick, nexine 0.5 μ thick; granular to baculate, about 0.8 μ high, crowded; size (25 specimens) 16.3 to 27.2 μ × 16.3 to 24.1 μ.

Holotype. Plate 8, figure 6; Slide RC 1830B-1/100.8 × 12.2; Lance Formation (Maestrichtian).

Differential Diagnosis. This species is distinguished from *Quercus granopollenites* Rouse, 1962, by its smaller size and somewhat thinner exine.

Stratigraphic Occurrence. Lance and Fort Union Formations.

Form Genus: *Fraxinoipollenites* Potonié, 1951

Fraxinoipollenites variabilis Stanley, 1965

(Pl. 8, figs. 8-10)

Fraxinoipollenites variabilis Stanley, Am. Paleontology Bull., v. 49, no. 222, p. 306, Pl. 45, figs. 29-35, 1965.

Stratigraphic Occurrence. Fort Union Formation.

Stanley (1965): Ludlow and Cannonball Members (early Paleocene) of the Fort Union Formation, South Dakota.

Fraxinoipollenites pachyexinous sp. nov.

(Pl. 9, figs. 1, 2)

Description. Tricolpate, colpi extending nearly to poles, colpi with margos, about 1 μ thick; shape prolate; exine 1.2 μ thick; very finely reticulate to finely punctate; muri as wide as or wider than lumina; lumina decrease in size adjacent to the colpi; grain size (15 specimens) 17 to 23 μ × 13 to 20 μ.

Holotype. Plate 9, figure 1; Slide RC 1831B-5/102.9 × 11.6; Fort Union Formation (Paleocene).

Differential Diagnosis. *Fraxinoipollenites pachyexinous* is distinguished from *F. staplini* Srivastava by its smaller size and from *F. variabilis* Stanley by its thicker exine, its wider muri, and its margos along the colpi.

Stratigraphic Occurrence. Fort Union Formation.

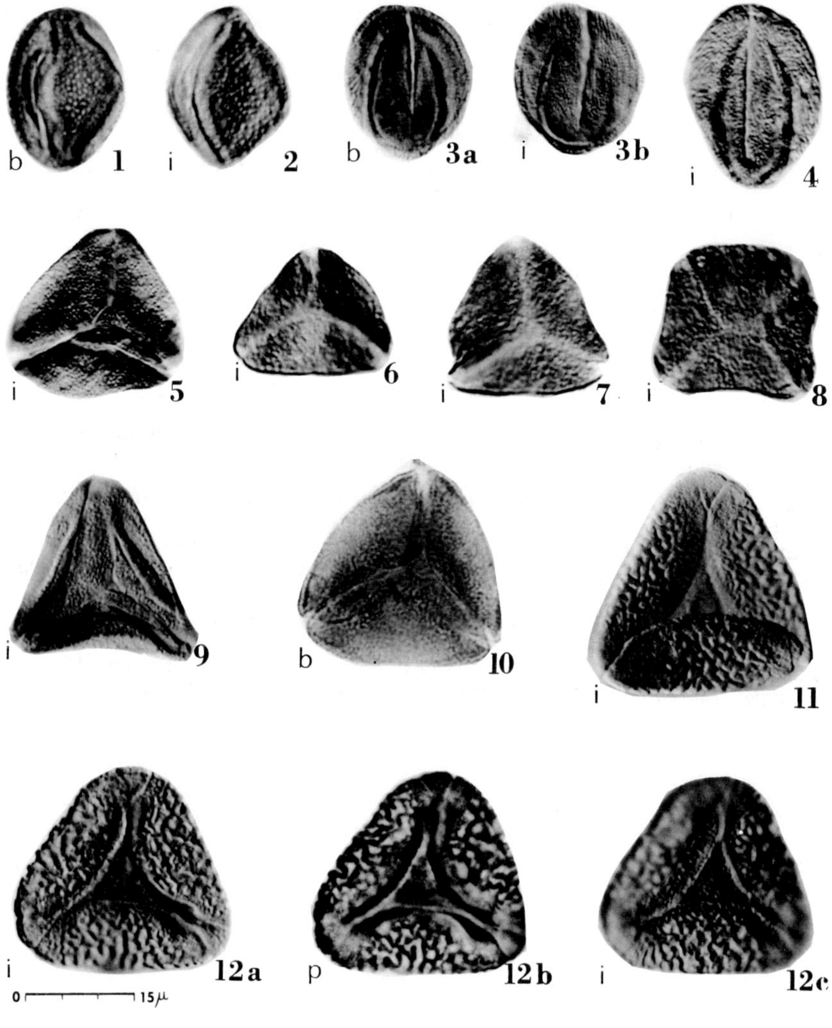

ILLUSTRATIONS OF PALYNOMORPHS

LEFFINGWELL, PLATE 9
Geological Society of America Special Paper 127

PLATE 9. ILLUSTRATIONS OF PALYNOMORPHS

All illustrations ×1000. Notations b, p, and i indicate photographs taken in bright field, phase contrast, and interference contrast, respectively.

Figure 1, 2. *Fraxinoipollenites pachyexinous* sp. nov. Figure 1, holotype. Page 45.

Figures 3, 4. *Striatopollis tectatus* sp. nov. Figure 3, holotype. Page 48.

Figures 5, 9, 10. *Cupanieidites inaequalis* sp. nov. Figure 10, holotype. Page 49.

Figures 6-8. *Syncolporites minimus* sp. nov. Figure 7, holotype. Figure 8, aberrant grain. Page 49.

Figures 11, 12. *Insulapollenites rugulatus* gen. nov., sp. nov. Figures 12, holotype; 12a illustrates sculpture on polar island. Page 48.

Form Genus: *Striatopollis* Krutzsch, 1959a
Striatopollis tectatus sp. nov.
(Pl. 9, figs. 3-4)

Description. Tricolpate, colpi extending nearly to the poles; shape prolate; exine 1.2 to 1.5 μ thick, tectate, columella layer about 0.5 μ high, tectum about 0.8 μ thick; striate sculpture upon tectum; size (20 specimens) 21 to 26 μ × 15.5 to 21 μ.

Holotype. Plate 9, figures 3a, 3b; Slide RC 1833B-2/89.7 × 13.7; Fort Union Formation (Paleocene).

Differential Diagnosis. *Striatopollis tectatus* is distinguished from *S. sarstedtensis* Krutzsch by its slightly larger size and its tectate exine. The striae are upon the tectum and therefore are not as deeply incised as those in Krutzsch's species, and the punctate-reticulate structure is of endosexinous origin rather than being extrapunctate and located on the ridge surface, as is *S. sarstedtensis*.

Stratigraphic Occurrence. Lance and Fort Union Formations.

Form Genus: *Insulapollenites* gen. nov.

Type Species. *Insulapollenites rugulatus* sp. nov.

Generic Diagnosis. Oblate, parasyncolpate pollen whose colpi bifurcate near the poles and near the equator to form equatorial and polar islands.

This genus is distinguished from *Duplopollis* Krutzsch, 1959a, and *Cupanieidites* Cookson and Pike, 1954, by being syncolpate rather than syncolporate and by the islands at the angles of the amb. The name *Insulapollenites* is derived from insula = islands, which refers to the polar and equatorial islands that characterize this genus.

Insulapollenites rugulatus sp. nov.

Unclassified pollen sp. 1 Groot and Groot, Palaeontographica, bd. 111, abt. B, p. 170, Pl. 31, figs. 10, 11, 1962.

Description. Parasyncolpate, colpi bifurcating near the poles and at the angles of the amb to form islands; polar islands triangular and approximately 10 μ in diameter; islands at the angles of the amb diamond-shaped, and in polar view one-half of each diamond-shaped island visible on each hemisphere; grain shape oblate, trangular in polar view, with rounded apices; exine 1.5 μ thick in the equatorial interapical areas, 0.8 μ thick at the angles of the amb, sexine thicker than nexine; coarsely rugulate in the interapical areas, becoming reticulate to very fine reticulate within 3 to 5 μ from the colpi; equatorial islands very finely reticulate, one polar island very finely reticulate, the opposite island smooth to scabrate; size (20 specimens) 26.5 to 34 μ.

Holotype. Plate 9, figures 12a-12c; Slide RC 1806-7/106.2 × 15.1; Fort Union Formation (Paleocene).

Differential Diagnosis. This species is distinguished from *Duplopollis* sp. B Drugg, 1967, by its more variable ornamentation, which is considerably smaller along the colpi and on the polar islands.

Stratigraphic Occurrence. Lebo Shale and Tongue River Members of the Fort Union Formation.

Groot and Groot (1962): Brightseat Formation (Paleocene), Maryland.

Form Genus: *Cupanieidites* Cookson and Pike, 1954

Cupanieidites inaequalis sp. nov.

(Pl. 9, figs. 5, 9, 10)

Cupanieidites aff. *C. major* Cookson and Pike, 1954, in Anderson, New Mexico Bur. Mines and Mineral Resources Mem. 6, p. 22, Pl. 6, fig. 6; Pl. 8, figs. 10-12; Pl. 10, fig. 16, 1960.

Description. Parasyncolporate, pores indistinct, colpi without margos, triangular polar islands at one or both poles; if present at both poles, usually of unequal size, the larger one 3 to 6 μ wide and the maximum size of the smaller one 3.5 μ; shape oblate, triangular to rounded triangular in polar view; exine 0.8 μ thick between apertures, thickening to about 2 μ at the pores, the nexine thickness varies from 0.5 μ between pores to 1 μ at the pores, sexine thickness varies from less than 0.5 μ between apertures to 1 μ at the pores; most of the exinous thickening occurs from 3 to 8 μ on either side of the apertures, and a partial tectum tends to develop at the pores; elsewhere, the exine clearly intectate; very finely reticulate, with lumina about 0.5 μ wide and approximately of equal size; size (25 specimens) 20 to 27.2 μ.

Holotype. Plate 9, figure 10; Slide RC 1982-3/104.3 × 12.3; Fort Union Formation (Paleocene).

Differential Diagnosis. This species is distinguished from *Cupanieidites major* Cookson by its smaller size and less distinct pores.

Stratigraphic Occurrence. Lance Formation and Tullock Member of the Fort Union Formation.

Anderson (1960): Ojo Alamo 2 florule (Late Cretaceous?) Nacimiento 2 florule (early Paleocene); rare in Lewis shale florule (Late Cretaceous? to early Paleocene).

Form Genus: *Syncolporites* van der Hammen, 1954

Syncolporites minimus sp. nov.

(Pl. 9, figs. 6-8)

Description. Syncolporate, occasionally slightly parasyncolporate, pores indistinct, colpi without margos; shape oblate, elongately triangular to rounded triangular in polar view; exine about 0.5 μ thick, finely granular; size (maximum dimension, 30 specimens) 15.5 to 23.3 μ.

Holotype. Plate 9, figure 7; Slide RC 1833B-2/88.0 × 8.7; Fort Union Formation (Paleocene).

Differential Diagnosis. This species is distinguished from *Syncolporites lisamae* van der Hammen, 1954, by its less distinct pores.

Stratigraphic Occurrence. Lance and Fort Union Formations.

Form Genus: *Kurtzipites* (Anderson, 1960) emend.

Kurtzipites Anderson, New Mexico Bur. Mines and Mineral Resources, Mem. 6, p. 24, 1960.

Type Species. *Kurtzipites trispissatus* Anderson, 1960, New Mexico Bur. Mines and Mineral Resources Mem. 6, p. 25, Pl. 2, fig. 15.

Emended Generic Diagnosis of Kurtzipites. Tricolporate pollen, although some species may be either indistinctly tricolporate or tricolpate(?); colpi short, meridionally oriented, extending from one-fourth to one-half of grain radius; pores atriate; grain shape, oblate, rounded triangular in polar view; exine scabrate; the nexine about the pore may be uniformly thick, of variable thickness, or not thickened at all; when the pore exine is thickened, a characteristic, more or less triangular thickening is often developed near the polar extremities of the colpi.

Discussion. The aperture of *Kurtzipites* is considered to be tricolporate (or possibly tricolpate in some species), rather than triporate as originally described, because the sexinous opening is slitlike and meridionally oriented and has a length-breadth ratio greater than 2. The pore is interpreted as being atriate. Consequently, the purpose of the emendation is to redescribe the apertures and to broaden the concept of the genus to include two species that morphologically are very similar to the type species, *K. trispissatus*. The first species is similar in all respects to the type except that its nexine is not perceptibly thickened anywhere along the pore margin. The second species was originally described by Norton (1963) in his Ph.D. dissertation. In this species, the pore margin is uniformly thickened about the pore and at the polar extremities of the colpi (Pl. 10, figs. 6, 7). Morphologically, the pore structure of *K. trispissatus* is intermediate between these two previously unpublished species in that its nexine is thickened at the equatorial pore margin and at the polar extremities of the colpi. The stratigraphic ranges of all three species are similar, but the types with nexinous thickenings at the pore generally occur more abundantly in rocks of Late Cretaceous age, and those with no perceptible nexinous thickening occur more frequently in rocks of early Paleocene age.

Kurtzipites trispissatus Anderson, 1960
(Pl. 10, figs. 1a, 1b)

Kurtzipites trispissatus Anderson, New Mexico Bur. Mines and Mineral Resources Mem. 6, p. 25, Pl. 2, figs. 15-17, 1960.

Stratigraphic Occurrence. Lance Formation and Tullock Member of the Fort Union Formation.
Anderson (1960): Kirkland Shale florule (Late Cretaceous), New Mexico.

Kurtzipites simplex sp. nov.
(Pl. 10, figs. 2-4b)

Description. Tricolporate(?) or tricolpate, colpi short, from one-quarter to less than one-half of grain radius; shape oblate, rounded triangular in polar view; exine about 0.5 μ thick, becoming slightly thicker at the colpi margins, scabrate, formed of very fine granules about 0.2 μ high; size (25 specimens) 19.5 to 26.5 μ.

Holotype. Plate 10, figures 4a, 4b; Slide RC 1777-3/91.8 x 10.5; Fort Union Formation (Paleocene).

Differential Diagnosis. *K. simplex* is distinguished from *K. trispissatus* by the absence of distinct nexinous thickenings at the polar extremities of the colpi or at the equatorial pore margin.

Stratigraphic Occurrence. Lance Formation and Tullock Member of the Fort Union Formation.

Kurtzipites sp.
(Pl. 10, figs. 6, 7)

This species was originally described by Norton (1963); it is characterized by a more or less uniform nexinous thickening about the pore. It is not described at this time because the amount of variation of the nexinous thickening requires additional study.

Stratigraphic Occurrence. Lance Formation and Tullock Member of the Fort Union Formation.

Form Genus: *Triporopollenites* (Pflug) Thomson and Pflug, 1953
Triporopollenites plektosus Anderson, 1960
(Pl. 10, figs. 5a, 5b)

Triporopollenites plektosus Anderson, New Mexico Bur. Mines and Mineral Resources Mem. 6, p. 27, Pl. 4, fig. 14; Pl. 8, fig. 16, 1960.

Stratigraphic Occurrence. Lance Formation and Tullock and Lebo Shale Members of the Fort Union Formation.

Anderson (1960): Ojo Alamo florule (Late Cretaceous?); Nacimiento florule (early Paleocene), New Mexico.

Form Genus: *Triatriopollenites* (Pflug) Thomson and Pflug, 1953
Genus: *Triatriopollenites granulatus* (Simpson) n. comb.
(Pl. 10, figs. 8a-9b)

Engelhardtia granulata Simpson, Roy. Soc. Edinburgh Trans., v. 64, no. 16, p. 445, Pl. 14, figs. 3, 4, 1961.

The specimens of *T. granulatus* encountered in this study are slightly smaller than those described by Simpson, their size being (20 specimens) 15.5 to 24 μ. The exine is 1.2 μ thick between the apertures, thinning to 0.8 μ at the pores; the nexine is uniformly thin, less than 0.5 μ, stopping 1.6 to 2.5 μ from the pores; a discernible endosexinous columnella layer between the apertures becomes imperceptible toward the pores; and the ectosexine thickens from 0.5 μ between the apertures

PLATE 10. ILLUSTRATIONS OF PALYNOMORPHS

All illustrations ×1000. Notations b, p, and i indicate photographs taken in bright field, phase contrast, and interference contrast, respectively.

Figure 1. *Kurtzipites trispissatus* Anderson. Page 50.
Figures 2-4. *Kurtzipites simplex* sp. nov. Figure 4, holotype. Page 51.
Figure 5. *Triporopollenites plektosus* Anderson. Page 51.
Figures 6, 7. *Kurtzipites* sp. Page 51.
Figures 8, 9. *Triatriopollenites granulatus* (Simpson). Page 51.
Figure 10. *Wodehouseia spinata* Stanley. Page 54.

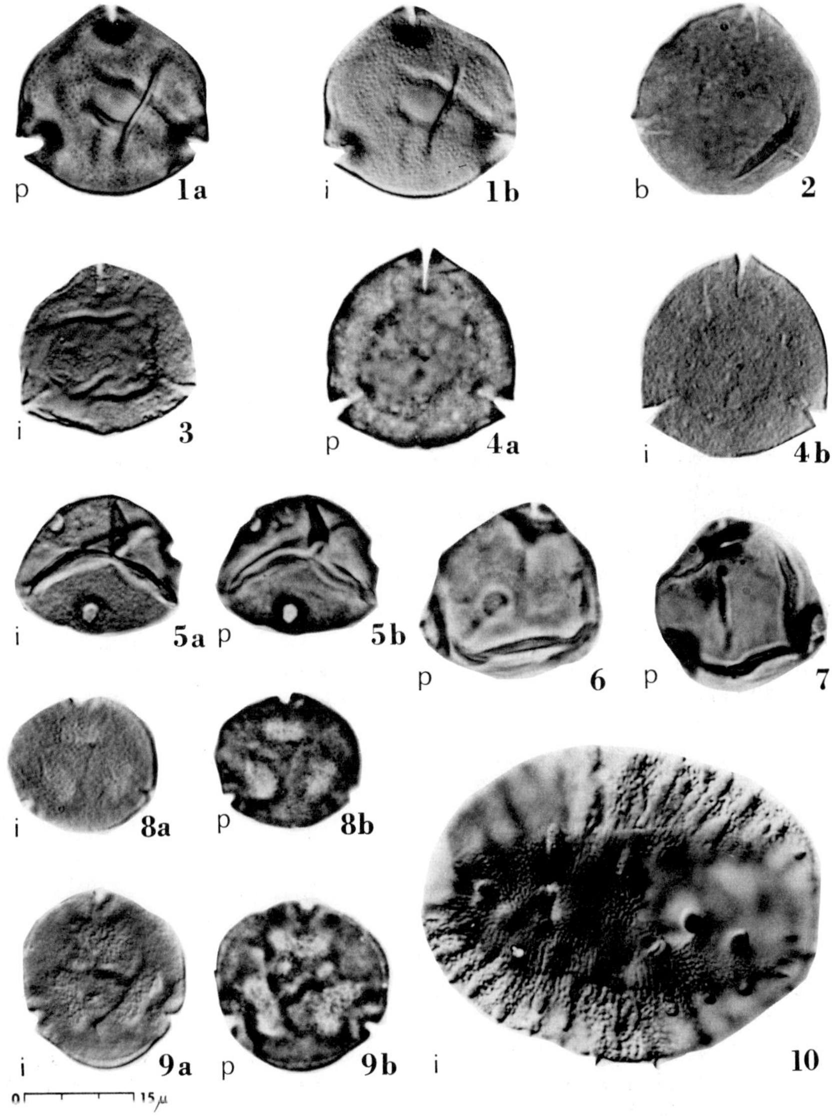

ILLUSTRATIONS OF PALYNOMORPHS

to 0.8 μ at the pores. The thinned exinous areas are more coarsely scabrate than the remainder of the grain.

Differential Diagnosis. The endosexinous columella layer between the pores and the crescent-shaped, thinned exinous areas distinguish this species from others in this genus.

Stratigraphic Occurrence. Uppermost Lance and Fort Union Formations.

Form Genus: *Proteacidites* Cookson, 1950
Proteacidites spp.

Most of the fossil pollen species of *Proteacidites* described from western North America are characterized by distinct exinous thickenings at the pore. This type of pore structure is similar to that found in some extant proteaceous (for example, *Lomatia*) pollen and in some symplocaceous pollen.

Stratigraphic Occurrence. Lance Formation and basal Tullock Member of the Fort Union Formation.

Rouse (1957, as *Beaupreadites*): Comax Formation (Late Cretaceous), British Columbia, Canada.

Sarmiento (1957, as Triporitae types 1, 2): Mancos Shale (Late Cretaceous), Utah.

Rouse (1962): Burrard Formation (Late Cretaceous), Canada.

Anderson (1960): Kirkland and Lewis Shales (Late Cretaecous), New Mexico.

Stanley (1965): Hell Creek Formation (Late Cretaceous), South Dakota.

Newman (1964): Mancos Shale, Iles Formation, Williams Fork Formation, Lewis Shale, Lance Formation (Late Cretaceous), Colorado.

Srivastava (1966): Edmonton Formation (Late Cretaceous), Alberta, Canada.

Form Genus: *Wodehouseia* Stanley, 1961
Wodehouseia spinata Stanley, 1961
(Pl. 10, fig. 10)

Wodehouseia spinata Stanley, Pollen et Spores, v. 3, p. 157, Pl. 1, figs. 1-12, 1961.

Kryshtofoviana vera Samoilovitch, *in* Samoilovitch and Mtchedlishvili, Vses. Neft. Nauchno-Issled. Geol-Razved. Inst. Trudy, v. 177, p. 234, Pl. 75, figs. 1a-3c, 1961.

Deplexipollis oculatus Chlonova, Geophysics Acad. Nauk SSSR Geol. Inst. Trudy No. 7, p. 81, Pl. 13, fig. 96, 1961.

Stratigraphic Occurrence. Lance Formation and basal Tullock Member, Fort Union Formation.

Other reported occurrences in Rocky Mountain area:

Stanley (1961, 1965): Hell Creek Formation (Late Cretaceous), South Dakota.

Srivastava (1966, 1967): Edmonton Formation (Late Cretaceous), Alberta, Canada.

Norton and Hall (1967): Hell Creek Formation (Late Cretaceous), Montana.

ACKNOWLEDGMENTS

The study of the upper Lance and Fort Union Formations in the type Lance area was completed in 1962 while I was employed by the Jersey Production Research Company. I gratefully acknowledge the field assistance and helpful discussions rendered during this period by Allen Dewees, consultant, formerly of the Humble Oil Company, and Jack D. Burgess of the Gulf Oil Technical Services Center. The type Lance and adjacent areas were resampled from 1965 to 1967, and I wish to acknowledge the field support of the Union Oil Company of California for visits to the type Lance and Cannonball areas. I also thank the Esso Production Research Company for providing cuts of the original samples. I am particularly indebted to Drs. W. A. Clemens, Erling Dorf, Steven Fox, R. H. Tschudy, Roger Y. Anderson, Karl R. Newman, Robert Sloan, Alan Cvancera, Norman Denson, W. C. Elsik, and Estella B. Leopold, to Wayne Chisholm and Mrs. Bernadine Tschudy for offering critical data, samples, and/or slides during various phases of the project, and to the Chevron Research Laboratory for use of various modern pollen preparations. I am also grateful to Dr. G. K. Guennel, Dr. Norman Norton, and Jack Burgess for critically reading the manuscript and making many helpful suggestions.

REFERENCES CITED

Anderson, R. Y., 1960, Cretaceous-Tertiary palynology, eastern side of the San Juan Basin, New Mexico: New Mexico Bur. Mines and Mineral Resources Mem. 6, 58 p., 11 pls.

Bell, W. A., 1949, Uppermost Cretaceous and Paleocene floras of western Alberta: Canadian Geol. Survey Bull. 13, 231 p, 67 pls.

Belsky, C. Y., Boltenhagen, E., and Potonié, R., 1965, Sporae dispersae der Oberen Kreide von Gabun, Aequatoriales Afrika: Paläont. Zeitschr., v. 39, pt. 1/2, p. 72-83, Pls. 12, 13.

Bratzeva, G. M., 1967, The problem of the Tsagaiansk flora with regard to spore-and-pollen analytical data: Rev. Paleobotany and Palynology, v. 2, p. 119-126, Pl. 1.

Brown, R. W., 1952, Tertiary strata in eastern Montana and western North and South Dakota: Billings Geol. Soc. Guidebook, 3d Ann. Field Conf., p. 89-92.

——1958, Fort Union Formation in the Powder River Basin, Wyoming: Wyoming Geol. Assoc. Guidebook, 13th Ann. Field Conf., p. 111-113.

——1962, Paleocene flora of the Rocky Mountains and Great Plains: U.S. Geol. Survey Prof. Paper 375, 119 p., 69 pls.

Calvert, W. R., 1912, Geology of certain lignite fields in eastern Montana: U.S. Geol. Survey Bull. 471, p. 187-201.

Chlonova, A. F., 1961, Spores and pollen from the upper half of the Upper Cretaceous of eastern Chosti, western Siberian lowlands: Akad. Nauk SSSR Geol. Inst. ser. Geophys., no. 7, 138 p., 17 pls.

Clemens, W. A., Jr., 1960, Fossil mammals of the type Lance Formation, Wyoming: California Univ., Berkeley, Ph.D. dissert., 510 p.

——1964, Fossil mammals of the type Lance Formation, Wyoming. Pt. I. Introduction and Multituberculata: California Univ. Pubs. Geol. Sci., v. 48, 105 p.

Cobban, W. A., 1958, Late Cretaceous fossil zones of the Powder River Basin, Wyoming and Montana: Wyoming Geol. Assoc. Guidebook, 13th Ann. Field Conf., p. 114-119.

Cookson, I. C., 1950, Fossil pollen grains of proteaceous type from Tertiary deposits in Australia: Australian Jour. Sci., ser. B, v. 3, no. 2, p. 166-177, 3 pls.

Cookson, I. C., and Pike, K. M., 1954, Some dicotyledonous pollen types from Cainozoic deposits in the Australian region: Australian Jour. Botany, v. 2, no. 3, p. 197-219, 2 pls.

Couper, R. A., 1953, Upper Mesozoic and Cainozoic spores and pollen grains from New Zealand: New Zealand Geol. Survey Paleontology Bull. 22, 77 p., 9 pls.

——1960, New Zealand Mesozoic and Cainozoic plant microfossils: New Zealand Geol. Survey Paleontology Bull. 32, 87 p., 12 pls.

Dettman, M. E., 1963, Upper Mesozoic microfloras from southeastern Australia: Royal Soc. Victoria Proc., new ser., v. 77, pt. 1, p. 1-148.

Dobbin, C. E., and Barnett, V. H., 1927, The Gillette coal field, northeastern Wyoming: U.S. Geol. Survey Bull. 796-A, 64 p.. 13 pls.

Dorf, E., 1940, Relationship between floras of the type Lance and Fort Union Formations: Geol. Soc. America Bull., v. 51, p. 213-236.

——1942, Upper Cretaceous floras of the Rocky Mountain region, II. Flora of the Lance Formation of its type locality, Niobrara County, Wyoming: Carnegie Inst. Washington Pub. 508, p. 79-168, 17 pls.

Drugg, W. S., 1967, Palynology of the Upper Moreno Formation (Late Cretaceous-Paleocene) Escarpado Canyon, California: Palaeontographica, bd. 120, abt. B, p. 1-71, Pls. 1-9.

Elsik, W. C., 1968, Palynology of a Paleocene Rockdale lignite, Milam County, Texas. I. Morphology and taxonomy: Pollen et Spores, v. 10, no. 2, p. 263-314, 15 pls.

Erdtman, G., 1952, Pollen morphology and plant taxonomy, angiosperms: Waltham, Massachusetts, Chronica Botanica, 539 p.

Funkhouser, J. W., 1961, Pollen of the genus *Aquilapollenites*: Micropaleontology, v. 7, no. 2, p. 193-198, Pls. 1-2.

Funkhouser, J. W., and Evitt, W. R., 1959, Preparation techniques for acid-insoluble microfossils: Micropaleontology, v. 5, no. 3, p. 369-375.

Groot, J. J., and Groot, C. R., 1962, Some plant microfossils from the Brightseat Formation (Paleocene) of Maryland: Palaeontographica, bd. 111, abt. B, p. 161-171.

Hail, W. J., Jr. and Leopold, E. B., 1960, Paleocene and Eocene age of the Coalmont Formation, North Park, Colorado: U.S. Geol. Survey Prof. Paper 400-B, p. B260-B261.

Hall, J. W., and Norton, N. J., 1967, Palynological evidence of floristic change across the Cretaceous-Tertiary boundary in eastern Montana: Palaeogeography, Palaeoclimatology, and Palaeoecology, v. 3, p. 121-132.

Hancock, E. T., 1921, The New Salem lignite field, Morton County, North Dakota: U. S. Geol. Survey Bull. 726-A, 39 p.

Hatcher, J. B., 1893, The *Ceratops* Beds of Converse County, Wyoming: Am. Jour. Sci., ser. 3, v. 14, p. 135-144.

——1896, Some localities for Laramie mammals and horned dinosaurs: Am. Naturalist, v. 30, p. 112-120.

Ibrahim, A., 1933, Sporenformen des Aegirhorizonts des Ruhr-Reviers: Wurzburg, Konrad Triltsch, 47 p.

Jeletzky, J. A., 1960, Youngest marine rocks in western interior of North America and the age of the *Triceratops*-beds; with remarks on comparable dinosaur-bearing beds outside North America: Copenhagen, Internat. Geol. Congr., Rept. XXIst Sess. Norden, pt. V, The Cretaceous-Tertiary boundary, p. 25-40.

——1962, The allegedly Danian dinosaur-bearing rocks of the globe and the problem of the Mesozoic-Cenozoic boundary: Jour. Paleontology, v. 36, no. 5, p. 1005-1018, 1 pl.

Jeletzky, J. A., and Clemens, W. A., 1965, Comments on Cretaceous *Eutheria,* Lance *Scaphites,* and *Inoceramus?* ex gr. *Tegulatus*: Jour. Paleontology, v. 39, no. 5, p. 952-959.

Jepsen, G. L., 1940, Paleocene faunas of the Polecat Bench Formation, Park County, Wyoming: Am. Philos. Soc. Proc., v. 83, p. 217-340.

Knowlton, F. H., 1909, The stratigraphic relations and paleontology of the "Hell Creek Beds," "*Ceratops* beds" and equivalents, and their reference to the Fort Union Formation: Washington Acad. Sci. Proc., v. 11, p. 179-238.

Knowlton, F. H., 1919, A catalogue of the Mesozioc and Cenozoic plants of North America: U.S. Geol. Survey Bull. 696, 815 p.

——1930, The flora of the Denver and associated formations of Colorado: U.S. Geol. Survey Prof. Paper 155, 142 p.

Lloyd, E. R., 1914, The Cannonball River lignite field, North Dakota: U.S. Geol. Survey Bull. 541-G, 51 p., 14 pls.

Love, J .D., Weitz, J. L., and Hose, R. K., 1955, Geologic map of Wyoming: U.S. Geol. Survey, scale 1:500,000.

Lull, R. S., 1915, The mammals and horned dinosaurs of the Lance Formation of Niobrara County, Wyoming: Am. Jour. Sci., ser. 4, v. 40, p. 319-348.

Marsh, O. C., 1889, Skull of the gigantic Ceratopsidae: Am. Jour. Sci., ser. 3, v. 38, p. 501-506, Pl. 12.

Newman, K. R., 1964, Palynologic correlations of Late Cretaceous and Paleocene Formations, northwestern Colorado, p. 169-180, 1 pl., *in* Cross, A. T., *Editor*, Palynology in oil exploration: Soc. Econ. Paleontologists and Mineralogists Spec. Pub. 11.

——1965, Upper Cretaceous-Paleocene guide palynomorphs from northwestern Colorado: Colorado Univ. Studies Ser. Earth Sci., v. 2, p. 1-21, 1 pl.

Norton, N. J., 1963, Palynology of the Upper Cretaceous and lower Tertiary in the type locality of the Hell Creek Formation: Minneapolis, Minnesota Univ., Ph.D. dissert., 175 p., 30 pls.

——1965, Three new species of *Aquilapollenites* from the Hell Creek Formation, Garfield County, Montana: Pollen et Spores, v. 7, no. 1, p. 135-143, 4 pls.

Norton, N. J., and Hall, J. W., 1967, Guide spormorphae in the Upper Cretaceous-lower Tertiary of eastern Montana (U.S.A.): Rev. Palaeobotany and Palynology, v. 2, p. 99-110, Pl. 1.

——in press, Palynology of the Upper Cretaceous and lower Tertiary in the type locality of the Hell Creek Formation, Montana, U.S.A.: Palaeontographica.

Olive, W. W., 1957, The Spotted Horse coalfield, Sheridan and Campbell Counties, Wyoming: U.S. Geol. Survey Bull. 1050, 83 p., 13 pls.

Pacltova, B., 1961, On some plant microfossils from fresh-water sediments of the Upper Cretaceous (Senonian) in the south Bohemian basins: Sbornik U.U.G., v. 26, p. 47-102, 24 pls.

Pipiringos, G. N., Chisholm, W. A., and Kepferle, R. C., 1965, Geology and uranium deposits in the Cave Hills area, Harding County, South Dakota: U.S. Geol. Survey Prof. Paper 476A, 64 p., 5 pls.

Potonié, R., 1951, Pollen-und sporenformen als Leitfossilien des Tertiärs: Mikroscopie, v. 6, p. 272-383, 2 pls.

——1960, Synopsis der gattungen der sporae dispersae, pt. 3: Beihefte Geol. Jahrb., v. 39, 189 p., 9 pls.

Potonié, R., Thompson, P., and Thiergart, F., 1950, Zur nomenklatur und klassifikation der neogenen sporomorphae (pollen und sporen): Geol. Jahrb., v. 65, p. 35-70, 3 pls.

Radforth, N., and Rouse, G., 1956, Floral transgressions of major geological time zones: Royal Soc. Canada Trans., 3d ser., sec. 5, v. L, p. 17-26, 1 pl.

Rouse, G. G., 1957, The application of a new nomenclatural approach to Upper Cretaceous plant microfossils from western Canada: Canadian Jour. Botany, v. 35, no. 347, p. 359-375, 3 pls.

——1962, Plant microfossils from the Burrard Formation of western British Columbia: Micropaleontology, v. 8, no. 2, p. 187-218, Pls. 1-5.

Russell, L. S., 1930, Upper Cretaceous dinosaur faunas of North America: Am. Philos. Soc. Proc., v. 69, p. 133-159.

Samoilovich, S. R., 1967, Tentative botanico-geographical subdivision of northern Asia in Late Cretaceous time: Rev. Palaeobotany and Palynology, v. 2, p. 127-139, Pls. 1-3.

Samoilovich, S. R., and Mtchedlishvili, N. D., 1961, Pollen and spores of western Siberia: Vses. Neft. Nauchno-Issled. Geol.-Razved. Inst. Trudy, v. 177, 659 p.

Sarmiento, R., 1957, Microfossil zonation of Mancos Group: Am. Assoc. Petroleum Geologists Bull., v. 41, no. 8, p. 1683-1693, 1 pl.

Selling, O., 1947, Studies in Hawaiian pollen statistics; Pt. 2, The pollens of Hawaiian phanerogams: Bishop Mus. Spec. Pub., 430 p., 58 pls.

Shafiqullah, M., Folinsbee, R. E., Baadsgaard, H., Cumming, G. L., and Lerbekmo, J. F., 1964, Geochronology of Cretaceous-Tertiary boundary, Alberta, Canada: Rept. 22nd Internat. Geol. Congr., Pt. III, Proceedings, p. 1-20.

Simpson, J. B., 1961, The Tertiary pollen-flora of Mull and Ardnamurchan: Roy. Soc. Edinburgh Trans., v. 64, no. 16, p. 421-468.

Srivastava, S. K., 1966, Upper Cretaceous microflora (Maestrichtian) from Scollard, Alberta, Canada: Pollen et Spores, v. 8, no. 3, p. 497-522, Pls. 78-88.

——1967, Palynology of the Late Cretaceous mammal beds, Scollard, Alberta (Canada): Palaeogeography, Palaeoclimatology, Palaeoecology, v. 3, p. 133-150, 3 pls.

Stanley, E. A., 1961, The fossil pollen genus *Aquilapollenites*: Pollen et Spores, v. 3, no. 2, p. 329-352, 8 pls.

——1965, Upper Cretaceous and Paleocene plant microfossils and Paleocene dinoflagellates and hystrichosphaerids from northwestern South Dakota: Am. Paleontology Bull., v. 49, no. 222, 384 p., Pls. 19-49.

Stanton, T. W., and Knowlton, F. H., 1897, Stratigraphy and paleontology of the Laramie and related formations in Wyoming: Geol. Soc. America Bull., v. 8, p. 127-156.

Tschudy, R. H., 1961, Palynomorphs as indicators of facies environments in Upper Cretaceous and lower Tertiary strata, Colorado and Wyoming: Wyoming Geol. Assoc. Guidebook, 16th Ann. Field Conf. p. 53-59.

van der Hammen, T., 1954, El desarrollo de la flora Columbiana en los periodos geologicos, I, Maestrichtiano hasta Terciario mas inferior: Bol. Geologico, v. 2, no. 1, p. 49-106, Pls. 1-21.

Wilson, L. R., and Webster, R., 1946, Plant microfossils from a Fort Union coal in Montana: Am. Jour. Botany, v. 33, p. 271-278, 2 pls.

Wodehouse, R. P., 1933, Tertiary pollen; Pt. 2, The oil shales of the Eocene Green River Formation: Torrey Botanical Club Bull., v. 60, p. 479-524, 56 figs.

Wood, H. E., Chaney, R. W., Clark, J., Colbert, E. H., Jepsen, G. L., Reeside, J. B., and Stock, C., 1941, Nomenclature and correlation of the North American continental Tertiary: Geol. Soc. America Bull., v. 52, p. 1-48, 1 pl.

APPENDIX TO REFERENCES

Several important publications were not considered in the taxonomic section because they were published after submission of the manuscript. Four of these are listed below:

Elsik, W. C., 1968, Palynology of a Paleocene Rockdale lignite, Milam County, Texas. II. Morphology and taxonomy: Pollen et Spores, v. 10, no. 3, p. 599-664, Pls. 16-44.

Norton, N. J., and Hall, J. W., 1969, (cited as 'in press' above): Palaeontographica, bd. 125, abt. B, p1-64, Pls. 1-8.

Snead, R. G., 1969, Microfloral diagnosis of the Cretaceous-Tertiary boundary, central Alberta: Res. Counc. Alberta Bull. No. 25, 155 p., 7 pls.

Srivastava, S. K., 1969, Assorted angiosperm pollen from the Edmonton Formation (Maestrichtian), Alberta, Canada: Canadian Jour. Botany, v. 47, no. 6, p. 975-989, Pls. 1-3.

APPENDIX 1: DESCRIPTIONS OF SAMPLE LOCALITIES AND LITHOLOGIES

Sample Number	Locality	Lithology
RC 1971	Claystone bed immediately overlying Smith Coal at Olive (1957) locality 456, SE¼, sec. 14, T. 56 N., R. 77 W.	Brownish-gray claystone
RC 1973	Claystone bed immediately overlying Roland Coal at Dobbin and Barnett	Brownish-gray claystone with

Sample Number	Locality	Lithology
	(1927) locality 378, SW¼, sec. 6, T. 45 N., R. 69 W.	carbonized plant fragments
RC 1978	Claystone bed overlying Coal E, locality same as RC 1973	Light-olive-gray, bentonitic claystone
RC 1977	Basal 2 inches of 1-foot, 8-inch thick Coal E at Dobbin and Barnett (1927) locality 92, SE¼, sec. 32, T. 40 N., R. 70 W.	Black lignite
RC 1964	Claystone bed immediately overlying Dietz Coal at Olive (1957) locality 630, NW¼, sec. 7, T. 56 N., R. 76 W.	Dark-gray claystone
RC 1966	Upper 6 inches of Canyon Coal, Olive (1957) locality 629, NE¼, sec. 7, T. 56 N., R. 76 W.	Black lignite
RC 1803	Channel sample of 8-inch impure lignite, 6 feet above Coal I at Dobbin and Barnett (1927) locality 21, SW¼, sec. 33, T. 39 N., R. 68 W.	Black, impure lignite
RC 1802	Claystone bed immediately overlying RC 1803, at same locality as RC 1803	Light-olive-gray, bentonitic claystone
RC 1807	Claystone bed immediately overlying Coal K, ¼ mile southeast of Dobbin and Barnett (1927) locality 10, approximately 200 feet from south side of road, SW¼, sec. 36, T. 39 N., R. 68 W.	Light-yellow-gray, bentonitic claystone
RC 1806	Channel sample, 30 to 36 inches below top of 5-foot-thick K Coal bed of Dobbin and Barnett (1927)	Black lignite
RC 1822	Claystone bed 5 feet above Coal L, near Dobbin and Barnett (1927) locality 192, east side of road just below top of sandstone-capped escarpment, SW¼, sec. 31, T. 42 N., R. 68 W.	Dark-gray claystone

Sample Number	Locality	Lithology
RC 1819	Channel sample of 18-inch thick Coal P, southeast of Dobbin and Barnett (1927) locality 62, SW¼, sec. 10, T. 40 N., R. 67 W.	Black lignite
RC 1777	Claystone bed immediately overlying Coal R, near Dobbin and Barnett (1927) locality 576, west side of road, NE¼, sec. 10, T. 49 N., R. 68 W.	Brownish-gray claystone with plant fragments
RC 1781	Channel sample of upper 6 inches of Coal R, locality same as RC 1777	Black lignite
RC 1782	Claystone 14 feet below top of Coal R, locality same as RC 1777	Medium-dark-gray claystone
RC 1786	6-inch channel sample from middle of 3-foot-thick Coal S bed, south side of Coal Draw, in vicinity of Dorf's Paleocene leaf locality, SE¼, sec. 33, T. 48 N., R. 67 W.	Black lignite
RC 1850 to RC 1826	Clemens' (1960) measured section B-B′; top of section at top of northeast facing escarpment on the J. N. Peterson Ranch, NE¼, sec. 8, T. 38 N., R. 65 W. (*see* Fig. 4 for stratigraphic position of samples in Clemens' section). Lithologies as follows:	
RC 1850	Claystone bed at base of Clemens' 2-foot gypsiferous shale bed, 22 to 24 feet from top of section	Light-olive-gray claystone with plant fragments
RC 1848C	Claystone at base of Clemens' 6-foot gypsiferous black shale bed, 24.5 to 30.5 feet from top of section	Medium-gray claystone with carbonized leaf fragments
RC 1846	Claystone at base of Clemens' 1-foot lignite bed, 52 to 53 feet from top of section	Light-gray bentonitic claystone with carbonized plant fragments

Sample Number	Locality	Lithology
RC 1843	Claystone immediately overlying Clemens' 2-foot lignite bed, 57 feet 9 inches from top of section	Pale-brown claystone with plant fragments
RC 1840	Channel sample of basal 6 inches of Clemens' 2-foot lignite bed, 57 to 59 feet from top of section	Black lignite
RC 1839	Channel sample of 1-inch underclay of Clemens' 2-foot lignite bed, 57 to 59 feet from top of section	Dark-gray silty claystone with carbonized leaf fragments
RC 1836	Channel sample from basal 1 foot of Clemens' 2-foot black shale bed, 64 to 66 feet from top of section	Medium-light-gray claystone with plant fragments
RC 1834	Channel sample from 2-inch lignite at base of Clemens' 2-foot black shale bed, 64 to 66 feet from top of section	Black lignite
RC 1833B	Claystone from Clemens' 1-foot brown carbonaceous shale bed, 78 to 79 feet from top of section	Brownish-gray claystone with plant fragments
RC 1833A	Claystone from 1-foot-thick bed underlying RC 1833B	Medium-light-gray claystone with plant fragments
RC 1832B	Channel sample, 4-inch lignite bed, 2 feet below top of Clemens' 14-foot alternating dark-brown and black shale bed, 84 to 98 feet from top of section	Black lignite
RC 1832A	Claystone 2.5 feet below top of Clemens' 14-foot alternating dark-brown and black shale bed, 84 to 98 feet from top of section	Medium-light-gray silty claystone with plant fragments
RC 1831B	Claystone, 4.5 feet below top of Clemens' 14-foot alternating dark-brown and black shale bed, 84 to 98 feet from top of section	Brownish-gray claystone with plant fragments
RC 1831A	Channel sample of 7-inch lignite, 5 feet below top of Clemens' 14-foot	Brownish-black lignite

Sample Number	Locality	Lithology
	alternating dark-brown and black shale bed, 84 to 98 feet from top of section	
RC 1830A	Claystone, 9 feet below top of Clemens' 14-foot alternating dark-brown and black shale bed, 84 to 98 feet from top of section	Medium-light-gray claystone with plant fragments
RC 1829	Claystone 3.5 feet below top of Clemens' 14-foot dark-gray shale bed, 111 to 125 feet from top of section	Medium-light-gray silty claystone
RC 1828	Claystone 8 feet below top of Clemens' 14-foot dark-gray shale bed, 111 to 125 feet from top of section	Medium-light-gray silty claystone with carbonized plant fragments
RC 1826	Claystone 4 feet below top of Clemens' 11-foot dark-gray shale bed, 145 to 156 feet below top of section	Brownish-gray claystone with plant fragments
RC 1645	Dark carbonaceous claystone from Clemens' vertebrate locality V5620, sample from approximately 1000 feet north of Clemens' monument marker, on the west side of gulley and approximately 175 yards north of fence crossing gulley, NE¼, sec. 9, T. 36 N., R. 64 W.	Grayish-black, silty, carbonaceous claystone
RC 1636	Clayey siltstone from Dorf's Lance fossil leaf locality P 3858, 10 feet above base of stream bed, on steep eastern bank of Lance Creek, NE¼, sec. 24, T. 37 N., R. 65 W.	Light-olive-gray clayey siltstone with carbonaceous plant fragments
RC 1824 and RC 1825	Shelly siltstone from Dorf's Lance fossil leaf locality P 3857, middle steep embankment along eastern bank of Lance Creek, from bed of thin-shelled, fragile fresh-water invertebrates, NE¼, sec. 25, T. 37 N., R. 65 W.	Yellow-gray siltstone with abundant thin-shelled, fragile fresh-water invertebrates

Palynology of the Cretaceous-Tertiary Boundary in the Northern Rocky Mountain and Mississippi Embayment Regions

ROBERT H. TSCHUDY
U.S. Geological Survey, Denver, Colorado

ABSTRACT

In both the Rocky Mountain and the Mississippi Embayment regions it is often difficult to distinguish Upper Cretaceous from lower Tertiary rocks on the basis of physical characteristics; the transition can be recognized with relative ease, however, on the basis of abrupt qualitative changes in plant microfossils. Many Cretaceous species vanish and new species appear in the Paleocene. In both regions the Paleocene yields fewer species than does the underlying Cretaceous.

A pronounced difference exists between the Late Cretaceous pollen and spore floras of the two regions. The spore-pollen assemblage from the Rocky Mountains is partially characterized by the presence of *Aquilapollenites, Proteacidites, Wodehouseia,* and an unnamed species of *Tricolpites.* These taxa are not found in the Late Cretaceous of the Mississippi Embayment region. The embayment region yields *Rugubivesiculites,* an unnamed species of *Araucariacites*, and several genera belonging to the Normapolles group that are not found in the Rocky Mountain region.

The Mississippi Embayment Cretaceous and Paleocene pollen floras show greater similarities to pollen floras from Europe than to those from the northern Rocky Mountains. The Rocky Mountain pollen floras exhibit closer similarities to the Alaskan and Siberian floras. It is suggested that the floral dissimilarity between the Mississippi Embayment and the northern Rocky Mountain regions may be accounted for by the separation of the Rocky Mountain province from the Mississippi Embayment province during Late Cretaceous time by the great north-trending Cretaceous epeiric sea.

It is further suggested that the floral changes across the Cretaceous-Tertiary transition may have been caused by slight climatic changes brought about by uplift or by withdrawal of the tempering influence of the epeiric sea.

CONTENTS

INTRODUCTION

Consistent recognition of the Cretaceous-Tertiary boundary in both the Mississippi Embayment and the Rocky Mountain regions has proven difficult. Animal fossils that differentiate the Cretaceous from the Tertiary do not occur with sufficient consistency to be used for recognizing the boundary. This difficulty occurs because animal fossils are chiefly confined to marine rocks and in many instances nonmarine beds lie on either side of the boundary. It is also not possible in either area to recognize the boundary on the basis of a pronounced change in lithology.

The present work was initiated because of the need for a reliable means of identifying the Cretaceous-Tertiary boundary. This investigation was stimulated by the continuing interest of paleontologists in the general floral and faunal changes occurring during the passage from the Cretaceous to the Tertiary.

The prime objective of this investigation was to determine the value of palynomorphs to consistently delineate the Cretaceous-

Tertiary boundary in both the Mississippi Embayment and the Rocky Mountain regions. To accomplish this objective it was necessary to collect control samples from the previously defined Upper Cretaceous and lower Tertiary rocks of the two regions, prepare these samples, and record any changes in the palynomorph assemblages derived from the samples. The characteristic palynomorphs occurring in the defined Upper Cretaceous and lower Tertiary rocks of both regions have been established, and these document the floral changes that occur at or near the defined boundaries. It is now feasible to consistently distinguish the defined Cretaceous-Tertiary boundary in the Mississippi Embayment and Rocky Mountain regions on the basis of palynomorph content. Although the bulk of the data presented is from Montana and Kentucky, a sufficient number of additional samples from the northern Rocky Mountains and the upper Mississippi Embayment have been examined to give conclusions a regional validity.

In the Hell Creek area of Montana, dinosaurs have been found only below the lowest definite lignite. Barnum Brown (1907, p. 835) said, "It is a most remarkable and significant fact that in no instance has a fragment of dinosaur bones been found in or above the lignite series by any of our party during five years' work in this region." R. W. Brown (1952, p. 92, 1962) gave a complete discussion of the Cretaceous-Tertiary contact controversy and provided a formula for locating this contact. In essence, the basal Paleocene Tullock Member of the Fort Union Formation is recognized by the presence of the first lignite bed above dinosaur bone-bearing rocks of the Hell Creek Formation.

In the northern Mississippi Embayment region no dinosaurs have been found. The Cretaceous-Tertiary contact, established by marine animal fossils, is located between the Paleocene Clayton and the Cretaceous McNairy or Owl Creek Formations. In Kentucky this contact cannot be distinguished if the basal fossiliferous limestone of the Clayton is lacking. As a consequence, rocks near the boundary are mapped under the term Clayton and McNairy Formations undifferentiated (W. W. Olive, 1964, oral commun.).

Sohl (1960, p. 25) reported, "The Owl Creek is not recognizable at all outcrops where the contact between the Cretaceous and Paleocene is exposed . . . and is so similar to the overlying sands of the Clayton Formation that where the basal limestone is not present, the contact is difficult to distinguish."

Furthermore, in neither region is there a pronounced and universally recognizable hiatus marked by an unconformity. The rocks are much the same on either side of the boundary, and where they do not crop out in large exposures they are very difficult to distinguish. The rocks consist mainly of continental sands and clays and contain few animal remains.

HELL CREEK SECTIONS, MONTANA

Collecting Localities

The Hell Creek localities studied are in the northeastern part of Montana, south of the Missouri River and north of Miles City (Fig. 1). Barnum Brown (1907, p. 829), in describing the Hell Creek Formation, said, "These beds, with overlying and underlying deposits are typically exposed on Hell Creek and nearby tributaries of the Missouri River; hence they have been designated the Hell Creek beds." Exposures are more continuous, however, in the

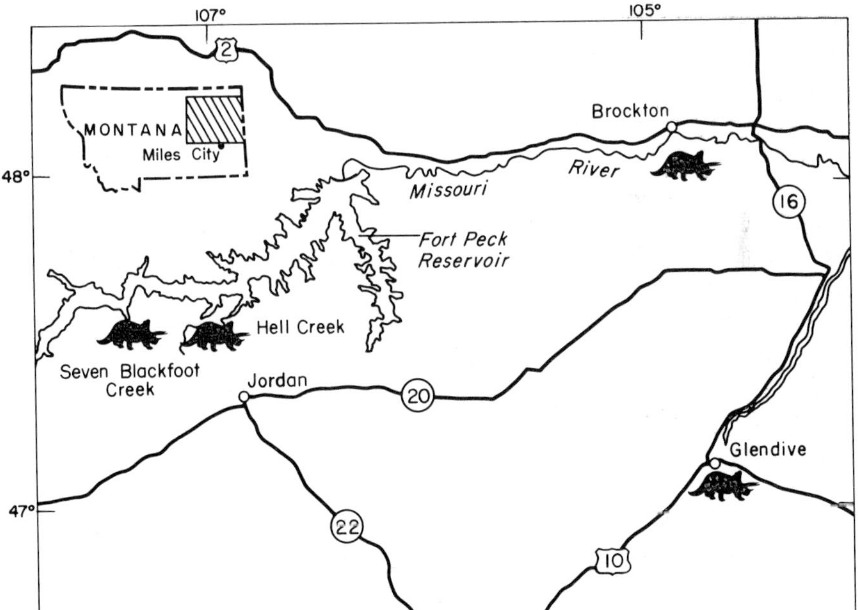

Figure 1. Index map of Hell Creek collecting localities in Montana. Seven Blackfoot Creek, Hell Creek, Brockton, and Glendive.

valley of Seven Blackfoot Creek to the west and are easier to measure and sample than on Hell Creek itself; therefore, the complete section sampled was taken from the valley of Seven Blackfoot Creek.

Measured section on eastern side of Seven Blackfoot Creek, sec. 29, T. 21 N., R. 34 E., Garfield County, Montana

(Section begins near creek bottom in the Colgate Member of the Fox Hills Sandstone and continues upward through the Hell Creek Formation and into the Tullock Member of the Fort Union Formation. Samples at localities D3472-A to D3472-O were collected in 1964 and those at localities D3754-A to D3754-E were collected in 1965.)

	Thickness (feet)	Locality Nos. of pollen samples
Tullock Member of Fort Union Formation:		
Coal, lignitic, very woody	3	D3472-P
Sandstone, light-tan, where silty has faint yellow darker banding	9	
Shale, carbonaceous, black, with some woody lignite at top	2	D3472-O
Sandstone, light-tan, silty with some shale partings	8	
Shale, thin, black, carbonaceous	1/6	D3472-N
Sandstone, massive, light-gray with faint ferruginous streaks; weathers in characteristic fluted form with many protruding log-sized concretions	37	
Coal, thin, lignitic with recognizable wood-fragments	1/3	D3472-L D3472-M
Siltstone and shale, light-tan to gray	8	
Coal, lignitic, and carbonaceous shale; much woody material at top	1	D3472-K D3754-D
Conformable contact		
Hell Creek Formation:		
Shale and siltstone, gray, slightly bentonitic	16	

Shale, dark-gray to black, friable, highly carbonaceous	1/3	D3472-J D3754-E
Sandstone, "salt and pepper" light gray; weathers in fluted form	30	
Siltstone, brown, carbonaceous, with shale partings	1/6	D3472-I
Sandstone and siltstone, alternating light-gray sandstone and light-gray-green bentonitic siltstone and shale	30	
Shale, gray, weathers purple, bentonitic	1/6	D3472-H
Siltstone and shale, light-gray to tan	15	
Shale, dark-gray, weathers purple, bentonitic (first dark-colored layer above massive sandstone)	1/6	D3472-G
Sandstone, massive, tan, well-cemented	135	
Shale, gray to black, bentonitic	1/4	D3472-F
Siltstone and shale, light-gray, banded	29	
Shale, black, carbonaceous	1/6	D3472-E
Sandstone and siltstone, light-gray to tan, banded	39	
Shale and siltstone, black and gray, bentonitic, carbonaceous	1	D3472-D
Shale and siltstone, light-gray and tan, banded	24	
Shale, gray, bentonitic (first dark layer above sandstone)	2/3	D3472-C
Sandstone, very friable, tan, with brown organic streaks	15	D3754-A D3472-B
Total Hell Creek Formation	336	

Conformable contact

Colgate Member of Fox Hills Sandstone:

Sandstone, white, poorly cemented	1/3	
Shale, carbonaceous, black	1/3	D3754-C
Sandstone, white, poorly cemented	11 1/3	
Shale, gray, base of Colgate	1/3	D3754-B

Measured section through the Cretaceous-Tertiary boundary on Hell Creek drainage, in gully 0.25 mile east of Brownie Butte, secs. 31 and 32, T. 21 N., R. 37 E., Garfield County, Montana

	Thickness (feet)	Locality Nos. of pollen samples
Tullock Member of Fort Union Formation:		
Sandstone, light-yellow, well-cemented with faint darker-yellow banding	50+	
Coal, lignitic, very woody	3	D3014-I
Shale, highly carbonaceous	1/3	
Shale, gray, slightly bentonitic	3	
Shale, laminated dark-gray with plant fragments	4	
Shale, gray, slightly bentonitic	3	
Coal, lignitic, woody, oxidized on surface to sooty consistency	3	D3014-H
Shale and siltstone, banded buff and gray	13	
Shale, dark-gray with plant fragments	3	
Shale, gray-brown, slightly carbonaceous	3	D3014-G
Shale, laminated light-gray-brown, bentonitic	1	
Coal, lignitic, silty, with much woody material	1	D3014-F
Shale, light-gray, slightly bentonitic	13	
Coal, lignitic, mostly broken leaves	1/12	D3014-E
Shale, light-gray, slightly bentonitic	3	
Sandstone, light-tan, faintly banded; minor ledge of 1-inch ferruginous concretions near top	13	
Shale, gray-brown, friable, carbonaceous	1	D3014-D
Coal, lignitic, with much silt and shale	1	D3014-C
Sandstone, white, friable, with minor layers of shale	5	
Shale, dark-gray to black, with plant fragments	3/4	
Coal, lignitic, very soft, with loosely consolidated plant fragments	1/6	D3014-B
Conformable contact		

Hell Creek Formation:		
Sandstone, gray-green, fine-grained; occasional layers of gray shale that weathers purple	27	
Sandstone, tan, laminated, with concretions that form a ledge	3	
Sandstone, light-gray-green, cross-bedded, poorly consolidated	10	
Shale, gray, silty; contains plant stems; dinosaur bones found at this level	1	D3014-A
Covered		

Measured section through the Cretaceous-Tertiary boundary south-southwest of Brockton, NE¼ sec. 4, T. 26 N., R. 52 E., Richland County, Montana

(South of Missouri River, along dirt road; section crosses Mortarstone Bluff)

	Thickness (feet)	Locality Nos. of pollen samples
Tullock Member of Fort Union Formation:		
Coal, lignitic with 2-foot clay parting, parting narrows to inches within 50 feet	3	D3625-G
Sandstone, tan, silty	5 1/2	
Coal, lignitic, hard, brittle, very woody	1 1/2	D3625-F
Sandstone and siltstone, light-gray, friable, with gray shale layers; a few small concretions in lower part	25	
Shale, dark-gray to brown, carbonaceous, some woody lignite at base	1 1/6	D3625-E
Sandstone, whitish, cross-bedded, friable, with large tan concretions	19	
Shale, dark-gray to black, carbonaceous, with definite lignite at center	2	D3516-B
Shale, silty, gray, bentonitic	17 1/2	
Coal, lignitic, shaly layers near top	3	D3516-A
Conformable contact		

Hell Creek Formation:		
Shale, gray, bentonitic, and friable gray siltstone	5	
Shale and siltstone, gray with darker carbonaceous partings; weathers darker than underlying sandstone	10	D3625-D
Sandstone, gray, friable, with gray bentonitic shale partings	20 1/2	
Shale, dark-gray, carbonaceous	1/2	D3625-C
Sandstone, light-gray, friable, weathers tan; some tan sandstone concretions	57	
Shale, dark-gray, bentonitic, carbonaceous	2	D3625-B
Sandstone, friable, tan, weathers slightly purple	11	
Shale, dark-gray to black, carbonaceous, slight amount of lignite in center	2/3	
(Remainder of Hell Creek covered at this locality)		

Measured section (reference section) near Glendive, secs. 2 and 14, T. 141 N., R. 55 E., Dawson County, Montana

(Section begins in bluffs east of intersection of railroad crossing and dirt road, sec. 14. Localities D3690-A to D3690-H are at the type locality of the Colgate Member of the Fox Hills sandstone 3.7 miles south of the intersection of Towne and Merrill Streets. Localities D3730-A to D3730-L are in bluffs in sec. 2, directly east of the city of Glendive. Water reservoir and radio towers are on this bluff.)

	Thickness (feet)	Locality Nos. of pollen samples
Tullock Member of Fort Union Formation:		
Sandstone, light-tan, well-cemented, extending to top of bluff	30 +	
Coal, lignitic, blocky, containing much woody material	1	D3730-L
Sandstone, light-tan, well-cemented, with faint brown banding	22	
Coal, lignitic, and carbonaceous shale	1	D3730-K

Sandstone, light-tan, friable, with faint brown banding	18	
Coal, lignitic, very woody	2	D3730-J
Conformable contact		
Hell Creek Formation:		
Siltstone, light-gray-green, slightly bentonitic	11 1/2	
Shale and siltstone, dark-gray to black, carbonaceous	2	D3730-I
Shale and siltstone, light-gray-green; tan ferruginous concretions at top	7	
Shale and siltstone, brown, carbonaceous, slightly bentonitic	2 1/2	D3730-H
Shale and siltstone, light-gray-green, bentonitic, with light-yellow sandstone concretions	14	
Shale, black, carbonaceous, bentonitic	3	D3730-G
Shale, gray-green, bentonitic	6	
Sandstone, "salt and pepper" light-gray, poorly cemented near top, with light-tan layer	11 1/3	
Shale, dark-gray to brown, carbonaceous	2/3	D3730-F
Sandstone, "salt and pepper" light-gray, poorly cemented	5	
Shale, thin, dark-gray, carbonaceous	1/6	D3730-E
Sandstone, "salt and pepper" light-gray, soft, poorly cemented, with log-sized tan concretions	39	
Shale, light-gray-green, bentonitic	5	
Shale, dark-gray to black, carbonaceous, bentonitic	1 1/4	D3730-D
Shale and siltstone, banded, bentonitic, light-gray-green and light-gray	24	
Shale and siltstone, highly carbonaceous, black	3/4	D3730-C
Sandstone, friable, light-tan, with thin layers of light-gray-green bentonitic shale	12	
Shale, dark-gray-green, bentonitic	7	

Shale, gray to black, carbonaceous, bentonitic	4	D3730-B
Sandstone, light-gray to tan, friable	7	
Shale, black, carbonaceous, bentonitic	1/2	D3690-H
Sandstone, well-cemented, light-gray, with bands of dark-gray-green bentonitic shale	11 1/2	
Shale, dark-gray to black, carbonaceous, bentonitic; dinosaur bones found at this level	1/2	D3730-A
Sandstone, well-cemented, light-gray, with bands of dark-gray bentonitic shale	5	
Siltstone, light-gray, bentonitic	1/2	D3690-G
Sandstone, poorly cemented, tan, with thin layers of gray to light-purple shale	11	
Sandstone, friable, cross-bedded in places; large log-sized concretions at top	58	
Shale, black, carbonaceous, bentonitic	1	D3690-F
Sandstone, friable, light-gray; some darker gray thin beds of shale	10	
Shale, black, carbonaceous	1	D3690-E
Shale, dark-gray to black, carbonaceous, bentonitic	13	
Shale, black, carbonaceous	1	D3690-D
Shale, variegated gray and black, highly bentonitic	5	
Sandstone, white, poorly cemented	11	
Shale, purplish to dark-gray, bentonitic	3	D3690-C
Total Hell Creek Formation	295	

Conformable contact

	Thickness (feet)	Locality Nos. of pollen samples
Colgate Member of Fox Hills Sandstone:		
Sandstone, massive, white, friable	13	
Coal, lignitic, interbedded with black carbonaceous shale	0-2	D3690-B
Sandstone, massive, white, friable; locally eroded in a fluted pattern	26	

Shale, dark-brown, carbonaceous	1/4	D3690-A
Sandstone, white, friable	2	
Covered		

Additional samples from another Cretaceous-Paleocene section on the eastern rim of Seven Blackfoot Creek were examined, as were samples from the basal part of the Hell Creek Formation, which were collected near the Hell Creek Recreation Camp. These samples are not listed as they are not specifically pertinent to the discussion.

Stratigraphy

The Hell Creek Formation of Late Cretaceous age lies between the Colgate Member of the Fox Hills Sandstone and the Paleocene Tullock Member of the Fort Union Formation. In many areas the Hell Creek directly overlies the main body of the Fox Hills and the Colgate Member is not recognized. The contact between the Hell Creek Formation and the underlying white Colgate Member of the Fox Hills Sandstone is particularly conspicuous in the Seven Blackfoot Creek drainage and is clearly displayed at the type locality of the Colgate at Colgate Station southwest of Glendive, Montana. R. W. Brown (1939, p. 240) referred to the sandstone at the type locality of the Colgate as follows: "Although this band of white sandstone appears to be a distinct unit, it is in reality not so sharply demarked from the preceding and succeeding strata as the conspicuous difference in color might at first suggest. Neither the base nor the top of the sandstone is separated from adjacent strata by other than minor local erosional discordancies." In Seven Blackfoot Creek the contact between the Fox Hills and the Hell Creek is disconformable (Jensen and Varnes, 1964, p. F-21) and is channeled in many places. In areas in eastern Montana where the Colgate is present, the base of the Hell Creek Formation is easily recognized.

The Hell Creek Formation is made up of alternating discontinuous sandstone and mostly dark-colored bentonitic clays and siltstones. Some of the sands are poorly cemented and weather in a characteristic fluted form. Large log-sized concretions in the sandstones are characteristic of the formation. The clays and siltstone are light tan to brown, with streaks or layers of darker organic material. Where the organic material is dominant, the beds take on a dark gray-blue "somber" color that is the basis for the reference to the Hell Creek Formation as the "somber beds."

The contact between the Hell Creek Formation (Pl. 1) and the overlying Tullock Member of the Fort Union Formation is conformable and gradational.

The Fort Union originally was subdivided into light and dark members. "These, beginning at the base are the light Tullock, dark Lebo, light Tongue River and dark Sentinel Butte" (R. W. Brown, 1952, p. 91). Outside of the original local areas this succession is difficult or impossible to trace. "For example, the Tullock and Lebo southeastward from Miles City, Mont. become gradually so similar in color and lithologic features that they were lumped as the Ludlow member" (R. W. Brown, 1952, p. 91). The relationship of the Hell Creek Formation to the underlying and overlying formations is shown on Figure 2.

Jeletzky (1960, p. 27) has indicated that the contacts of the Pierre, Fox Hills, Hell Creek, and Fort Union Formations are conformable and, for the most part, gradational. Apparent unconformities are local and are caused by channeling, cross-bedding,

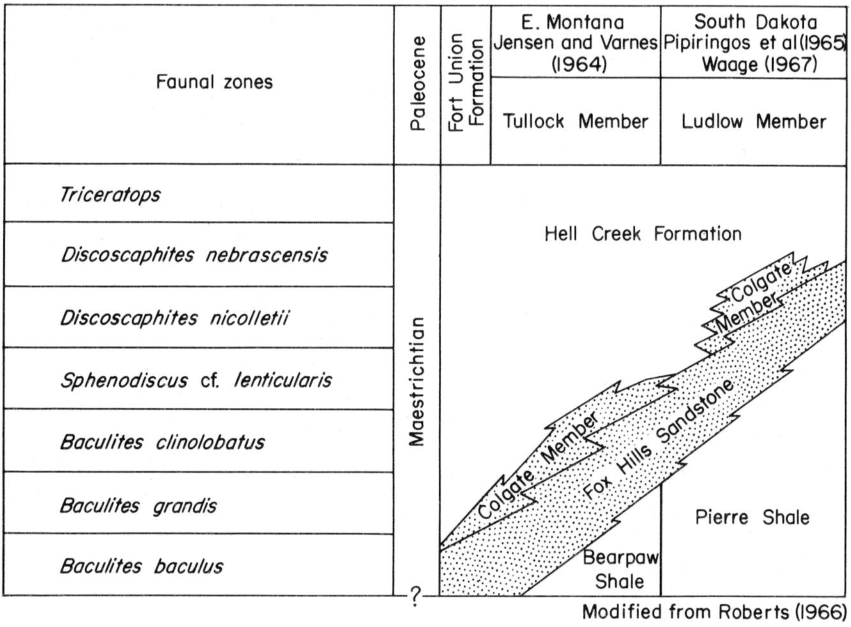

Figure 2. Correlation chart of Upper Cretaceous and lower Paleocene rocks in eastern Montana and South Dakota.

HELL CREEK LITHOTYPES EXPOSED NEAR GLENDIVE, MONTANA

Sugary fluted sandstone at top, banded drab "somber" clay and siltstone below.

TSCHUDY, PLATE 1
Geological Society of America Special Paper 127

or (rarely) faulting. Dating of these formations is discussed in detail by Jeletzky (1960, 1962) with the conclusion that the Hell Creek Formation is of Maestrichtian age and the lower part of the Fort Union is early Paleocene in age. A potassium-argon date for the early Paleocene near Fort Peck, Montana (the Big Dirty coal seam), is given as 63 m.y. by Folinsbee and others (1963, p. 77-78).

Seven Blackfoot Creek and Glendive Sections

Samples for pollen analysis were taken throughout the Hell Creek section, including the underlying Colgate Member of the Fox Hills Sandstone and the overlying Paleocene Tullock Member of the Fort Union Formation at Seven Blackfoot Creek and Glendive, Montana. The upper part of the Hell Creek sections, which include the Cretaceous-Paleocene transition, were also sampled on Hell Creek, on the eastern rim of Seven Blackfoot Creek, and south of Brockton, Montana. These localities are shown on Figure 1.

The section on Seven Blackfoot Creek was measured by Jacob's staff from the top of the Colgate Member of the Fox Hills Sandstone to the first definite lignitic horizon (Pl. 2). This section consisted of 336 feet of alternating sands and bentonitic shales and clays. The measured thickness compares favorably with the 319 feet reported by Barnum Brown (1907). Jensen and Varnes (1964) erroneously attribute to Brown a report of 510 feet of Hell Creek beds at the type locality on Hell Creek. Collier and Knechtel (1939) reported a thickness of 135 feet for the Hell Creek Formation in McCone County, Montana, just east of Seven Blackfoot Creek and Hell Creek. Jensen and Varnes (1964, p. F-18) reported that a written communication from R. B. Colton stated that the Hell Creek was "from 220 to 280 feet in total thickness in the bluffs south of the Missouri River, about 45 miles east of Frazer between Poplar and Brockton."

Most of the massive sands were devoid of organic matter, but the shales contained organic darker bands, which were sampled. Nine productive samples were obtained from the Hell Creek sequence, two from the underlying Colgate, and five from the overlying Tullock.

Examination of samples across the defined Cretaceous-Tertiary boundary (below and above the first definite lignite) demonstrated that a marked palynologic change occurs at the level of the first definite lignite (not lignitic shale but definite ignitable lignite).

EXPOSURE OF CRETACEOUS-TERTIARY BOUNDARY IN SEVEN BLACKFOOT CREEK

Base of coal horizon about the middle of the picture marks the beginning of Paleocene deposition.

TSCHUDY, PLATE 2
Geological Society of America Special Paper 127

CRETACEOUS AND TERTIARY STRATA EXPOSED NEAR GLENDIVE, MONTANA

Arrow points to coal bed at base of Tullock Formation. Hell Creek Formation below.

In a section measured at Glendive, Montana, the Hell Creek Formation between the top of the Colgate and the first definite lignite bed of the Tullock (Pl. 3) is about 295 feet thick. Thirteen productive Hell Creek samples were obtained from this section together with two productive samples from the underlying Colgate and two from the overlying Tullock Member of the Fort Union Formation.

The Seven Blackfoot Creek and Glendive sections yielded nearly the same sequence of fossils. Selected groups of genera and species from these sections were plotted in relation to their occurrences in the rocks. In both sections some groups, including *Pemphixipollenites, Pterocarya,* and *Schizosporis complexus* Stanley, were limited to the lower part of the Hell Creek Formation. Groups such as *Aquilapollenites, Proteacidites,* and *Tricolpites interangulus* Newman ranged throughout the Cretaceous sections but were not found in the overlying Tullock. A species of *Polycolpites* and a new palynomorph genus were limited to the upper parts of both Hell Creek sections. Several genera and species ranged throughout the Hell Creek and basal Tullock, and many species occurred in such low frequencies that their ranges could not be determined with any degree of confidence. In both sections the Cretaceous-Tertiary boundary, coinciding with the boundary as defined by R. W. Brown (1952, p. 92), was recognized by the disappearance of many Cretaceous species and the appearance of species of *Ostryoipollenites, Tricolpopollenites,* and *Triporites.* The distribution of these groups is shown on Figures 3 and 4.

The relationship of the two sections is shown on Figure 5. The correlation horizon separating the lower from the upper part of the Hell Creek appears to migrate upsection toward the east (Glendive). This may indicate that the Glendive section is somewhat younger than the Seven Blackfoot Creek section. This apparent difference in the two sections may be spurious, however. The thick sandstone layers are conspicuously lenticular, as are the intervening more somber organic clay beds. The lower stratigraphic position of the horizon in the Seven Blackfoot Creek section may be a reflection of lesser deposition of sands in the basal part and a greater thickness of sand in the middle part.

Characteristic Hell Creek and Tullock Species

Many of the Hell Creek and Tullock species have not yet been named. Rather than include the complete flora found, I have shown

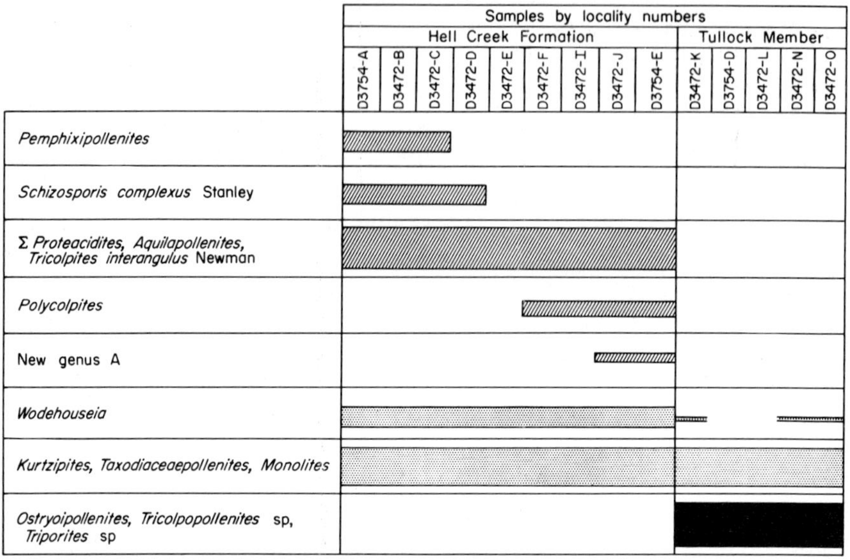

Figure 3. Stratigraphic occurrence of selected groups of palynomorphs from the Seven Blackfoot Creek section. Width of bars represents a rough index of abundance.

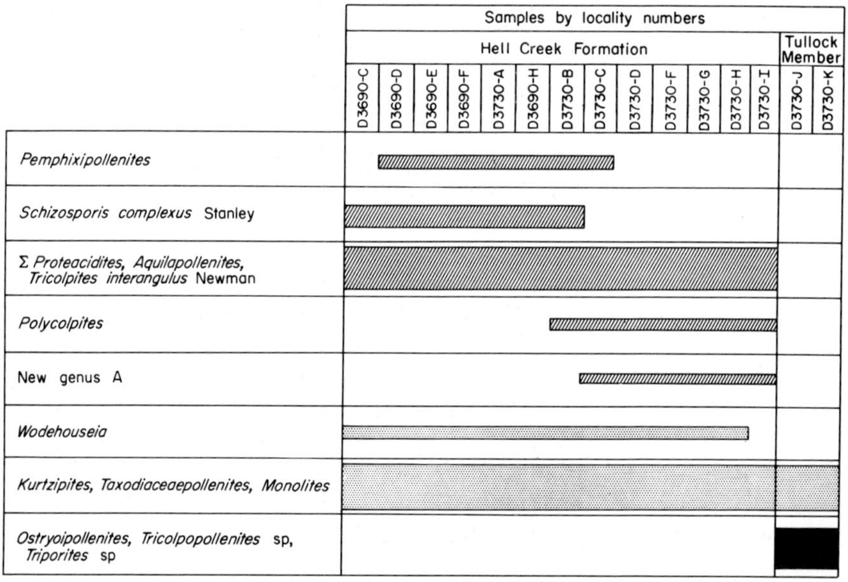

Figure 4. Stratigraphic occurrence of selected groups of palynomorphs from the Glendive section. Width of bars represents a rough index of abundance.

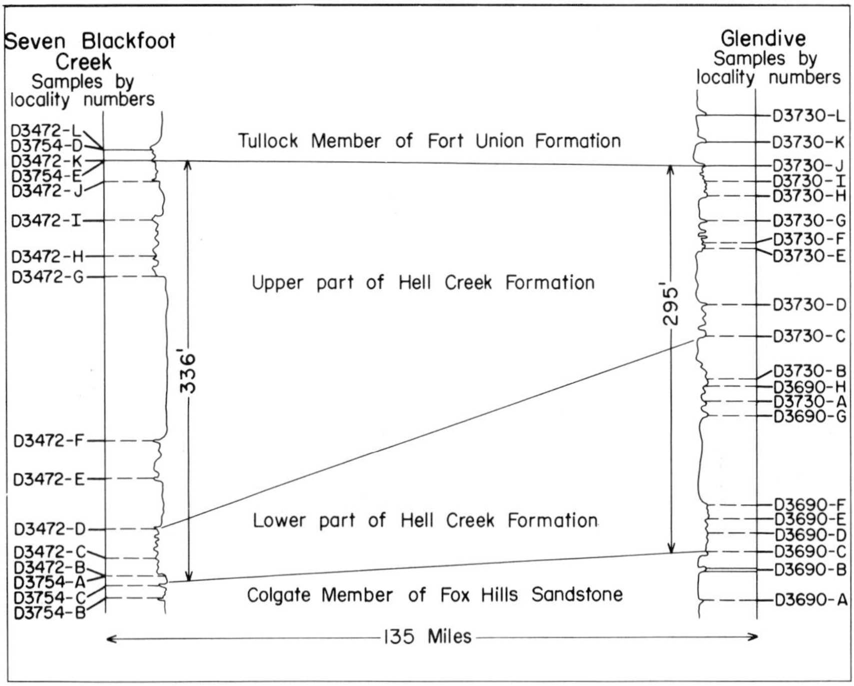

Figure 5. Correlation horizons established between the Seven Blackfoot Creek and Glendive sections.

the most characteristic and commonly occurring species on Plate 4. Specimens shown above the line are from the Tullock Member of the Fort Union Formation; those below are from the Hell Creek Formation.

MISSISSIPPI EMBAYMENT SAMPLES

Collecting Localities

The principal localities from which control and Cretaceous-Tertiary boundary samples were taken are all in the upper part of the Mississippi Embayment region in northern Mississippi and in Tennessee, Missouri, and Kentucky (Fig. 6). No exposures are known that include all of the uppermost Cretaceous and lowermost Tertiary rocks in the Mississippi Embayment region. In contrast to the excellent exposures in the badlands of eastern

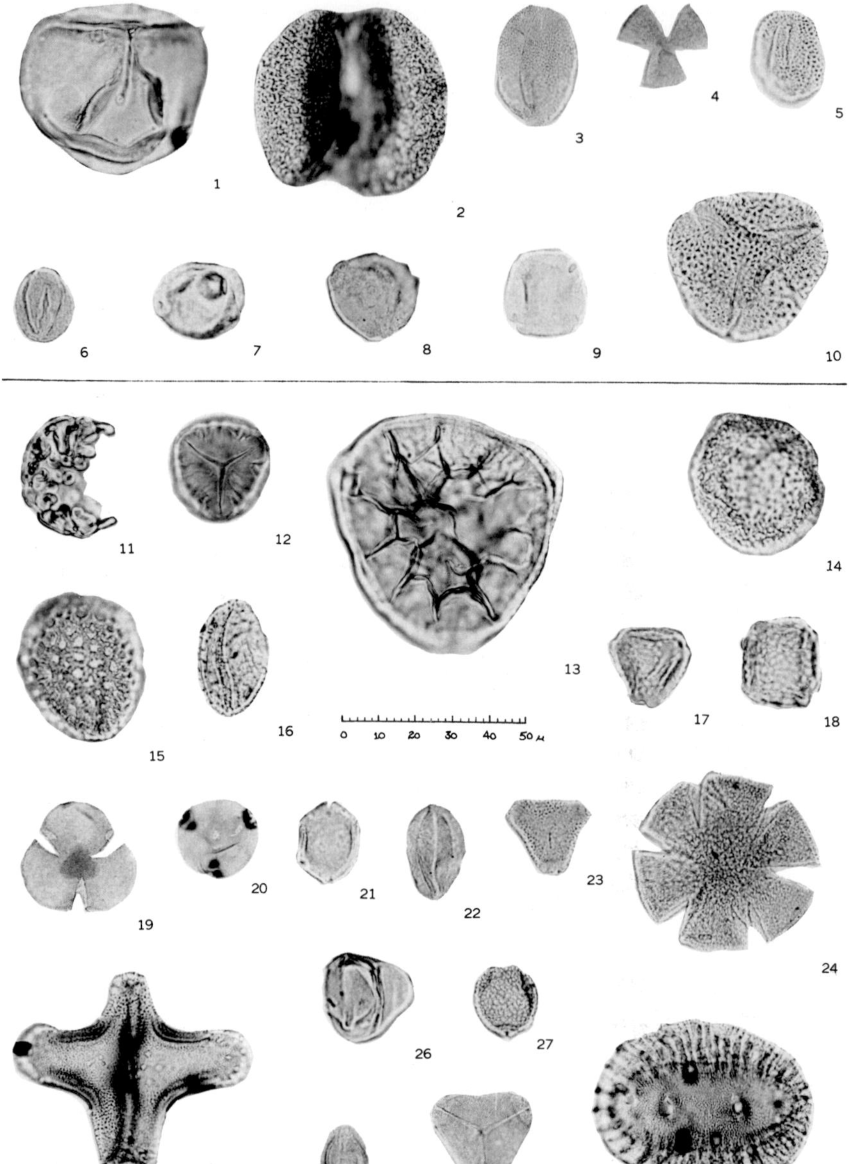

CHARACTERISTIC SPECIES OF TULLOCK MEMBER OF FORT UNION FORMATION AND HELL CREEK FORMATION, MONTANA

TSCHUDY, PLATE 4
Geological Society of America Special Paper 127

PLATE 4. CHARACTERISTIC SPECIES OF TULLOCK MEMBER OF FORT UNION FORMATION AND HELL CREEK FORMATION, MONTANA

All figures ×500. Figures 1-10 are Tullock Member. Figures 11-30 (below line) are Hell Creek Formation.

Figure
1. *Leiotriletes* sp.
2. cf. *Pinus semicircularis* Stanley.
3. *Monosulcites* sp.
4. *Tricolpopollenites* sp.
5. *Tricolpopollenites* sp.
6. *Tricolpopollenites* sp.
7. *Ostryoipollenites* sp.
8. *Myricaceoipollenites* sp.
9. cf. *Ulmipollenites* sp.
10. *Tricolpopollenites.*
11. Incertae sedis new genus A.
12. *Hamulatisporis* sp.
13. *Zlivisporis* cf. *Z. blauensis* Pacltova.
14. *Osmundacidites* sp.
15. *Schizosporis complexus* Stanley.
16. *Monosulcites* sp.
17. 18. *Ulmipollenites* sp.
19. *Tricolpites interangulus* Newman.
20. *Kurtzipites* cf. *K. trispissatus* Anderson.
21. *Pterocarya* cf. *P. levis* Stanley.
22. *Tricolpopollenites* sp.
23. *Proteacidites* cf. *P. retusus* Anderson.
24. *Polycolpites* sp.
25. *Aquilapollenites attenuatus* Funkhouser.
26. *Orbiculapollis* cf. *O. lucidus* Chlonova.
27. *Proteacidites* sp.
28. *Tricolpopollenites* sp.
29. *Pemphixipollenites* sp.
30. *Wodehouseia spinata* Stanley.

Figure 1. Exposure in Seven Blackfoot Creek. Arrow indicates coal horizon. Hell Creek Formation below, Tullock Member of Fort Union Formation above.

Figure 2. 6.5 foot outcrop of Clayton Formation and McNairy Sand (below line), Calloway Co., Kentucky. Arrow indicates location of Cretaceous-Tertiary boundary. This is the same section that is shown on Figure 9.

COMPARISON OF EXPOSURES OF CRETACEOUS AND TERTIARY ROCKS IN MONTANA AND KENTUCKY

TSCHUDY, PLATE 5
Geological Society of America Special Paper 127

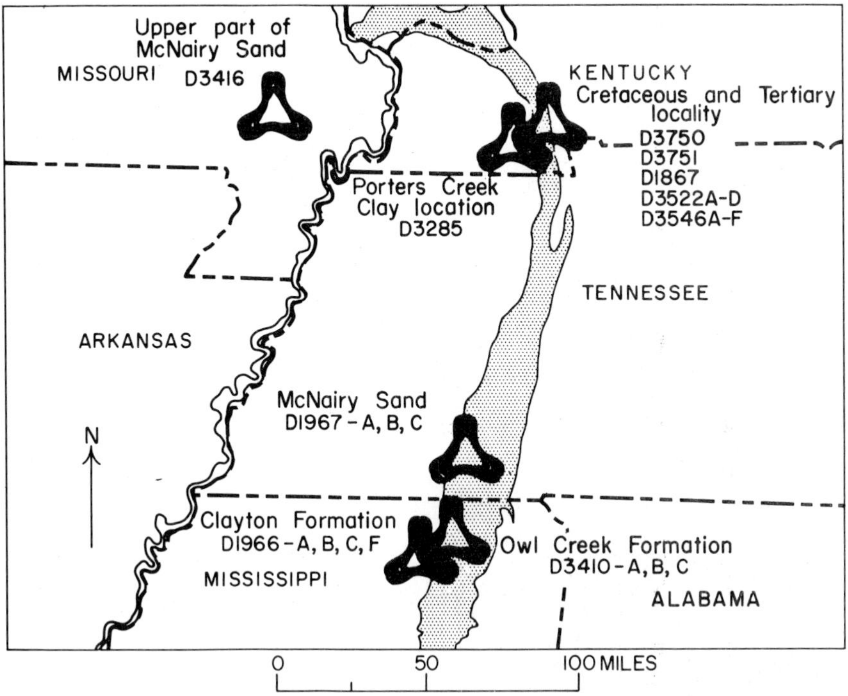

Figure 6. Index map of collecting localities in the upper Mississippi Embayment region. Shaded area indicates Cretaceous outcrops.

Montana (Pl. 5, fig. 1), exposures are limited and discontinuous (Pl. 5, fig. 2). The lack of adequately exposed sequences made it necessary that palynological control samples be obtained from type or well-documented exposures. Late Cretaceous control samples were obtained from the McNairy Sand and from the Owl Creek Formation. Paleocene control was obtained from the Clayton Formation and from the Porters Creek Clay. Later, after control had been established, a series of samples was obtained from an outcrop of the Clayton (?) and McNairy Formations undifferentiated from southwestern Kentucky. The distinctive assemblages found in the Late Cretaceous and Paleocene control samples were such that the Cretaceous-Tertiary boundary could be located within inches on the basis of the change in the palynomorphs recovered from successive samples.

Section on channel of McCullough Fork, 300 feet north of the Kentucky-Tennessee state line, Kentucky State coordinates, south zone, E. 1,263,500 feet, N. 70, 250 feet, Hazel quadrangle, Calloway County, Kentucky

	Thickness (feet)	Locality Nos. of pollen samples
Alluvium		
Unconformity		
Porters Creek Clay:		
Clay, thin-bedded, gray-green, silty, kaolinitic, micaceous, slightly glauconitic with conchoidal fracture	5-8	D3285
Covered		

Section on southwestern wall of tributary to Guyton Creek, along County road 1.5 miles south-southwest of Cotton plant, S½NW¼ sec. 8, T. 6 S., R. 3 E., Tippah County, Mississippi

(*From* Conant, 1942, p. 37)

	Thickness (feet)	Locality Nos. of pollen samples
Clayton Formation:		
Clay, bluish-gray, laminated, and abundant thin laminae of sand; upper part contains purer clay layers, some of which resemble Porters Creek Clay; lower part more carbonaceous and contains much comminuted plant material (corresponding beds exposed in road ditch 0.25 mile to south are cross-bedded)	22	D1966-F, 12-15 feet above base; D1966-D, 5 feet above base; D1966-C, 3 feet above base
Sand, stratified, yellow	1	
Clay, bluish-gray; thin sand laminae; a few poorly preserved plant remains	6	D1966-B, near top
Clay, slightly laminated, glauconitic, bluish-brown and greenish-brown, highly fossiliferous; lower 1 foot indurated, almost the composition of limestone	6	D1966-A, 2 feet above base
Total Clayton Formation	35	

Type section from right bank of Owl Creek, 2.5 miles northeast of Ripley, E½ sec. 7, T. 4 S., R. 4 E., Tippah County, Mississippi

(*From* Stephenson and Monroe, 1940, p. 232)

	Thickness (feet)	Locality Nos. of pollen samples
Clayton Formation		
Unconformity (sharp contact)		
Owl Creek Formation:		
Sand, dark-greenish-gray, fine, compact, tough, massive, micaceous, calcareous, argillaceous, slightly glauconitic; contains many well-preserved shells from many genera and species	22	D3410-C, 22 feet above base D3410-B, 14 feet above base D3410-A, 1 foot above base

Type section from western end of big cut of the Southern Railway east of Dismal Swamp and Tuscumbia River, south-central part of the NW¼ SW¼ of the Chewala quadrangle, McNairy County, Tennessee

Base is approximately the same elevation as the tracks and is 40 feet northwest of railroad crossing. [*See* Stephenson and Monroe, 1940, p. 178.]

	Thickness (feet)	Locality Nos. of pollen samples
McNairy Sand:		
Sand		
Clay, dark-brown-gray; hackly fractures; grading upward into zone of bleached clay several inches thick	5/6	D1967-C
Clay, dark-brown-gray; hackly fractures; thinner bedded than below; with micaceous sand partings	1	D1967-B
Clay, dark-brown-gray; hackly fractures; beds at base as much as 2 to 3 inches thick; sandy partings at top. Comminuted plant fragments on clay partings	1 1/6	D1967-A
Covered		

Section along road in southeast-facing slope of Crowleys Ridge, 0.35 mile northwest of Ardeola, in NW¼NW¼ sec. 10, T. 27 N., R. 11 E., Stoddard County, Missouri

	Thickness (feet)	Locality Nos. of pollen samples
Clayton Formation		
Unconformity		
Owl Creek Formation	11	
Unconformity (sharp contact)		
McNairy Sand:		
Clay and sand, brown, laminated, cross-bedded, with parting planes of muscovite mica; contains abundant comminuted plant remains; borings in upper 8 to 10 inches are filled with sandy matrix from overlying Owl Creek Formation	11	D3416, 6 feet below unconformity

Stratigraphy

The uppermost Cretaceous rocks in this area consist of the McNairy Sand and, where present, the Owl Creek Formation. The Owl Creek is separated from the McNairy by an unconformity. In Mississippi the Owl Creek is restricted, in comparison to the Ripley Formation, and intertongues with the Prairie Bluff Chalk (Sohl, 1960). The basal Paleocene Clayton Formation is also separated from Cretaceous rocks by an unconformity. Paleocene animal fossils, which identify the Clayton, are for the most part obtained from the bed of limestone at the base of the Clayton. Where the limestone is absent, the Clayton, Owl Creek, and McNairy rocks are so similar that the contacts are difficult to recognize. The widespread marine Porters Creek Clay overlies the Clayton in much of the area. The relationships of these formations, except for the Prairie Bluff Chalk, are shown in Figure 7.

Control Samples

Study of the palynomorph assemblages from the control samples obtained from the McNairy, Owl Creek, Clayton, and Porters Creek Formations shows a distinct and easily recognizable change between the uppermost Cretaceous samples and samples from the

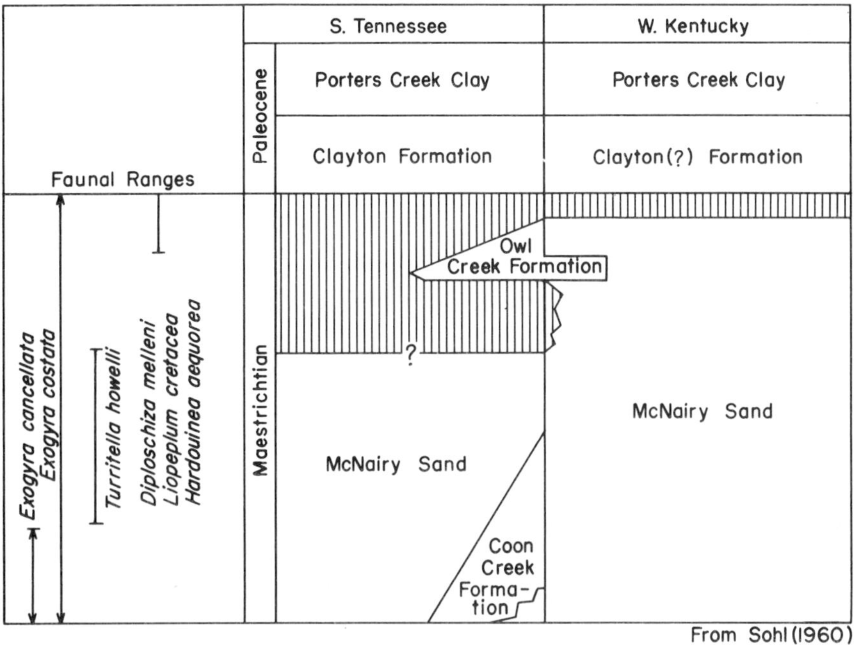

Figure 7. Correlation of Upper Cretaceous and Paleocene rocks in the upper part of the Mississippi Embayment region.

Paleocene Clayton Formation. The changes may be briefly characterized by the facts that some genera, notably *Rugubivesiculites*, and some species of *Araucariacites* are limited to the Cretaceous, and that, above the Cretaceous-Tertiary boundary, new species, notably species of *Momipites* and *Triatriopollenites*, appear. Some species are limited to the McNairy and were not found in the Owl Creek samples, supporting the postulate that a hiatus exists at least in some areas between the McNairy and the Owl Creek.

Some genera, notably members of the Normapolles group of Pflug (1953), are common in the Mississippi Embayment samples and are extremely rare in Rocky Mountain samples. Members of this pollen group are numerous in Upper Cretaceous rocks and persist into the Tertiary.

The relationships of groups of palynomorphs that serve to distinguish the contrasting Cretaceous and Tertiary assemblages are shown on Figure 8. The Porters Creek Clay can commonly be recognized by its distinctive lithology in combination with the presence of palynomorphs of marine origin (acritarchs and dinoflagellates) and also by the absence of *Latipollis* and the presence of a species of *Interporopollenites*.

The Cretaceous and Tertiary control samples yielded an abundant and diverse flora of more than 430 species. These samples have provided a firm basis on which authentic boundary determinations may be made. As additional samples are examined new species are continually being added, augmenting the diverse flora and strengthening the basis for separating Cretaceous from Tertiary rocks.

The Cretaceous-Tertiary Boundary in Kentucky

Because of the general absence of marine fossils and the similarity of the Cretaceous and Tertiary rocks in the boundary area, this boundary is difficult to recognize in Kentucky. A 6-foot-thick outcrop of dark-gray to black organic clay in Calloway County, Kentucky, was sampled, and the upper part yielded a palynomorph assemblage similar to the one found in the control Clayton

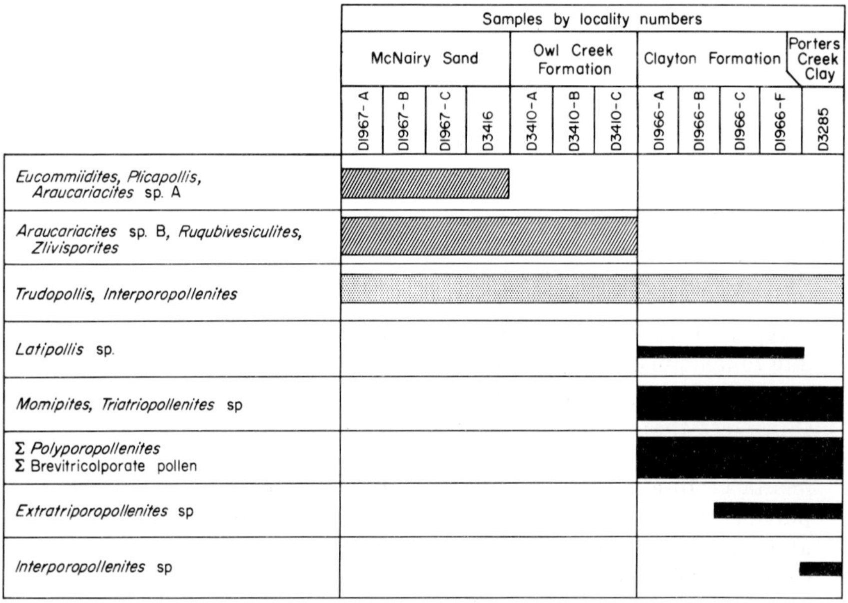

Figure 8. Stratigraphic occurrence of selected groups of palynomorphs from the Upper Cretaceous and Paleocene formations of the upper Mississippi Embayment.

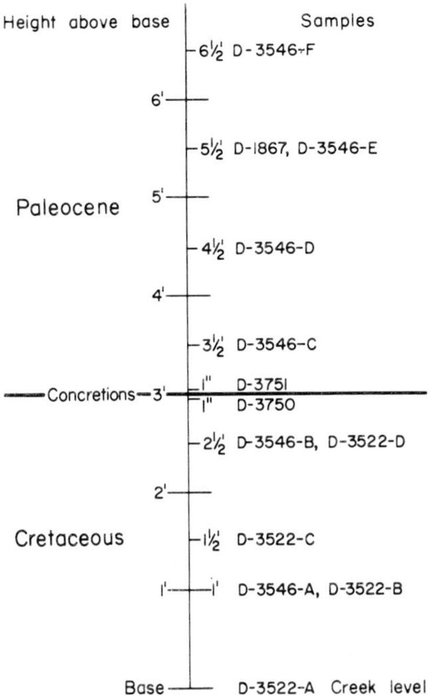

Figure 9. Location of the Cretaceous-Tertiary boundary based on palynomorphs in a 6.5-foot outcrop section in Kentucky. *See* also Plate 5, figure 2.

samples, whereas the lower part yielded an assemblage similar to that found in the McNairy control samples. Subsequent close sampling throughout this 6-foot sequence served to locate the horizon at which the Cretaceous assemblage terminated and the Paleocene assemblage began. The positions of the samples are shown on Figure 9 and the outcrop is shown on Plate 5, figure 2. W. W. Olive (1966, written commun.) supplied the following measured section at this locality.

Measured section in Calloway County, Kentucky, 600 feet southeast of Independence School, Kentucky coordinates, south zone E. 1,268,350 feet, N. 145,100 feet, Hico quadrangle, Kentucky

	Thickness (feet)
Gravel, reddish-brown, with pebbles as much as 5 inches across; sand, orange, silty, fine-grained, micaceous, interlaminated with silty light gray clay, intergrades with unit below. Light color is due to leaching and oxidation	10
Clay, dark-gray to black, sandy, micaceous, contains thin (less than 0.5 inch) interlaminae and lenses of fine white micaceous sand. Also contains fragments of lignitized wood and irregularly shaped platy ferrous concretions as much as 6 inches long and 1.5 inches wide; many are concentrated along Cretaceous-Tertiary boundary	3½
Clay, dark-brownish-gray, sandy, structureless, lignitic	2½
Sand, yellowish-orange, fine to medium	½+

A sample taken 1 inch above the level of concretions yielded a Paleocene palynomorph assemblage, and a sample at the very top of the structureless lignitic clay, 1 inch below the concretionary level, yielded a Cretaceous palynomorph assemblage. Rainwater (1960, p. 100) said that throughout much of the Gulf Coastal Plain there is a great faunal change at the Cretaceous-Tertiary boundary, often accompanied by the presence of abundant phosphatic nodules and glauconite just above the unconformity. Sohl (1960) also mentioned silty pebbles found at the unconformity, and Stephenson and Monroe (1940, p. 223) showed a picture of the Cretaceous-Paleocene unconformity "obscured by weathering and marked only by the line of clay boulders."

Data obtained from type Upper Cretaceous and lower Paleocene areas indicate that a recognizable change in the palynomorph content of the rocks coincides with the change previously observed in animal fossils. For practical purposes, the Cretaceous-Tertiary boundary can be located by the pollen and spores in the rocks. In this Calloway County locality, at least, the boundary also coincides with a horizon of concretions that may represent the unconformity.

Characteristic Upper Cretaceous and Lower Paleocene Species

It is as true in the Mississippi Embayment area as in the Rocky Mountains that many of the palynomorph species that have been found have not yet been named. Rather than all of the palynomorphs, only the most characteristic and commonly occurring species will be mentioned; these are shown on Plate 6. Those above the line are from the Paleocene, and those below the line are from the Upper Cretaceous.

COMPARISON OF ASSEMBLAGES FROM MONTANA AND THE UPPER MISSISSIPPI EMBAYMENT

The two areas differ in three major aspects: (1) the general palynomorph content of the rocks, (2) dominant species present, and (3) representation of the Normapolles group.

In both areas it is now possible to distinguish palynologically the uppermost Cretaceous rocks from the lowermost Tertiary rocks. The differences between essentially synchronous assemblages from the two regions is particularly striking (*compare* Pls. 4 and

	Montana		Miss. Emb.
	Seven Blackfoot Creek	Glendive	
In Cretaceous	132	143	314
In Paleocene	39	45	176
Limited to Cretaceous	119	123	262
Limited to Paleocene	26	25	124
In both	13	20	52
Total species	158	168	438

Figure 10. Comparison of the number of palynomorph species, exclusive of marine forms, found in Cretaceous and Paleocene samples from Montana and the Mississippi Embayment region.

6). The number of palynomorph species in the Cretaceous and Paleocene, exclusive of marine forms, is shown on Figure 10. From these data two facts emerge. One is that the pollen and spore floras of the Mississippi Embayment were more diverse in species than were the Rocky Mountain floras. This diversity is accentuated by the fact that the Mississippi Embayment floras were obtained from fewer samples than were the western floras. The second fact is that the Paleocene floras of both areas are impoverished in relation to the floras of the underlying Cretaceous.

Available plant megafossil evidence, mostly from fossil leaves, parallels the data from pollen and spores. Dorf (1942, p. 105-112) collected 70 plant megafossil species from the type Lance Formation and reported 64 species from the Medicine Bow Formation and 41 species from the Hell Creek Formation, a total of 100 Late Cretaceous species from the three formations. (These formations are Hell Creek equivalents.) This figure may be contrasted with 33 plant species reported from the basal Paleocene Tullock and Ludlow Members of the Fort Union (Dorf, 1942, p. 119). Thus plant megafossil evidence from the northern Rocky Mountain area provides supporting evidence for an impoverished flora in the basal Paleocene of this region.

It is not possible to obtain similar megafossil evidence from the Mississippi Embayment region. Although Berry (1925) records a Ripley flora of 135 species, no more than 66 have been recorded from any single locality. This is a slightly greater number of megafossil species than have been recorded from synchronous rocks in the Rocky Mountain region. This assemblage cannot be compared with Mississippi Embayment Paleocene collections as the latter are almost nonexistent. Berry (1916, p. 9) states that "All the determinable Midway(?) plants have come from a single outcrop near Earle, in Bexar County, Tex., . . ." Only 10 species have been recorded from that locality.

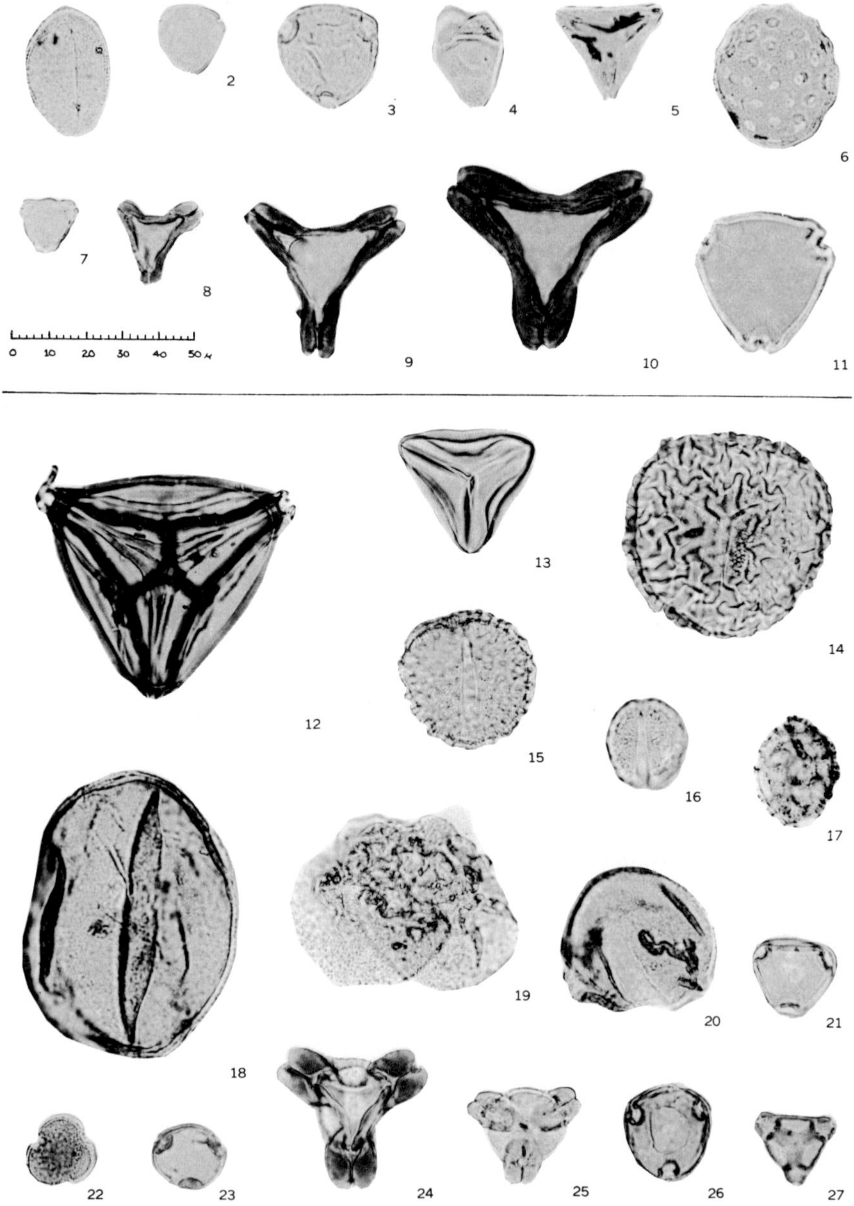

CHARACTERISTIC SPECIES OF UPPER CRETACEOUS AND LOWER PALEOCENE, MISSISSIPPI EMBAYMENT

TSCHUDY, PLATE 6
Geological Society of America Special Paper 127

PLATE 6. CHARACTERISTIC SPECIES OF UPPER CRETACEOUS AND LOWER PALEOCENE, MISSISSIPPI EMBAYMENT

All figures ×500. Figures 1-11 are Paleocene. Figures 12-27 (below line) are Upper Cretaceous.

Figure
1. *Monosulcites* sp.
2. *Momipites tenuipolus* Anderson.
3. *Interporopollenites* sp.
4. *Latipollis* sp.
5. *Sporopollis?* sp.
6. *Polyporopollenites* sp.
7. *Triatriopollenites* sp.
8. *Extratriporopollenites* sp.
9. *Extratriporopollenites* sp.
10. *Extratriporopollenites* sp.
11. *Nyssapollenites* sp.
12. Fern spore, new genus B.
13. *Gleicheniidites* sp.
14. *Hamulatisporis* sp.
15. *Monosulcites (Schizosporis?)* sp.
16. *Monosulcites (Schizosporis?)* sp.
17. *Monosulcites* sp.
18. *Araucariacites* sp. B.
19. *Rugubivesiculites* sp.
20. *Granabivesiculites?* sp.
21. *Interporopollenites* sp.
22. *Triporopollenites* sp.
23. *Triporopollenites* sp.
24. *Oculopollis* sp.
25. *Oculopollis* sp.
26. *Interporopollenites* sp.
27. *Plicapollis* sp.

This paucity of Paleocene leaf localities in the Mississippi Embayment region is probably due to accidents of preservation, in light of the facts that Paleocene leaf floras have been assiduously searched for and that an abundant Paleocene flora existed in the region as shown by the 176 pollen and spore species recovered from basal Paleocene rocks.

The dominant or most commonly occurring species recovered from the two regions are nearly mutually exclusive. This floral provincialism is particularly true of the Cretaceous, less consistently true of the Paleocene. Species commonly found in the two regions are shown on Plates 4 and 6.

The Normapolles group is a group of genera erected by Pflug (1953, p. 95) to include genera possessing, in Pflug's terms, so-called bizarre structural elements. Most, if not all, of these genera are now extinct. They had their origin in the Late Cretaceous and died out in mid-Tertiary time. Pflug included the following 13 genera in this group: *Sporopollis, Plicapollis, Trudopollis, Vacuopollis, Conclavipollis, Extratriporopollenites, Nudopollis, Basopollis, Oculopollis, Papillopollis, Interporopollenites, Tetrapollis,* and *Pentapollis.*

Krutzsch (1959) described several new genera and species from middle European Upper Cretaceous and Tertiary rocks. The following genera erected by Krutzsch can also be included in the Normapolles group: *Latipollis, Monstruosipollis, Turonipollis, Complexiopollis, Pflugipollis, Emscheripollis, Minorpollis,* and *Quedlinburgipollis.*

Species belonging to these genera are extremely rare in the western part of the United States. I have found one species of *Plicapollis* and one species of *Interporopollenites* in the Hell Creek Formation, but no Normapolles species in the Paleocene. Specimens of these species are extremely rare. They are so rare in Hell Creek samples that one can expect to find perhaps one Normapolles specimen in every 10 samples examined. Drugg (1967) reported *Nudopollis* from the lower Tertiary and *Vacuopollis* from the Upper Cretaceous of California. Newman (1964) also found specimens to be very rare in the Mancos and Iles Formations (Campanian) in the Rocky Mountains. He reported *Conclavipollis, Extratriporopollenites, Trudopollis,* and *Sporopollis.*

In striking contrast, Normapolles genera are common in samples from the Mississippi Embayment region. At least 10

Genera	Cretaceous		Paleo-cene	Eocene	
	Kpn	Kon	Tm	Tw	Tc
Extratriporopollenites	2	3	11	4	3
Trudopollis	2	3	2	1	2
Plicapollis	2	3			2
Latipollis	6	1	4	1	
Interporopollenites	1	2	5	1	
Oculopollis	1	2			
Minorpollis	1				
Conclavipollis	1	1			
Sporopollis	1		1?	1?	
New genus			1		

Figure 11. Number of species of several genera of the Normapolles group found in samples representing pre-McNairy (Kpn), McNairy, Owl Creek (Kon), Midway (Tm), Wilcox (Tw), and Claiborne (Tc) rocks of the Mississippi Embayment. No Normapolles species have been found in upper Claiborne or Jackson samples.

Normapolles genera and many species have been found in Upper Cretaceous and Paleocene rocks. The recovery of Normapolles species in the Mississippi Embayment region is shown in Figure 11. Some of the genera, such as *Oculopollis,* and most of the species have not been reported from the Rocky Mountain region. Species of the Normapolles group are particularly common in Upper Cretaceous and lower Tertiary rocks in the Mississippi Embayment region but become progressively fewer in the early and middle Eocene and apparently all die out by late Eocene time.

WORLD DISTRIBUTION OF SOME LATE CRETACEOUS POLLEN GENERA

The pronounced differences between palynomorph assemblages in the two regions led me to examine the world-wide distribution of a few of the more characteristic and distinctive Late Cretaceous genera. The genera chosen were *Aquilapollenites, Wodehouseia, Proteacidites, Pemphixipollenites,* and the Normapolles group.

Aquilapollenites and *Wodehouseia* are characteristic genera of the Late Cretaceous in the western United States, western Canada, Alaska, Siberia, and West Africa. *Proteacidites* has been seen in samples from the Mississippi Embayment, the Rocky Mountains, western Canada, Alaska, Siberia, West Africa, and Australia. *Pemphixipollenites* is present in samples from the western United States, Siberia, and West Africa. The Normapolles group, except for a few specimens, has been reported only from Europe and the United States east of the Mississippi River. The distribution of these genera, shown on Figure 12, suggests the presence of several

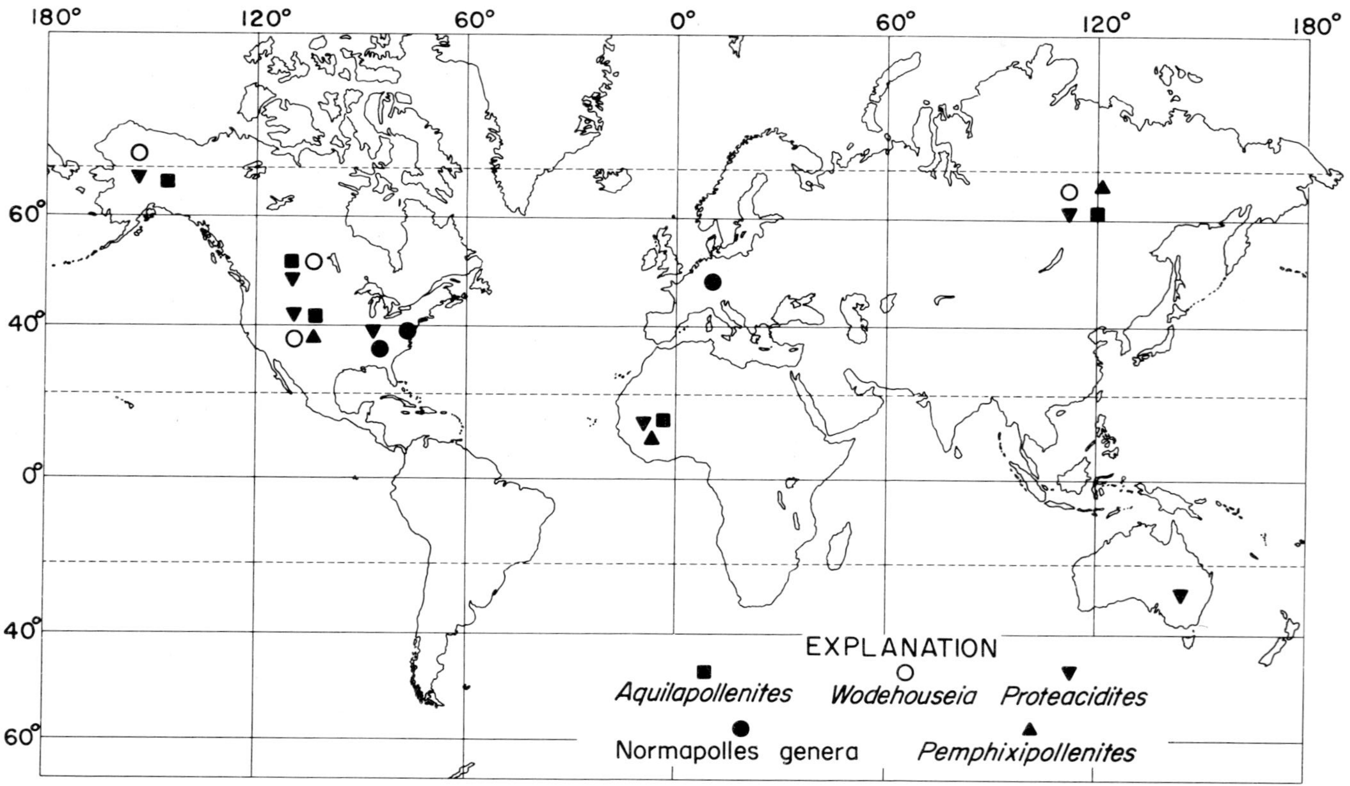

Figure 12. Presently known world distribution of some Late Cretaceous pollen genera.

floral provinces. Certainly at least two Late Cretaceous floral provinces are represented in the continental United States.

Late Cretaceous assemblages from Europe (Pflug, 1953; Krutzsch, 1957; Krutzsch and others, 1960) are distinctly different from those of comparable age reported from Siberia (Chlonova, 1960, 1961; Samoilovitch and others, 1961; Mtchedlishvili, 1965). The western United States region has yielded assemblages that compare directly with those from Siberia. Some of the genera reported from West Africa are found in Siberia and the western United States and have not been reported from Europe. Jardiné and Magloire (1965, p. 194) have recorded *Aquilapollenites* from the upper Maestrichtian of Senegal and *Proteacidites* from the basal Senonian through the Maestrichtian of the same region. Stover (1963) reported the genus *Pemphixipollenites* from middle Cretaceous rocks of Senegal and the Ivory Coast. He suggested affinities for this genus in the Proteaceae, Loranthaceae, and *Aquilapollenites*, even suggesting that *Pemphixipollenites* might be the progenitor of *Aquilapollenites*. Spore and pollen genera and species from the eastern United States are very similar to genera and species reported from Europe. Takahashi (1964), using the known distribution of *Cicatricosisporites dorogensis* R. Pot. and Gelletich, *Corrugatisporites solidus* (R. Pot.) Thoms. and Pflug, *Laevigatosporites ovatus* Wilson and Webster, *Schizosporis, Phyllocladidites, Aquilapollenites,* and *Pentapollenites,* demonstrated floral dissimilarities between Europe, Asia, Australia, and eastern and western North America.

At least a partial explanation of the Late Cretaceous dissimilarities in the pollen and spore floras from eastern and western North America may found in the separation of these floral provinces during much of Cretaceous time by the great north-trending Cretaceous epeiric sea (Fig. 13). This great epicontinental sea, in places as much as 1000 miles wide, probably was an effective barrier to east-west or west-east migration of plants during the Late Cretaceous.

The western province may have had a land connection to Asia across the Bering Strait. The stratigraphic distribution of species of *Aquilapollenites* reported by Mtchedlishvili (1965) is from the lower Senonian through the Danian. One Siberian species, *Aquilapollenites quadrilobus,* is recorded from the Aptian through the remainder of the Late Cretaceous. No records of this genus from the North American continent are known that are earlier than the

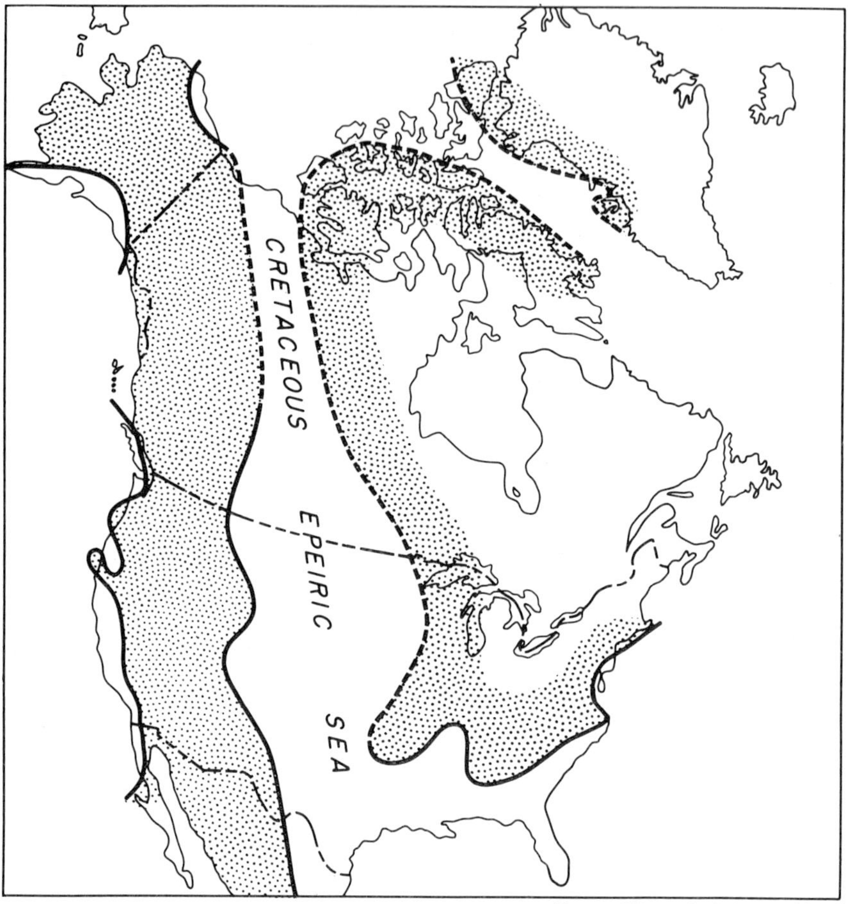

Figure 13. Probable distribution of land and sea in North America during late Campanian time.

Santonian (Rouse, 1957, p. 371). These facts suggest that *Aquilapollenites* originated in Asia and that the intermittent land bridge across the Bering Strait was inoperative during Late Cretaceous time until the Santonian. During and after the Santonian, genera such as *Aquilapollenites* were able to migrate to the North American continent.

The similarity of pollen assemblages from the Late Cretaceous of the eastern United States and Europe may have a comparable explanation. Although a land bridge, connecting Europe and the

North American continent by way of Greenland and the Arctic Islands, has been postulated on the basis of the continental drift theory, little concrete botanical evidence is available to support such a theory. Lindroth (1957), in discussing the faunal connections between Europe and North America, carefully documents the evidence pro and con for the existence of a pre-Pleistocene land connection. There is little doubt that certain biological species indigenous to Europe have arrived in Greenland, Iceland, and eastern North America. Lindroth rejects the possibility that the European character of certain faunas could be the result of introduction of the species by man. He concludes that the "main part of the fauna, especially of the soil bound members, reached Iceland and Greenland over a land connection now submerged." One need not postulate a complete land connection but merely islands close enough to provide for effective island jumping by species of plants. If a land-based fauna was able to find a migration corridor, then probably seeds and spores could have gradually spread plants as well. The climate of Greenland and Iceland during Cretaceous time was certainly not as rigorous as it is at present. It was mild enough so that swamps, indicated by present-day coal deposits, were present in Greenland (Seward, 1926; Miner, 1935). Whatever the explanation, the fact is clear that some of the Late Cretaceous plants of eastern North America had their closest affinities with European plants. Whether migration was from west to east or east to west is not at present known.

PLANT CHANGE AT THE END OF THE CRETACEOUS

The easily recognized floral change at the beginning of the Tertiary in eastern Montana and elsewhere in the Rocky Mountain region coincides with the Cretaceous-Tertiary boundary as defined by R. W. Brown (1952, p. 92). The flora existing during latest Cretaceous time lost many species at the boundary, and new ones superseded them. Coincident with this floral change is the slow but recognizable change in the general characteristics of the lower Paleocene Tullock rocks as contrasted with those of the Hell Creek. Tullock rocks in general are more sandy and more resistant to erosion, and the grain size is slightly larger, which indicates a greater transport energy. This energy change is probably a reflection of a slight and perhaps continuing uplift during Tullock time.

The floral change observed in plant megafossils is also pronounced. Dorf (1942, p. 120) says "that of a total of over 250 Paleocene species there are only 30 . . . which are reported as extending from the Paleocene into the Lance or from the Lance into the Paleocene." (The Lance in Dorf's usage includes Hell Creek, Medicine Bow, and other Late Cretaceous Rocky Mountain floras.)

In the Mississippi Embayment area, the Cretaceous-Tertiary floral change also coincides with the lithologic boundary and unconformity separating the Cretaceous McNairy and Owl Creek Formations and the Paleocene Clayton Formation. In the Mississippi Embayment this change may have been due in part to environmental changes during the time represented by the unconformities. Nevertheless, the occurrence of palynomorphs is similar to that observed in the west. Few species persist across the boundary. Whatever the cause of this palynomorph change, the results in the two areas are similar. A great many Cretaceous species died out, leaving unfilled biological niches that were gradually filled by new species evolving from genera already present. The Paleocene pollen and spore floras of the two regions have yielded fewer species than the Cretaceous—yet an examination of upper Paleocene and Eocene rocks discloses an increasing number and diversity of species with a more or less modern aspect.

Dorf (1942, p. 120) suggests as possible causes of floral change across the boundary in the west:

> (1) there may exist an unrecognized hiatus between the Lancian and Paleocene deposits; (2) there have as yet not been a sufficient number of specimens collected from the upper parts of the Lancian formations or the lower parts of the Paleocene to present a true picture of the floral changes; (3) the late Lancian orogeny in the Rocky Mountain region may have been of sufficient intensity to cause rather sudden and widespread environmental changes and concurrent floral changes; (4) the sudden floral change in the early Paleocene may have been a result of a rather rapid invasion of the Rocky Mountain region, possibly from eastern North America by way of the newly opened route around the north end of the early Paleocene Cannonball sea.

This last hypothesis finds some support from pollen evidence. The Paleocene pollen assemblages from the Rocky Mountain region and the Mississippi Embayment have more similar elements than do the Cretaceous assemblages. One example is the species *Momi-*

pites tenuipolus Anderson. This species makes its first appearance in the Mississippi Embayment in basal Paleocene rocks of the Clayton Formation. It is not present in the lower part of the Tullock Member of the Fort Union Formation in the Hell Creek area. It first appears at about 150 feet above the base of the Tullock at about the level of the Big Dirty coal horizon. Above this level it is a characteristic member of Rocky Mountain Paleocene assemblages.

The available evidence from pollen and leaf floras from the two regions does not suggest a profound climatic change such as is in evidence at the onset of the Pleistocene glaciations. Dorf (1942, p. 103) said that characteristic Lance species were perhaps more nearly warm-temperate than subtropical. R. W. Brown (1962, p. 95) saw in the western Paleocene an intermingling of temperate species with relicts from a warmer subtropical flora. Wolfe and Hopkins (1967, p. 72) indicated a warm-temperate to subtropical but deteriorating climate for the Paleocene of the northwestern United States. Berry (1925, p. 13, 14), in discussing the Ripley flora from the Mississippi Embayment, indicated that in his opinion the flora was lowland coastal and the climate was uniform warm-temperate. The pollen and spore flora of this region also suggest a warm-temperate to subtropical climate during the latest Cretaceous. No profound climate change at the Cretaceous-Tertiary transition is evident, however. As evidence, *Gleichenia,* a tropical and subtropical fern genus, is represented in the latest Cretaceous of the Mississippi Embayment by five palynomorph species and in the lower Paleocene by two species.

In the northern Rocky Mountain province, one may postulate that a slight climatic change caused by the withdrawal of the epeiric sea, or by the slow rising of the land surface, or both, might have caused a climatic change such as the lowering of the frost line or the inception of more pronounced seasonal changes. Such a slight change could have eliminated species that were not frost-tolerant and could have thus provided unoccupied niches for new species to invade. Indeed, Russell (1966, p. 203) postulated a slight climatic change as having been responsible for the extinction of the dinosaurs in North America. The advent of seasonal change, even of cold winters mild by present standards, is the type of change that very well may have been responsible for the floral change in the Rocky Mountain region. In the Mississippi Embayment region, the absence of the tempering influence of the epicontinental seas to the west might have given rise to similar changes, although whether or not frost can be postulated in this region during Paleocene time is open to question. Certainly as more data accumulate our ideas concerning similarities and differences in the synchronous provincial land floras, concerning origin, evolution, and migration, and even concerning the abrupt floral changes at the Cretaceous-Tertiary boundary will be placed on much firmer foundations.

ACKNOWLEDGMENTS

I thank Bernadine D. Tschudy and Sharon D. Van Loenen for assistance in collecting samples from the Rocky Mountain localities. W. W. Olive, U.S. Geological Survey, Paducah, Kentucky, and Elliott Cushing, U.S. Geological Survey, Memphis, Tennessee, supplied samples from the embayment region. Mrs. Van Loenen helped also in the preparation of samples and photography of specimens.

REFERENCES CITED

Berry, E. W., 1916, The lower Eocene floras of southeastern North America: U.S. Geol. Survey Prof. Paper 91, 481 p.

——1925, The flora of the Ripley formation: U.S. Geol. Survey Prof. Paper 136, 94 p.

Brown, Barnum, 1907, The Hell Creek beds of the upper Cretaceous of Montana; their relation to contiguous deposits, with faunal and floral lists, and a discussion of their correlation: Am. Mus. Nat. History Bull. 23, p. 823-845.

Brown, R. W., 1939, Fossil plants from the Colgate member of the Fox Hills sandstone and adjacent strata: U.S. Geol. Survey Prof. Paper 189-I, p. 239-275.

——1952, Tertiary strata in eastern Montana and western North and South Dakota, p. 89-92 *in* Billings Geol. Soc. Guidebook 3d Ann. Field Conf., Black Hills-Williston Basin, 1952.

——1962, Paleocene flora of the Rocky Mountains and Great Plains: U.S. Geol. Survey Prof. Paper 375, 119 p.

Chlonova, A. F., 1960, Vidovoi sostav pyl'tsy i spor v otlozheniiakh verkhnego mela Tchulymo-Enlseiskoi vpadiny: Akad. Nauk SSSR Sibirskoye Otdeleniye, Inst. Geologii i Geofiziki Trudy, issue 3, 104 p.

——1961, Spory i pyl'tsa verkhnei poloviny verkhnego mela vostochnoi chasti Zapadno-Sibirskoi nizmennosti: Akad. Nauk SSSR Sibirskoye Otdeleniye, Inst. Geologii i Geofiziki Trudy, issue 7, 140 p.

Collier, A. J., and Knechtel, M. M., 1939, The coal resources of McCone County, Montana: U.S. Geol. Survey Bull. 905, 80 p.

Conant, L. C., 1942, Union County Mineral Resources: Mississippi Geol. Survey Bull. 45, p. 1-100.

Dorf, Erling, 1942, Upper Cretaceous floras of the Rocky Mountain region; 1, Stratigraphy and paleontology of the Fox Hills and Lower Medicine Bow formations of southern Wyoming and northwestern Colorado; 2, Flora of the Lance formation at its type locality, Niobrara County, Wyoming: Carnegie Inst. Washington Pub. 508, Contr. Paleontology, 168 p.

Drugg, W. S., 1967, Palynology of the Upper Moreno Formation (Late Cretaceous-Paleocene), Escarpado Canyon, California: Palaeontographica, abt. B, v. 120, no. 1-4, p. 1-71.

Folinsbee, R. E., Baadsgaard, H., and Cumming, G. L., 1963, Dating of volcanic ash beds (bentonites) by the K-Ar method, p. 70-82 *in* Nuclear Geophysics, Nuclear Science Ser., Rept. 38: Natl. Acad. Sci.-Natl. Research Council Pub. 1075.

Gill, J. R., and Cobban, W. A., 1966, The Red Bird section of the Upper Cretaceous Pierre Shale in Wyoming: U.S. Geol. Survey Prof. Paper 393-A, p. A-1-A-71.

Jardiné, Serge, and Magloire, Lili, 1965, Palynologie et Stratigraphie du Crétacé des Bassins du Sénégal, et de Côte d'Ivoir, p. 187-245 *in* Colloque International de Micropaléontologie, Dakar 1963: Bur. Recherches Géol. et Minières Mem. 32, 269 .

Jeletzky, J. A., 1960, Youngest marine rocks in western interior of North America and the age of the Triceratops-beds; with remarks on comparable dinosaur-bearing beds outside North America: Rept. 21st Internat. Geol. Congress, Copenhagen, 1960, pt. 5, p. 25-40.

——1962, The allegedly Danian dinosaur-bearing rocks of the globe and the problem of the Mesozoic-Cenozoic boundary: Jour. Paleontology, v. 36, no. 5, p. 1005-1018.

Jensen, F. S., and Varnes, H. D., 1964, Geology of the Fort Peck area, Garfield, McCone and Valley Counties, Montana: U.S. Geol. Survey Prof. Paper 414-F, p. F-1-F-49.

Krutzsch, Wilfried, 1957, Sporen- und Pollengruppen aus der Oberkreide und dem Tertiär Mitteleuropas und ihre stratigraphische Verteilung: Zeitschr. Angew. Geologie, v. 3, no. 11-12, p. 509-548.

——1959, Einige neue Formgattungen und -arten von Sporen und Pollen aus der mitteleuropäischen Oberkreide und dem Tertiär: Palaeontographica, abt. B, v. 105, no. 5-6, p. 125-155.

Krutzsch, Wilfried, Pchalek, J., and Spiegler, D., 1960, Tieferes Pälaozän (?Montien) in Westbrandenburg: Rept. 21st Internat. Geol. Congress, Copenhagen, 1960, pt. 6, p. 135-143.

Lindroth, C. H., 1957, Faunal connections between Europe and North America: New York, John Wiley & Sons, Inc., 344 p.

Miner, E. L., 1935, Paleobotanical examinations of Cretaceous and Tertiary coals; 1, Cretaceous coals from Greenland: Am. Midland Naturalist, v. 16, no. 4, p. 585-625.

Mtchedlishvili, N. D., 1965, Znachenie pokrytose mennykh rastenii dlia stratigrafii verkhnemelovykh otlozhenii, p. 5-34 *in* Samoilovitch, S. R., *Editor,* Paleofitologicheskii sbornik: Vses. Neft. Nauchno-Issled. Geol.-Razved. Inst. Trudy, new ser., issue 239, 508 p.

Newman, K. R., 1964, Palynologic correlations of late Cretaceous and Paleocene formations, northwestern Colorado, p. 169-180 *in* Palynology in oil exploration—A symposium, San Francisco, California, 1962: Soc. Econ. Paleontologists and Mineralogists Spec. Pub. 11.

Pflug, H. D., 1953, Zur Entstehung und Entwicklung des angiospermiden Pollens in der Erdgeschichte: Palaeontographica, abt. B, v. 95, no. 4-6, p. 60-171.

Pipiringos, G. N., Chrisholm, W. A., and Kepferle, R. C., 1965, Geology and uranium deposits in the Cave Hills area, Harding County, South Dakota: U.S. Geol. Survey Prof. Paper 476-A, p. A-1-A-64.

Rainwater, E. H., 1960, Paleocene of the Gulf Coastal Plain of the United States of America: Rept. 21st Internat. Geol. Congress, Copenhagen, 1960, pt. 5, p. 97-116.

Roberts, A. E., 1966, Geology and coal resources of the Livingston coal field, Gallatin and Park Counties, Montana: U.S. Geol. Survey Prof. Paper 526-A, p. 1-56.

Rouse, G. E., 1957, The application of a new nomenclatural approach to Upper Cretaceous plant microfossils from western Canada: Canadian Jour. Botany, v. 35, no. 3, p. 349-375.

Russell, L. S., 1966, The changing environment of the dinosaurs in North America: Adv. Sci., v. 23, no. 110, p. 197-204.

Samoilovitch, S. R., Mtchedlishvili, N. D., Rusakova, L. I͡A., and Iashchurzhinskaia, A. B., *Editors,* 1961, Pyl'tsa i spory zapadnoi Sibiri, uira-paleotsen: Vses. Neft. Nauchno-Issled. Geol.-Razved. Inst. Trudy, issue 177, 660 p.

Seward, A. C., 1926, The Cretaceous plant-bearing rocks of western Greenland: Royal Soc. London Philos. Trans., ser. B., v. 215, p. 57-175.

Sohl, N. F., 1960, Archeogastropoda, Mesogastropoda, and stratigraphy of the Ripley, Owl Creek, and Prairie Bluff formations: U.S. Geol. Survey Prof. Paper 331-A, p. 1-151.

Stephenson, L. W., and Monroe, W. H., 1940, The Upper Cretaceous deposits: Mississippi Geol. Survey Bull. 40, 296 p.

Stover, L. E., 1963, The Cretaceous pollen Genus *Pemphixipollenites*: Pollen et Spores, v. 5, no. 1, p. 161-168,

Takahashi, Kiyoshi, 1964, Sporen und Pollen der oberkretazeischen Hakobuchi-Schichtengruppe, Hokkaido: Kyushu Univ. Fac. Sci. Mem., ser. D, v. 14, no. 3, p. 159-271.

Waagé, K. M., 1967, Cretaceous transitional environments and faunas in central South Dakota, p. 237-266 *in* Paleoenvironments of the Cretaceous seaway in the western interior—A symposium: Colorado School Mines preprints of papers presented at Geol. Soc. America, Rocky Mountain Sec., 20th Ann. Meeting, 266 p.

Wolfe, J. A., and Hopkins, D. M., 1967, Climatic changes recorded by Tertiary land floras in northwestern North America, p. 67-76 *in* Tertiary correlations and climatic changes in the Pacific: Japan, Ed. K. Hatai Sasaki Co.

PUBLICATION AUTHORIZED BY THE DIRECTOR, U.S. GEOLOGICAL SURVEY

Aquilapollenites (Rouse) Funkhouser — Selected Rocky Mountain Taxa and Their Stratigraphic Ranges

BERNADINE D. TSCHUDY
U.S. Geological Survey, Denver, Colorado

ESTELLA B. LEOPOLD
U.S. Geological Survey, Denver, Colorado

ABSTRACT

The genus *Aquilapollenites* and its type species, *A. quadrilobus*, are redescribed, and the latter is illustrated. This report summarizes our present information on stratigraphic ranges of selected Rocky Mountain species and varieties of *Aquilapollenites*. Of the 17 taxa included in this report, 10 are previously named species, 4 are new species, and 3 are new varieties. One species is transferred from the genus *Mancicorpus* and another from the genus *Parviprojectus*. The stratigraphic data concerning *Aquilapollenites* incorporate available information from U.S. Geological Survey pollen collections.

CONTENTS

INTRODUCTION

The genus *Aquilapollenites* Rouse, 1957, first recognized in Canadian strata of Late Cretaceous age, has since been reported from a wide range of geographic localities. Rocky Mountain records of the genus include the works of: Funkhouser (1961), who emended the genus and described 7 species from Wyoming; Stanley (1961), who named 5 species from South Dakota; Norton (1965), who described 3 species from Montana; and Srivastava (1966, 1968a, 1968b, 1969), who named 35 new species of *Aquilapollenites* and 5 of *"Mancicorpus"* from Alberta, Canada. These records are mainly of Late Cretaceous age but also include localities of Paleocene (and Eocene?) age. In the United States east of the Great Plains, *Aquilapollenites* has been reported from the Porters Creek Clay of Paleocene age from Arkansas (Jones, 1962) and Tennessee (Herrick and Tschudy, 1967). Ames and Riegel (1962, oral commun.) reported the genus from the Paleocene or Eocene of Alaska. Unpublished records from U.S. Geological Survey collections and from collections by Dale Wiggins (Standard Oil Company of California, 1966, oral commun.) indicate that the genus was widespread in Alaska during the Late Cretaceous; apparently it occurred also in the late Oligocene in Alaska (Estella B. Leopold, unpublished data).

Old World records center largely around central and eastern Asia. In Russia, *Aquilapollenites* species have been named and described from Upper Cretaceous and (or) Paleocene sediments by Bolkhovitina (1959), Mtchedlishvili (*in* Samoilovitch and others, 1961), Chlonova (1961), Samoilovitch (1965), and Bratzeva (1965). Some of the Russian species of *Aquilapollenites* that are similar or probably identical to Rocky Mountain species are discussed in this report. Varied forms of the genus *Aquilapollenites* from Upper Cretaceous deposits of the U.S.S.R., were illustrated as proteaceous pollen in Agranovskaia and others (1960). An isolated record from the lower Oligocene of the western Siberian lowlands of Russia was reported by Voitsel' and others (1956, p. 195). This pollen grain was described and illustrated but was not named; however, it undoubtedly belongs to the genus *Aquilapollenites.* In Japan, Sato (1961) and Takahashi (1964) reported and named several species of *Aquilapollenites* from Upper Cretaceous strata. Simpson (1961) and Martin (1968) reported several species of the genus from deposits of probable Paleocene age from Scotland; Simpson (1961) described the genus *Taurocephalus* as new, but it is undoubtedly the same as *Aquilapollenites.* Three species of *Aquilapollenites* have been named from Upper Cretaceous beds of northern Africa (Sénégal, about 13° N. lat) by Jardiné and Magloire (1965), and one of these species was reported in Upper Cretaceous sediments from Gabun, equatorial Africa (about 0° lat), by Belsky and others (1965).

Judging from the published records, the Late Cretaceous distribution of the genus *Aquilapollenites* includes the Rocky Mountain region of the United States and southwestern Canada, Alaska, Siberia, Japan and northwestern and west-central Africa. Early Tertiary records of the genus are from a few isolated localities: Paleocene and Eocene(?) of the Rocky Mountains, Paleocene of Arkansas and Tennessee, Paleocene or Eocene and Oligocene of Alaska, late Paleocene from the northern part of the western Siberian lowlands, and probable Paleocene of Scotland. Some comments on the worldwide distribution of this genus during the Late Cretaceous are made by R. H. Tschudy (this volume).

POLLEN LOCALITIES

Pollen collections studied, with their respective formations, and section, township, and range positions are listed in the appendix.

The numbers prefixed by the letter "D" are USGS paleobotany locality numbers from the Denver Pollen Laboratory. The collection localities are shown on the index map (Fig. 1).

STRATIGRAPHY

The stratigraphic ranges of selected taxa of *Aquilapollenites,* based entirely on U.S. Geological Survey pollen collections, are plotted in Figure 2. The taxa mentioned in Figure 2 are illustrated on Plate 1. Because the precise geologic ages for many of the samples were not known, age was inferred in the following way: the position of individual samples in relation to the local thickness of the formation was noted and compared with Cobban and Reeside's (1952) stratigraphic charts of Cretaceous formations of the western United States; this position was then plotted on our stratigraphic range chart (Fig. 2). For a few collections, more recent information concerning stratigraphic position was available and was used. Gill and Cobban (1966) discussed in detail the age of the Pierre Shale at Red Bird, Wyoming. The ages of the Miner Creek and Cokedale Formations and the Eagle Sandstone in Montana were summarized by Roberts (1963, 1965). The Irish Creek core penetrating the Pierre Shale and Fox Hills Sandstone in South Dakota was described lithologically by Tourtelot and Schultz (1961), and approximate ages of the samples from this core were determined by W. A. Cobban (1967, oral commun.). The age of the Meeteetse Formation in the Wind River Basin in Wyoming was discussed by Keefer (1965).

In addition to the Rocky Mountain stratigraphic units, the Colville Group of Alaska, whose age was reviewed by Robinson (1958), is referred to in this report.

Previously published occurrences of species of *Aquilapollenites* are not included on the chart of stratigraphic ranges (Fig. 2) because the physical positions in the formation at local exposures were not generally indicated by the various authors.

In the Rocky Mountain region, the genus *Aquilapollenites* ranges from lower Campanian strata "within the *Scaphites hippocrepis zone*" (W. A. Cobban, 1968, oral commun.), through middle and probably upper Paleocene (within the Sentinel Butte Member of the Fort Union Formation). To date we have found no occurrences of the genus in the lower Paleocene.

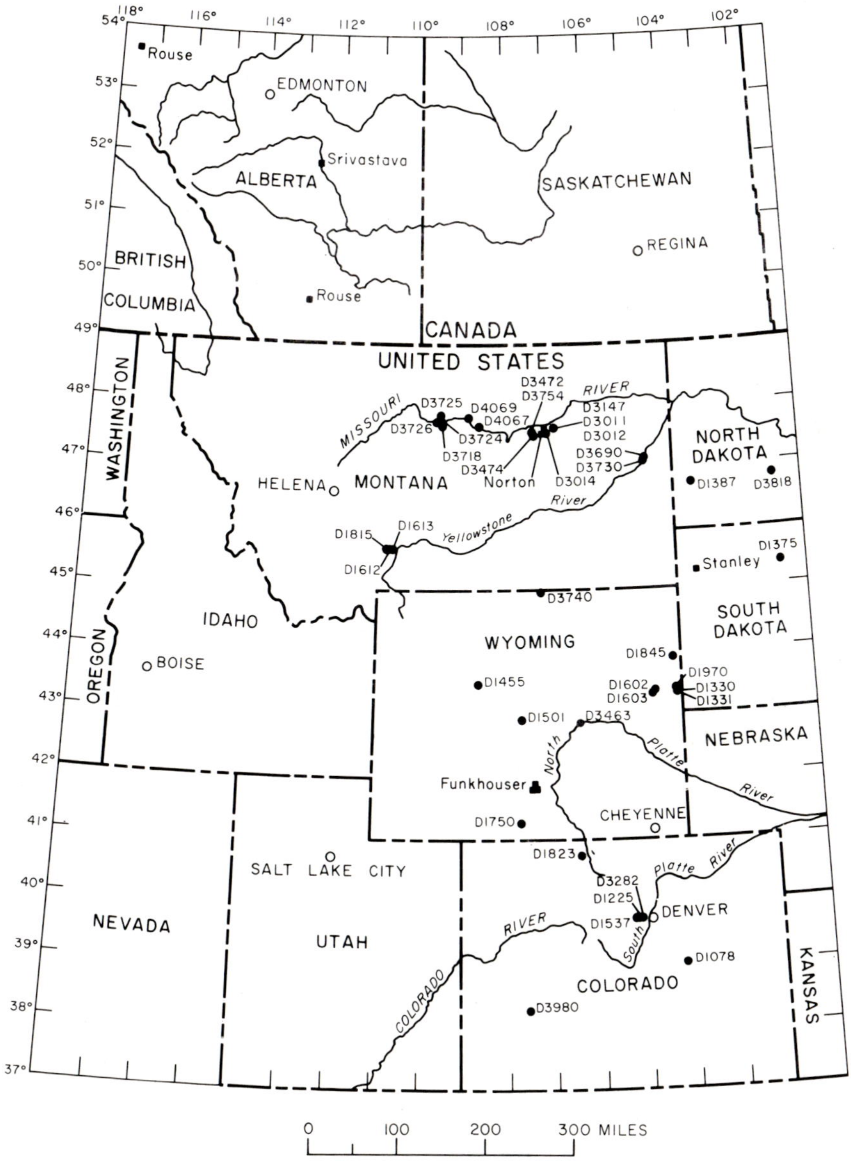

Figure 1. Index map showing U. S. Geological Survey Rocky Mountain pollen collection localities (circles) and localities of other authors (squares).

Please foldout here

PLATE 1. SELECTED TAXA OF *AQUILAPOLLENITES*

All magnifications ×500; taxa plotted on Figure 2.

Figure

1. *Aquilapollenites rigidus* n. sp. Equatorial view. USGS paleobotany locality D3718-I, slide 1, at 102.8 × 16.4.
2. *Aquilapollenites senonicus* (Mtchedlishvili) n. comb. Equatorial view. USGS paleobotany locality D3472-D, slide 1, at 76.7 × 15.5.
3. *Aquilapollenites trialatus* var. *variabilis* n. var. Equatorial view. USGS paleobotany locality D3718-I, slide 1, at 94.4 × 13.9.
4. *Aquilapollenites calvus* n. sp. Equatorial view of paratype. USGS paleobotany locality D1330-16, slide 6, at 114.5 × 10.5.

5A. *Aquilapollenites quadrilobus* Rouse. Equatorial view. USGS paleobotany locality D1603, slide 1, at 90.6 × 2.6.

5B. *Aquilapollenites quadrilobus* Rouse. Polar view. USGS paleobotany locality D3014-A, slide 2, at 103.5 × 5.2.

6. *Aquilapollenites turbidus* n. sp. Equatorial view. USGS paleobotany locality D1078, slide 1, at 102.9 × 21.8.
7. *Aquilapollenites trialatus* var. *uniformis* n. var. Equatorial view. USGS paleobotany locality D3725-C, slide 4, at 90.4 × 8.7.

8A. *Aquilapollenites tripodiformis* n. sp. Equatorial view of holotype. USGS paleobotany locality D1330-16, slide 4, at 78.9 × 15.2.

8B. *Aquilapollenites tripodiformis* n. sp. Equatorial view. USGS paleobotany locality D1375-14, slide 1, at 100.1 × 15.5.

9. *Aquilapollenites catenireticulatus* Srivastava. Equatorial view. USGS paleobotany locality D3147-B, slide 4, at 103.7 × 10.8.
10. *Aquilapollenites attenuatus* Funkhouser. Equatorial view. USGS paleobotany locality D1603, slide 4, at 92.9 × 11.9.

11A. *Aquilapollenites reductus* Norton. Equatorial view. USGS paleobotany locality D3012, slide 4, at 93.2 × 23.4.

11B. *Aquilapollenites reductus* Norton. Polar view. USGS paleobotany locality D3012, slide 2, at 76.8 × 21.9.

12A. *Aquilapollenites reticulatus* (Mtchedlishvili) n. comb. Equatorial view. USGS paleobotany locality D3147-B, slide 3, at 82.8 × 5.8.

12B. *Aquilapollenites reticulatus* (Mtchedlishvili) n. comb. Polar view. USGS paleobotany locality D3147-B, slide 5, at 115.7 × 6.1.

13. *Aquilapollenites delicatus* Stanley. Equatorial view of holotype. USNM 131311.
14. *Aquilapollenites delicatus* var. *collaris* n. var. Equatorial view of holotype, medium focus. USGS paleobotany locality D3147-B, slide 2, at 87.2 × 16.5.
15. *Aquilapollenites conatus* Norton. Equatorial view. USGS paleobotany locality D3147-B, slide 5, at 115.2 × 7.2.

16A. *Aquilapollenites bertillonites* Funkhouser. Holotype. USNM 41308.

16B. *Aquilapollenites bertillonites* Funkhouser. Polar view. USGS paleobotany locality D3472-C, slide 1, at 111.0 × 15.4.

17A. *Aquilapollenites spinulosus* Funkhouser. Equatorial view of holotype. USNM 41310.

17B. *Aquilapollenites spinulosus* Funkhouser. Polar view of paratype. USNM 41309.

AQUILAPOLLENITES (Rouse) Funkhouser, 1961

Aquilapollenites Rouse, 1957, p. 370-371.

Type species. *Aquilapollenites quadrilobus* Rouse, 1957.

Holotype. Designated by Rouse, 1957, p. 371; illustrated *in* Radforth and Rouse, 1954, Plate 1, figure 14, (*see* also Pl. 2, figs. 1A, 1B, in this report); Mc. U. P. C. (McMaster University Palaeobotanical Collection) Slide C_S-31; Alberta, Canada; Brazeau Formation, Upper Cretaceous.

Aquilapollenites (Rouse) Funkhouser, 1961 (April), p. 193-194.

Triprojectacites Mtchedlishvili, *in* Samoilovitch and others, 1961 (July), p. 203-204 (supergroup of Mtchedlishvili).

Triprojectus Mtchedlishvili, *in* Samoilovitch and others, 1961 (July), p. 204-205.

Aquilapollenites (Rouse) Mtchedlishvili, *in* Samoilovitch and others, 1961 (July), p. 209.

Integricorpus Mtchedlishvili, *in* Samoilovitch and others, 1961 (July), p. 217.

Mancicorpus Mtchedlishvili, *in* Samoilovitch and others, 1961 (July), p. 218-219.

Parviprojectus Mtchedlishvili, *in* Samoilovitch and others, 1961 (July), p. 225.

Projectoporites Mtchedlishvili, *in* Samoilovitch and others, 1961 (July), p. 227.

Triceropollis Chlonova, 1961 (August), p. 85.

Taurocephalus Simpson, 1961 (December), p. 440.

Aquilapollenites (Rouse) Funkhouser restr. Srivastava, 1968a, p. 668-670.

Mancicorpus (Mtchedlishvili) Srivastava, 1968a, p. 695-697.

Historical Background of Genus

Aquilapollenites as originally introduced by Rouse (1957) was described as bilaterally symmetrical; having two to four winglike projections extending from a broadly subrectangular body; having body ornamentation coarsely granular, weakly papillate or punctate; having ornamentation of the projections levigate to finely papillate; range of length 40 to 100 μ. Rouse assigned two species to the genus—the type species, *A. quadrilobus* (see p. 125-131), whose type specimen is now known to be heteropolar, and the second species, *A. trialatus,* which is isopolar (to subisopolar).

In April 1961, *Aquilapollenites* was emended by Funkhouser. In brief, he described the genus as follows: heteropolar or isopolar angiosperm pollen having one or both poles extended into polar protrusions;

having three equatorial protrusions; having three colpi or three pairs of "demicolpi" borne on the outer margins of the equatorial protrusions; in addition, some species having three colpi on and parallel to the equator between the equatorial protrusions; grains more or less smooth or variously sculptured.

Mtchedlishvili (*in* Samoilovitch and others, 1961) and Srivastava (1968a) proposed subdivisions of the genus *Aquilapollenites;* these revisions are not acceptable because (1) the genera are not clearly distinct from each other, and (2) *Aquilapollenites* would be deprived of its type species, and one genus—*Mancicorpus*—proposed by Mtchedlishvili and emended by Srivastava, would usurp the type species of *Aquilapollenites.*

Our interpretation of *Aquilapollenites* is presented in the description below. It is not intended as an emendation since it neither expands nor limits *Aquilapollenites* as circumscribed by Funkhouser.

Description

Angiosperm pollen grains, heteropolar, or isopolar to subisopolar; having three lateral equatorial protrusions; tricolpate to trisyncolpate. Heteropolar grains having a major and minor polar protrusion—the major one prominently protruding to nearly flat and the minor one slightly protruding to concave. Isopolar to subisopolar grains having two equal or nearly equal poles that protrude slightly or prominently. Equatorial protrusions of heteropolar species extending at right angles from polar axis of grain or declining at various angles away from major pole; equatorial protrusions of isopolar to subisopolar species extending virtually at right angles from the polar axis. In equatorial views of isopolar to subisopolar species the contact areas of the equatorial protrusions with the body (the axillae) are concave. *Colpi* extending total or part length of equatorial protrusions, in plane passing through and parallel to polar axis of grain; in some species colpi extending varying distances onto body or even reaching as far as and joining (syncolpate) at one or both poles. Some species also having three equatorial furrows along the equator, between the equatorial protrusions. *Exine* at least two-layered; axillary endexinal thickenings (costae) adjacent to or underlying colpi, extending varying distances onto body and onto equatorial protrusions; in some cases specimens having faintly visible costae or having light-colored axillary areas, which probably indicate areas from which costae have dissolved.[1] Ektexinal sculpture various. Equatorial protrusions often thinner-walled and with finer or different sculpture than body of grain.

---

[1]Preliminary experiments by Imogene Doher in this laboratory have shown that long Schulze treatment destroys endexinal thickenings in the species tested.

Discussion

Mtchedlishvili (*in* Samoilovitch and others, 1961, p. 203-204) proposed a supergroup, Triprojectacites, that as described appears to be indistinguishable from the genus *Aquilapollenites* (Rouse) Funkhouser. Mtchedlishvili restricted the genus *Aquilapollenites* and placed it under the supergroup Triprojectacites; in addition, she described five new genera that she also included under Triprojectacites (*see* Mtchedlishvili, *in* Samoilovitch and others, 1961, p. 204-227). We do not accept Mtchedlishvili's five new genera because they are not mutually exclusive; one of them would deprive *Aquilapollenites* of its type species, and as *Aquilapollenites* was circumscribed by Mtchedlishvili, its type species, *A. quadrilobus,* would be excluded. Mtchedlishvili's five new genera and her restriction of *Aquilapollenites* are discussed below.

(1) *Aquilapollenites.* Mtchedlishvili reserved *Aquilapollenites* for isopolar to subisopolar grains; this conception of *Aquilapollenites* is not acceptable because *A. quadrilobus,* the type species of *Aquilapollenites,* is heteropolar and would be excluded (*see A. quadrilobus,* p. 125-131).

(2) *Mancicorpus* was created exclusively for heteropolar species; according to Mtchedlishvili, the body is developed on one side and on the other side the body is absent or is represented by a small node. Perhaps *Mancicorpus* would not have been erected if it had been realized that *A. quadrilobus* was a heteropolar species. Misconceptions of *A. quadrilobus* are understandable since the species was originally incompletely described and its type specimen is difficult to interpret from the photograph (*in* Radforth and Rouse, 1954, Pl. I, fig. 14); *A. quadrilobus* was originally further confused because it was illustrated by three specimens belonging to at least two different species. It is obvious that Mtchedlishvili did not know that *A. quadrilobus* is a heteropolar species; she described it as isopolar to subisopolar and assigned to it grains that clearly are not *A. quadrilobus. Mancicorpus* is not acceptable because it would usurp *A. quadrilobus,* the type species of *Aquilapollenites.*[2]

(3) *Triprojectus.* Mtchedlishvili stated: "Pollen grain of *Triprojectus* differ from the pollen of other groups of this supergroup in a large body and rather long projections" (translation by I. Mittin, from Mtchedlishvili, *in* Samoilovitch and others, 1961, p. 205). The total range in body lengths of species that she placed in *Triprojectus* was 42.9 to 64.0 μ, the range in body widths was 14.1 to 35.2 μ, and the range of lengths of equatorial protrusions was 13.8 to 32.3 μ. These dimensions decidedly

---

[2]According to the International Code of Botanical Nomenclature, Article 52 (*see* Lanjouw, 1961): "When a particular species was originally designated as the type, the generic name must be retained for the genus including that species."

overlap those of species that Mtchedlishvili placed in *Aquilapollenites*—these species had body lengths ranging from 21.5 to 58.5 μ, body widths ranging from 10.1 to 19.4 μ and equatorial-protrusion lengths ranging from 10.0 to 28.2 μ. Hence, dimensions of species that Mtchedlishvili included in *Triprojectus* were not substantially different from those for species that she included under *Aquilapollenites,* even though size characters were supposed to differentiate *Triprojectus* from other genera.

(4) *Projectoporites,* according to Mtchedlishvili, in size and outline most closely resembles *Triprojectus* but differs conspicuously in absence of furrows and in smaller nexine thickenings. *Projectoporites* was described as a triporate genus; however, this character is not clear-cut since Mtchedlishvili stated: "In individual grains [of *Projectoporites spinulosus*] in which the equatorial projection is facing the slide one could observe a very fine furrow or slit on the body of the pollen grain as if a continuation of the pore opening and slightly resembling a narrow furrow." (translation by I. Mittin, from Mtchedlishvili, *in* Samoilovitch and others, 1961, p. 229). As illustrated by Mtchedlishvili, nexine thickenings in species of *Projectoporites* are not consistently smaller than those in her *Triprojectus* species.

(5) *Parviprojectus* was differentiated by Mtchedlishvili from pollen of all other genera of the supergroup Triprojectacites by its small equatorial projections and large body. This does not in fact distinguish it from *Triprojectus, Projectoporites,* or *Integricorpus* to which Mtchedlishvili assigned some species having proportionally large bodies and small equatorial protrusions (for example *Triprojectus unicus* (Chlonova) Mtch., *Triprojectus echinatus* Mtch., *Projectoporites magnus* Mtch., *Projectoporites spinulosus* Mtch., and *Integricorpus bellum* Mtch.).

(6) *Integricorpus,* with its reticulate exine and lack of surface spines or clavae, is distinguishable from Mtchedlishvili's isopolar to subisopolar genera *Triprojectus* and *Projectoporites;* it is not distinct from her genus *Parviprojectus,* which includes species with reticulate exine. The only species that she placed in *Integricorpus, I. bellum,* has small equatorial projections with a large body—the same characters that she used to differentiate *Parviprojectus* from all other genera of her supergroup.

Chlonova (1961, p. 85) described the genus *Tricerapollis* and distinguished it from *Aquilapollenites* as having one pole of the body convex, the other concave; however, this does not distinguish it from *Aquilapollenites* because Funkhouser (1961, p. 193) stated in his description of the genus that "one or both of the poles is extended into what may be called polar protrusions." The species *A. striatus* Funkhouser has one concave and one extended pole.

Srivastava (1968a, p. 668-670) restricted *Aquilapollenites* to isopolar and slightly heteropolar pollen. He (1968a, p. 695-696) emended

Mancicorpus and proposed to distinguish it from *Aquilapollenites* by an arbitrary ratio of polar projection lengths, whereby the genus *Mancicorpus* "should include only those heteropolar pollen whose length of the reduced pole (distance from the center of the equatorial axis to the tip of the polar projection or polar area) is less than half of the length of the developed polar projection (distance from the center of the equatorial axis to the tip of the developed polar projection)." We cannot accept Srivastava's proposal because *A. quadrilobus*, the type species of *Aquilapollenites*, would by his definition belong in the genus *Mancicorpus* instead of *Aquilapollenites* (*see* our discussion under *A. quadrilobus*, p. 127). As indicated in footnote 2, *A. quadrilobus* must remain in the genus *Aquilapollenites*. We consider *Mancicorpus* a part of *Aquilapollenites*.

SELECTED ROCKY MOUNTAIN TAXA AND THEIR OCCURRENCES

In this section four new species and three new varieties of *Aquilapollenites* are named and described. Four previously named species are redescribed and two of these are expanded to include the new varieties named here. Two species, *Mancicorpus senonicum* and *Parviprojectus reticulatus*, are transferred to *Aquilapollenites*. U.S. Geological Survey records of occurrence are listed for each of 17 taxa, and these are plotted in Figure 2.

Slides containing type specimens of new taxa, as well as slides containing nontype grains represented on the plates, are in the slide files at the U.S. Geological Survey Denver Pollen Laboratory. The type specimens are within black ink circles marked on the slides; they may also be located by the mechanical stage coordinates given in the plate explanations and with the individual holotype and paratype assignments. For conversion of our coordinates to those of another mechanical stage: our coordinate readings for the center point of a 1 x 3 standard microscope slide are 108.2 (horizontal axis) x 12.4 (vertical axis). When the slide label is placed on the left on the microscope stage our vertical coordinates decrease toward the lower edge of the slide and our horizontal coordinates decrease to the right on the slide. (*See* Traverse (1958) and Tschudy (1966) for conversion of coordinates.)

In the sections of this report, "USGS Records of Occurrence," formations, stages, and series are cited as tripartite; they are divided into upper, middle, and lower parts according to thickness—not into two parts as used by Gill and Cobban (1966).

Heteropolar Species

Aquilapollenites quadrilobus Rouse, 1957
(Pl. 1, fig. 5A, 5B; Pl. 2, fig. 1A-6)

Aquilapollenites quadrilobus Rouse, 1957, p. 371.

Holotype: Designated by Rouse, 1957, p. 371; illustrated in Radforth and Rouse, 1954, Plate 1, figure 14 (*see also* Pl. 2, figs. 1A, 1B, in this report); Mc. U. P. C. Slide C_s-31; Entrance, Alberta, Canada; Brazeau Formation, Upper Cretaceous.

Aquilapollenites polaris Funkhouser, 1961, p. 198, Pl. 1, figs. 1a-1d, 2; western flank of Rawlins uplift, Rawlins, Wyoming; Lance Formation, Maestrichtian.

Aquilapollenites pulcher Funkhouser, 1961, p. 198, Pl. 1, figs. 7a-7c; western flank of Rawlins uplift, Rawlins, Wyoming; Lance Formation. Maestrichtian.

Aquilapollenites pulvinus Stanley, 1961, p. 347-348, Pl. 7, figs. 1-12; Crow Butte, Harding County, South Dakota; Hell Creek Formation, Maestrichtian.

Aquilapollenites hirsutus Srivastava, 1969, p. 139, Pl. V, fig. 30; Alberta, Canada; Edmonton Formation, Maestrichtian.

USGS Records of Occurrence

Formation	*Stage or series*	*USGS paleobotany locality*	*Locality*
	Undated Upper Cretaceous	D1750-B	Wyoming
Lance (upper part)	Upper Maestrichtian	D3463 D1602	Wyoming
Hell Creek (upper part)	Upper Maestrichtian	D3014-A D3472-I, -J D3474-B D3730-G, -H, -I	Montana
Hell Creek (middle part)	Upper and middle Maestrichtian	D3730-B, -C, -E	Montana
Lance (lower part)	Middle Maestrichtian	D1603	Wyoming
Lance (lower part)	Maestrichtian	D1501-C	Wyoming
Hell Creek (lower part)	Middle Maestrichtian	D3472-D, -F D3690-C, -D	Montana
Fox Hill Sandstone, Colgate Member	Middle or lower Maestrichtian	D3690-B D3754-B	Montana

Formation	*Stage or series*	*USGS paleobotany locality*	*Locality*
Bearpaw Shale	Lower Maestrichtian or upper Campanian	D3012	Montana
Pierre Shale, Virgin Creek Member	Lower Maestrichtian or upper Campanian	D1375-9	South Dakota
Pierre Shale, lower unnamed shale member	Upper Campanian	D1330-19	Wyoming
Miner Creek	Upper Campanian	D1612	Montana
Bearpaw Shale (basal part) or Judith River (uppermost part)	Upper Campanian	D4067	Montana
Judith River (upper part)	Upper Campanian	D3725-A, -C	Montana
Fruitland (basal part)	Upper Campanian	D3980-A	Colorado
Judith River (lower part)	Middle Campanian	D3740-B	Wyoming
Pierre Shale, Mitten Black, Shale Member	Middle or lower Campanian	D1330-4	Wyoming

Description. The following description is based on more than 150 specimens (including the holotype) of *A. quadrilobus* found on the type slide and more than 40 specimens found on USGS paleobotany slides. Tricolpate, heteropolar pollen grains with three equatorial protrusions. *In equatorial view*: grains with a domelike major pole, a nodelike to nearly flat minor pole, and three equatorial protrusions drooping slightly toward minor pole; polar axis 28 to 44 μ, equatorial diameter (including equatorial protrusions) 36 to 65 μ; major polar protrusions usually larger than, but in some specimens equal to, or rarely smaller than, equatorial protrusions. *In polar view*: grains triangular, with extended corners; outer ends of equatorial protrusions slightly flattened in plane parallel to polar axis. *Apertures*: colpi extending entire length of equatorial protrusions, terminating approximately at junction of protrusions with body; in some specimens colpi appearing to anastomose at minor pole (Pl. 2, fig. 4), apertures usually not visible in equatorial views, best seen in oblique and polar views of grains. *Pollen wall*: exine two-layered, endexine (exclusive of axillary costae) thin; costae lens-shaped to bow-shaped, up to 3 μ thick in thickest part, occupying the axillae, usually terminating near base of polar protrusions, some extending a short distance onto base of major polar protrusions and a few appearing to anastamose on minor pole; costae often faintly visible or lacking (*see* thin lens-shaped areas, Pl. 1, fig. 5A, from which costae have apparently been dissolved during chemical treatment of grains); ektexine about 1 μ thick on major pole, often as much as 1.5 μ thick on minor pole and in axillae; in equatorial view of grains exine gradually thinning from base of equatorial protrusions outward toward their free ends; exine thin at colpal margins at ends of equatorial protrusions; ektexinal elements columnar; columellae fused at tips, forming partial tectum and resulting in a finely reticulate to foveolate surface sculpture; sculpture usually finer on equatorial protrusions than on polar protrusions, some grains having a collar of slightly coarser reticulum surrounding approximate lower half of major polar protrusion; on poorly preserved grains reticulate (or foveolate) sculpture often faintly visible or surface appearing smooth; spinae projecting as much as 2 μ above surface of muri, ornamenting polar protrusions; on polar areas spinae usually perpendicular to surface, on neck of major polar protrusion spinae tending to point toward major pole; spinae on equatorial protrusions usually smaller than on polar protrusions, tending to point toward polar axis of grain; along colpal margins at free ends of equatorial protrusions spinae usually retroflexed.

Remarks. The slide containing the type specimen of *A. quadrilobus* had been misplaced but was recently found by Dr. Wayne W. Brideaux of McMaster University, Ontario, Canada, and generously loaned to us. The type specimen was found on the slide and rephotographed (*see* Pl. 2, figs. 1A, 1B). *A. quadrilobus* was abundantly represented on the slide: more than 150 specimens were found and nearly 50 of them were photographed in polar, equatorial, and tilted orientations. The specimens varied, as in our description, in shape, size, ornamentation, and apparently in length of apertures; variables were gradational and the specimens were considered to be one natural group or species.

The holotype of *A. quadrilobus* in the original illustration (Radforth and Rouse, 1954, Pl. 1, fig. 14) is oriented with the major polar protrusion on the left and the minor polar protrusion on the right. The minor polar protrusion is not clear in the photographs because it is obscured by one of the equatorial protrusions. One equatorial protrusion points upward, the second one downward, and the third to the right—although the third is not clear in the photograph. The holotype is in a tilted equatorial view as can be seen in our photographs (Pl. 2, figs. 1A, 1B).

Srivastava's (1968a, p. 669, p. 696) ratio for the separation of *Mancicorpus* and *Aquilapollenites* was applied to 26 of the photographed specimens (the only ones in equatorial view) of *A. quadrilobus* from

PLATE 2. THE TYPE SPECIES AND A NEW TAXON OF *AQUILAPOLLENITES*

All magnifications ×750.

Figure

1A, 1B. *Aquilapollenites quadrilobus* Rouse. Oblique equatorial view of holotype, two focal levels. Mc. U.P.C. slide C_s-31 at USGS coordinates 110.4 × 20.6.

2. *Aquilapollenites quadrilobus* Rouse. Polar view. Mc. U.P.C. slide C_s-31 at USGS coordinates 119.0 × 6.7.

3. *Aquilapollenites quadrilobus* Rouse. Oblique equatorial view. Mc. U.P.C. slide C_s-31 at USGS coordinates 100.2 × 21.7.

4. *Aquilapollenites quadrilobus* Rouse. Polar view. Mc. U.P.C. slide C_s-31 at USGS coordinates 83.7 × 21.0.

5. *Aquilapollenites quadrilobus* Rouse. Equatorial view. Mc. U.P.C. slide C_s-31 at USGS coordinates 89.4 × 4.5.

6. *Aquilapollenites quadrilobus* Rouse. Equatorial view. USGS paleobotany locality D3740-B, slide 1, at 105.8 × 8.1.

7A, 7B. *Aquilapollenites turbidus* n. sp. Equatorial view of holotype, two focal levels. USGS paleobotany locality D1330-14, slide 3, at 79.0 × 15.3.

8A, 8B. *Aquilapollenites turbidus* n. sp. Oblique polar view. USGS paleobotany locality D1330-16, slide 4, at 107.7 × 14.2.

9. *Aquilapollenites turbidus* n. sp. Equatorial view. USGS paleobotany locality D1330-5, slide 3, at 76.1 × 20.8.

10. *Aquilapollenites turbidus* n. sp. Oblique equatorial view. USGS paleobotany locality D1375-13, slide 1, at 105.2 × 21.5.

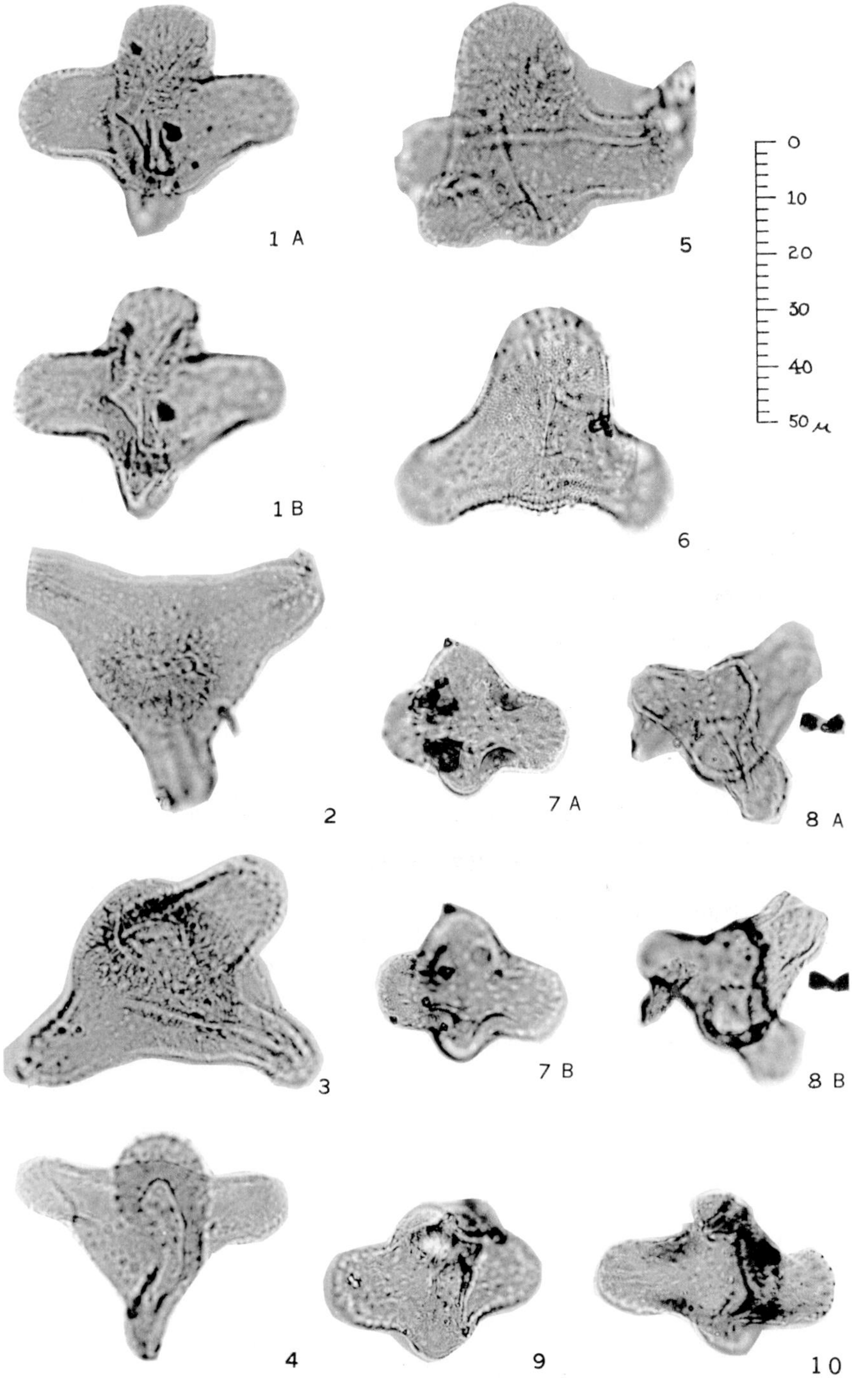

THE TYPE SPECIES AND A NEW TAXA OF *AQUILAPOLLENITES*

TSCHUDY AND LEOPOLD, PLATE 2
Geological Society of America Special Paper 127

Rouse's type slide. Nineteen of the specimens were found to conform to *Mancicorpus* (of Srivastava) and the remainder to *Aquilapollenites* (of Srivastava). Many of the grains were slightly tilted from true equatorial views, as was the holotype; consequently, measurements could not be accurately made. We were able to conclude, however, that most of these *A. quadrilobus* grains, according to Srivastava's definition, would belong in *Mancicorpus*. But *A. quadrilobus* cannot be transferred to *Mancicorpus* because it is the type species of *Aquilapollenites*.

It is surprising to us that Srivastava (1969, p. 139) placed *A. hirsutus* in *Aquilapollenites*, because, according to our understanding of his definition for separation of *Mancicorpus* and *Aquilapollenites*, *A. hirsutus* would belong in his genus *Mancicorpus*.

We have examined the type specimens of *A. polaris* Funkhouser and *A. pulcher* Funkhouser and believe them to be indistinguishable from each other and synonymous with *A. quadrilobus*. As *A. polaris* and *A. pulcher* were described by Funkhouser, the main distinction between them was that the "apertural areas" of *A. polaris* are covered with punctae and the "apertural areas" of *A. pulcher* are devoid of ornamentation. "Punctae" can be clearly seen on the "apertural areas" of the holotype of *A. polaris*, probably because the grain is fairly well preserved and translucent, whereas the heavily stained local wall thickenings (costae) of the holotype of *A. pulcher* may obscure sculptural detail in the "apertural areas." The presence or lack of punctae on the "apertural areas" is not a clear-cut feature and so does not seem adequate for the separation of species. Also, Funkhouser stated that *A. polaris* may be syncolpate at the reduced pole. We agree that it may be syncolpate at the reduced pole, but this feature is observable only in the polar view (paratype specimen). Funkhouser did not furnish a polar view of *A. pulcher*, and consequently colpal endings at the reduced pole in that species could not be observed. In the description of *A. pulcher*, he mentioned a band of "punctae" surrounding the long polar protrusions; however, the holotype of *A. polaris* also has such a band of slightly coarser reticulum in the same general area. The over-all surface sculpture of the *A. pulcher* holotype appears indistinct, perhaps due to poor preservation of the grain.

We have also placed *A. pulvinus* Stanley in synonymy with *A. quadrilobus*. We have examined the holotype of *A. pulvinus* and are of the opinion that it is not distinct from *A. quadrilobus*. The minor pole appears almost flat in the photographs of the holotype of *A. pulvinus*; however, similar grains were observed on the type slide of *A. quadrilobus* where an abundance of variable but intergrading forms was found.

We have not examined the holotype of *A. hirsutus* Srivastava but have placed it in synonymy with *A. quadrilobus* because Srivastava's (1969, Pl. V, fig. 30) illustrations of *A. hirsutus* so strikingly resembles many of the specimens of the specimens of *A. quadrilobus* that we found on the type slide of *A. quadrilobus* (for example, *see* our Pl. 2, fig. 5).

Srivastava's description of *A. hirsutus* fits in all important respects Rouse's type slide specimens of *A. quadrilobus*. Srivastava did state that the colpi of *A. hirsutus* are short and that they are restricted within the apical regions, whereas we stated in our description that the colpi of *A. quadrilobus* extend the entire length of the equatorial protrusions. In equatorial views of grains, such as that of the holotype of *A. hirsutus,* apertures are difficult to observe; however, our observations of colpi length are based on the analysis of a large number of specimens in many different orientations. In our experience colpi in the genus *Aquilapollenites* extend approximately as far as the poleward end of the endexinal costae. Srivastava's holotype specimen has costae extending to the base of the polar protrusions, indicating a similar extent of the colpi.

Mtchedlishvili (*in* Samoilovitch and others, 1961, p. 210-211) suggested the substitution of Rouse's (1957, Pl. II, fig. 9) specimen for the original designated type specimen of *A. quadrilobus.* Such a change of type specimen would be unacceptable because according to the International Code (Lanjouw and others, 1961, p. 64): "The choice made by the original author, if definitely expressed at the time of the original publication of the name of the taxon, is final."

Similar Species. *"Mancicorpus" solidum* Mtchedlishvili (*in* Samoilovitch and others, 1961, p. 223-224, Pl. 72, fig. 1a) from the Senonian-Maestrichtian of Siberia resembles *A. quadrilobus,* except that *"M." solidum* has spines mainly confined to the equatorial protrusions and the "distal" side of the body whereas *A. quadrilobus* has spines on the equatorial protrusions and on both sides of the body (on both polar protrusions). Mtchedlishvili (1965, p. 27) stated that *A. polaris* and *A. pulvinus* are similar to *M. solidum;* as noted in the synonymy above, we consider *A. polaris* and *A. pulvinus* equal to *A. quadrilobus.*

Aquilapollenites senonicus (Mtchedlishvili, 1961) n. comb.
(Pl. 1, fig. 2)

Mancicorpus senonicum Mtchedlishvili, 1961, *in* Samoilovitch and others, 1961, p. 224-225, Pl. 72, figs. 2a-3.

Holotype. Plate 72, figure 2a-2c, western Siberian lowlands, River Sym, locality Kolokol'nikov Yar; Lower Symian subsuite, Senonian.

Mancicorpus senonicus Mtchedlishvili, 1961, *in* Srivastava, 1968b, p. 1488, Pl. 1, figs. 9, 10; Alberta, Canada; Edmonton Formation, Maestrichtian.

USGS Records of Occurrence

Formation	*Stage or series*	*USGS paleobotany locality*	*Locality*
?	Undated Upper Cretaceous	D1750-B	Wyoming
?	Undated Upper Cretaceous	D1799	Alaska*

Formation	*Stage or series*	*USGS paleo-botany locality*	*Locality*
Colville Group	Upper Cretaceous	D3124-B, -G	Alaska*
Lance (upper part)	Upper Maestrichtian	D1602	Wyoming
Hell Creek (middle part)	Upper and middle Maestrichtian	D3730-A, -D	Montana
Hell Creek (lower part)	Middle Maestrichtian	D3147-B D3472-C, -D, -E, -F D3690-C, -D, -H	Montana
Fox Hills Sandstone	Lower Maestrichtian	D1331-2, -5	Wyoming
Lance (lower part)	Middle Maestrichtian	D1603	Wyoming
Fox Hills Sandstone, Colgate Member	Middle or lower Maestrichtian	D3690-B	Montana
Hell Creek (basal part)	Middle Maestrichtian	D3472-B D3754-A	Montana
Judith River (upper part)	Upper Campanian	D3725-A	Montana
Fruitland	Upper Campanian	D3980-E	Colorado
Judith River (lower part)	Middle Campanian	D3740-B	Wyoming
Pierre Shale, Mitten Black Shale Member	Middle or lower Campanian	D1330-5	Wyoming
Eagle Sandstone (middle part)	Lower Campanian	D3718-E	Montana

*Not plotted on stratigraphic range chart, since not a Rocky Mountain locality, but occurrence is included to show wide range.

Remarks. As explained earlier in this report (p. 122-124), we do not consider *Mancicorpus* a valid genus and therefore have transferred *Mancicorpus senonicum* to *Aquilapollenites.*

Aquilapollenites delicatus Stanley, 1961

Aquilapollenites delicatus Stanley, 1961, p. 346-347, Pl. 4, figs. 1-12.

Holotype. Stanley, 1961, Plate 4, figures 1-5; USNM 131311, slide SCB-11-20; SE¼ sec. 30 or NE¼ sec. 31, T. 15 N., R. 5 E., Crow Butte, Harding County, South Dakota; Hell Creek Formation, Maestrichtian.

Description. This description is based on an examination of Stanley's types and more than 60 of our own specimens. The circumscription of *A. delicatus* is expanded slightly to include a new variety, *A. delicatus* var. *collaris,* which is described in this report (*see* p. 135). Tricolpate, heteropolar pollen grains with three equatorial protrusions. *In equatorial view*: polar axis 28 to 45 μ; equatorial diameter (including equatorial protrusions) 38 to 67 μ; equatorial protrusions 13 to 30 μ long, 10 to 19 μ wide; length and width of equatorial protrusions usually approximately equal to corresponding dimensions of major polar protrusion; equatorial protrusions and major polar protrusion having broadly rounded ends; equatorial protrusions inclined slightly toward minor pole, or sometimes oriented at right angles to polar axis of grain; minor polar protrusion short, protruding very slightly. *In polar view*: grains triangular, with slightly to extremely concave sides; equatorial protrusions paddle-shaped, flattened in plane parallel to polar axis of grain. *Apertures*: colpi apparently spanning entire length of equatorial protrusions, in some cases appearing to extend a short distance onto base of major polar protrusion; in some specimens colpi agape at ends of equatorial protrusions—the open colpi thence continuous with a narrow unornamented line (or closed furrow), which terminates at approximate base of major and minor polar protrusions; distinct slightly open colpi, which reach full length of equatorial protrusions, visible in some specimens in polar view. *Pollen wall*: exine two-layered; endexine (exclusive of costae) as much as 0.5 μ thick on polar protrusions, thinner on equatorial protrusions; axillary endexinal costae (when present) as much as 3 μ thick in thickest part, bow-shaped to sausage-shaped, occupying inner one-half to two-thirds the length of the equatorial protrusions, extending slightly onto base of major polar protrusion, reaching almost to minor pole; some endexinal costae only faintly visible, some lacking (probably were removed by chemical processing during removal of grains from rock matrix); ektexine (exclusive of spinae) very thin to as much as 1.0 μ thick on equatorial protrusions, 0.5 to 1.5 μ thick on polar protrusions; on equatorial protrusions ektexinal granula and columellae closely spaced, surface sculpture granulate to finely foveolate-reticulate; on polar protrusions ektexinal columellae more dispersed, surface sculpture varying from fine to fairly coarse reticulate; lumina of reticulum as much as 2 μ in diameter, muri less than 0.5 μ wide; conical spinae projecting 0.5 to 1.5 μ above surface of muri, scattered over polar protrusions, at poles tending to point perpendicular to surface, on neck of major polar protrusions spinae usually slightly inclined toward major pole; smaller spinae scattered over surface of equatorial protrusions, tending to point toward polar axis of grain; colpal margins bordered by a row of closely spaced conical spinae, which are usually retroflexed.

Aquilapollenites delicatus Stanley var. *delicatus*

(Pl. 1, fig. 13)

Holotype. Same as for *Aquilapollenites delicatus* Stanley, 1961.

USGS Records of Occurrence

Formation	*Stage or series*	*USGS paleobotany locality*	*Locality*
?	Undated Upper Cretaceous	D1750-B	Wyoming
Laramie (lower part)	Middle Maestrichtian	D1537	Colorado
Hell Creek (lower part)	Middle Maestrichtian	D3147-A D3472-C, -E D3690-C, -E, -F, -H D3754-A	Montana
Fox Hills Sandstone, Colgate Member	Middle or lower Maestrichtian	D3690-B	Montana
Pierre Shale, Virgin Creek Member	Lower Maestrichtian or upper Campanian	D1375-9	South Dakota

Description. Same as for *A. delicatus* (described above, p. 132-133), except as follows: polar axis 28 to 39 μ, equatorial diameter (including equatorial protrusions) 38 to 50 μ, equatorial protrusions 13 to 22 μ long and 10 to 15 μ wide; reticulate sculpture of major polar protrusion nearly uniform from base of protrusion to pole or in some specimens distinctly finer on polar dome; ektexine of equatorial protrusions 0.5 to 1.0 μ thick, elements columnar and spaced at intervals about equal to their diameter.

Remarks. Our specimens that we have assigned to *A. delicatus* var. *delicatus* appear to be very similar to Stanley's holotype and paratype and to his designated polar view of that species.

A. delicatus var. *delicatus* resembles *Aquilapollenites senonicus* (Mtchedlishvili) n. comb.; however, the two taxa can easily be distinguished because the variety has prominent spines on the polar protrusions and *A. senonicus* lacks spines.

A. delicatus var. *delicatus* slightly resembles *A. quadrilobus* Rouse but can easily be distinguished by the much more coarsely reticulate sculpture on its major polar protrusion.

Aquilapollenites delicatus Stanley var. *collaris* n. var. (Pl. 1, fig. 14; Pl. 3, fig. 6)

Holotype. USGS paleobotany locality D3147-B, slide 2, at 87.2 × 16.5; Plate 3, figure 6; Plate 1, figure 14; almost directly west of recreation camp at Fort Peck State Park, Montana, sec. 31, T. 22 N., R. 38 E.; lower part of Hell Creek Formation (43.5 feet above contact of Hell Creek Formation and Fox Hills Sandstone), middle Maestrichtian.

USGS Records of Occurrence

Formation	*Stage*	*USGS paleobotany locality*	*Locality*
Hell Creek (upper part)	Upper Maestrichtian	D3472-I, -J D3730-G, -H, -I	Montana
Hell Creek (middle part)	Upper and middle Maestrichtian	D3730-C	Montana
Hell Creek (lower part)	Middle Maestrichtian	D3147-A, -B D3472-C D3690-F	Montana

Description. Same as for *Aquilapollenites delicatus* (described above, p. 132-133), except as follows: polar axis 40 to 45 μ, equatorial diameter (including equatorial protrusions) 51 to 67 μ, equatorial protrusions 25 to 30 μ long and 10 to 19 μ wide; a well-defined collarlike zone ($\pm$10 μ wide) of thinner ektexine and finer reticulate sculpture separates more coarsely reticulate polar dome from more coarsely reticulate basal half($\pm$) of major polar protrusion; many specimens having neck of major polar protrusions slightly constricted; ektexine of equatorial protrusions very thin, elements consisting of granules.

Remarks. In equatorial view, this variety is easily separable from *A. delicatus* var. *delicatus* by its collar of distinctly finer reticulate sculpture below the polar dome of the major polar protrusion. Polar views of specimens of the two varieties are often indistinguishable.

A. delicatus var. *collaris* is common in the upper and middle parts of the Hell Creek Formation of the Fort Peck and Glendive localities of northern Montana; it was also found in the lower part of the Hell Creek Formation, where it occurs with *A. delicatus* var. *delicatus*. We have not found *A. delicatus* var. *delicatus* in the upper or middle parts of the Hell Creek Formation at the sites mentioned above.

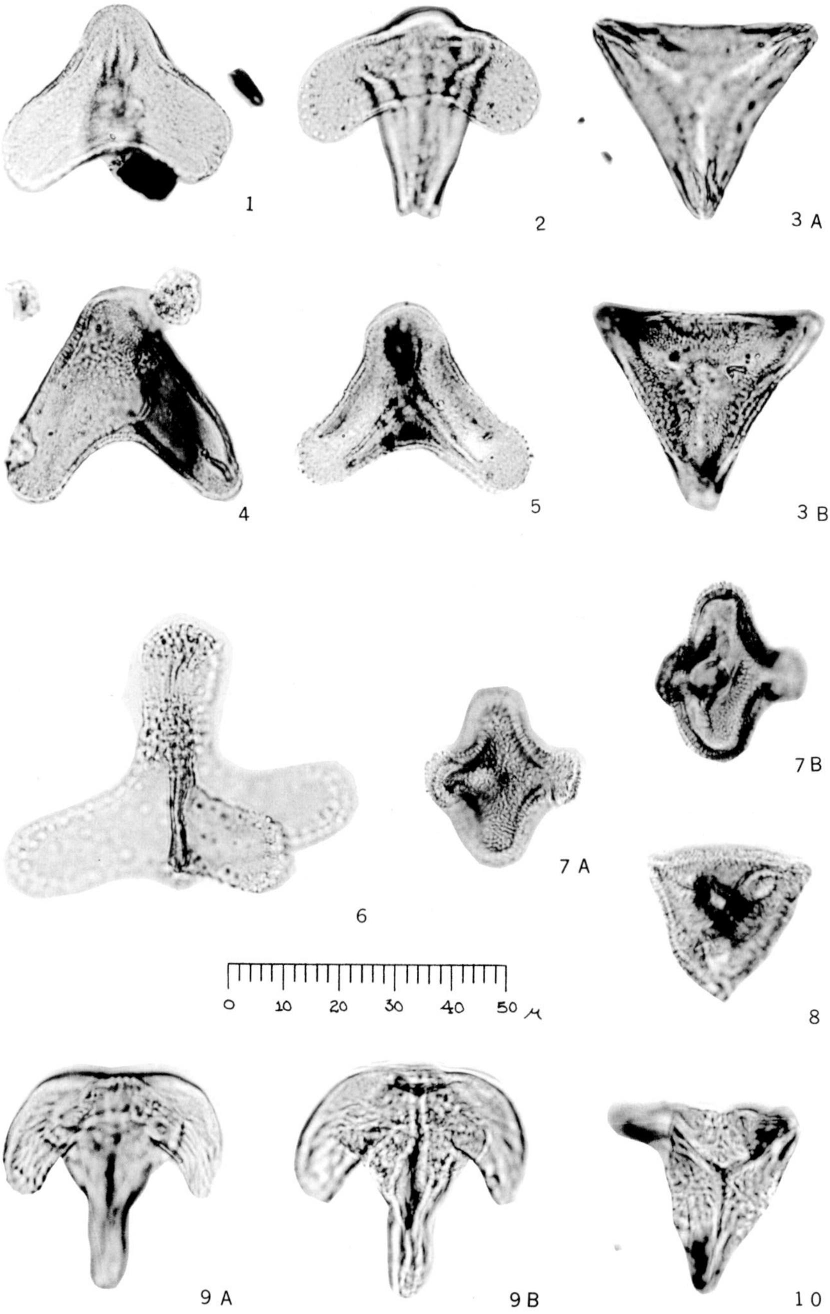

NEW AND EXISTING TAXA OF *AQUILAPOLLENITES*

TSCHUDY AND LEOPOLD, PLATE 3
Geological Society of America Special Paper 127

PLATE 3. NEW AND EXISTING TAXA OF *AQUILAPOLLENITES*

All magnifications ×750.

Figure

1. *Aquilapollenites calvus* n. sp. Equatorial view of holotype. USGS paleobotany locality D1330-16, slide 6, at 104.3 × 21.6.
2. *Aquilapollenites calvus* n. sp. Oblique equatorial view. USGS paleobotany locality D3725-C, slide 4, at 85.5 × 22.3.

3A, 3B. *Aquilapollenites calvus* n. sp. Polar view, two focal levels. USGS paleobotany locality D3725-C, slide 3, at 83.5 × 13.5.

4. *Aquilapollenites calvus* n. sp. Equatorial view. USGS paleobotany locality D3725-C, slide 3, at 96.4 × 16.9.
5. *Aquilapollenites calvus* n. sp. Equatorial view. USGS paleobotany locality D1375-13, slide 1, at 100.9 × 13.6.
6. *Aquilapollenites delicatus* var. *collaris* n. var. Equatorial view of holotype, high focus. USGS paleobotany locality D3147-B, slide 2, at 87.2 × 16.5.

7A, 7B. *Aquilapollenites catenireticulatus* Srivastava. Equatorial view, two focal levels. USGS paleobotany locality D3147-B, slide 4, at 103.7 × 10.8.

8. *Aquilapollenites catenireticulatus* Srivastava. Polar view. USGS paleobotany locality D3147-B, slide 2, at 98.6 × 15.1.

9A, 9B. *Aquilapollenites tripodiformis* n. sp. Equatorial view of holotype, two focal levels. USGS paleobotany locality D1330-16, slide 4, at 78.9 × 15.2.

10. *Aquilapollenites tripodiformis* n. sp. Polar view of paratype. USGS paleobotany locality D1330-16, slide 4, at 101.9 × 2.0.

The grain assigned by Drew and Tschudy (1968, Pls. 1, 2) to *A. delicatus* Stanley probably belongs to *A. delicatus* var. *collaris*.

A. delicatus var. *collaris* resembles *A. paplionis* Srivastava (1969, p. 140, Pl. V, figs. 33-36) but can be distinguished by its long major polar protrusion and by its well-defined collarlike zone of thinner ektexine with finer reticulate sculpture at the base of the major polar dome. These taxa are probably closely related to each other and to *A. delicatus* var. *delicatus*.

Aquilapollenites calvus n. sp.
(Pl. 1, fig. 4; Pl. 3, fig. 1-5)

Holotype. USGS paleobotany locality D1330-16, slide 6, at 104.3 × 21.6; Plate 3, figure 1; NW¼NE¼ sec. 23, T. 38 N., R. 62 W.; Black Hills area, Niobrara County, Wyoming; lower unnamed shale member, Pierre Shale, upper Campanian.

Paratype. USGS paleobotany locality D1330-16, slide 6, at 114.5 × 10.5; Plate 1, figure 4; same locality data as for holotype.

USGS Records of Occurrence

Formation	*Stage*	*USGS paleobotany locality*	*Locality*
Denver (lower part)	Upper Maestrichtian	D1225-K	Colorado
Laramie (lower part)	Middle Maestrichtian	D1537	Colorado
Pierre Shale, Virgin Creek Member	Lower Maestrichtian or upper Campanian	D1375-10	South Dakota
Pierre Shale, DeGrey Member	Upper Campanian	D1375-13	South Dakota
Bearpaw (basal part) or Judith River (uppermost part)	Upper Campanian	D4067	Montana
Judith River (upper part)	Upper Campanian	D3725-A, -C	Montana
Judith River (middle part)	Upper Campanian	D4069	Montana
Pierre Shale, lower unnamed shale member	Upper Campanian	D1330-16	Wyoming
Judith River (lower part)	Upper or middle Campanian	D3726-E	Montana

Formation	*Stage*	*USGS paleobotany locality*	*Locality*
Pierre Shale, Gregory Member	Middle Campanian	D1375-14	South Dakota
Pierre Shale, Mitten Black Shale Member	Middle or lower Campanian	D1330-5	Wyoming

We have also found this species in Rouse's slide Leth. No. 8, 1:5(i) GER from the Oldman Formation (Upper Cretaceous) from south-central Alberta, Canada.

Description. This description is based on the examination of more than 50 specimens in equatorial and oblique views and 5 specimens in polar view. Tricolpate, heteropolar pollen grains with three equatorial protrusions. *In equatorial view*: shape of grains tripodal, junction of major polar protrusion with equatorial protrusions forming plane to slightly concave surfaces; major pole domelike, minor pole not protruding; equatorial protrusions with broadly rounded ends, nearly uniform in width throughout their length, or some slightly broadened or narrowed near free ends; polar axis 16 to 31 μ; diameter (including equatorial protrusions) 28 to 47 μ (rarely as much as 59 μ); width of equatorial protrusions 14 to 20 μ. *In polar view*: grains triangular; with plane, slightly concave or slightly convex sides; with corners slightly extended or not extended. *Apertures*: colpi extending full length of equatorial protrusions, meeting in some specimens at minor pole, reaching nearly to major pole. *Pollen wall*: exine two-layered; endexine thin or as much as 0.5 μ thick; endexinal costae (and costal cavities as much as 3 μ thick in thickest part, extending onto equatorial protrusions for one-half to two-thirds the length of the protrusions, reaching nearly to major pole, reaching to and joining at minor pole; many specimens lacking endexinal thickenings but having light-colored axillary areas, which probably indicate cavities from which costal thickenings have dissolved during chemical processing of sample; ektexine as much as 2 μ thick, and total wall as much as 3 μ thick, both usually thickest below major polar dome in areas where equatorial protrusions merge with body; ektexine thin at colpal margins near free ends of equatorial protrusions; ektexinal elements columnar, in some specimens very distinct, in others only faintly visible even at 1000 magnification; elements fused at tips—resulting in a finely reticulate to rugulate surface sculpture; distinct verrucae and baculae as much as 1 μ long scattered over surface of grain, except usually lacking from or very sparse on major polar dome; a row of larger spines (1 to 2 μ long) present along colpal margins at rounded free ends of equatorial protrusions, spine tips usually pointing backward toward polar axis, bases of these retroflexed spines in some views appearing to be blunt projections at the colpal margins.

Remarks. This species was given the name *calvus* (=bald) because of its bald or virtually unornamented major polar dome.

Similar Species. *A. calvus* resembles *"Mancicorpus" anchoriforme* Mtchedlishvili (*in* Samoilovitch and others, 1961, p. 219-220, Pl. 70, figs. 3a-4b) from the Senonian and Maestrichtian-Danian from the eastern part of the western Siberian lowlands of the U.S.S.R.; however, the two species differ in the following respects: *"M." anchoriforme* (1) has equatorial protrusions not forming angles at contact with the body, (2) has dull spines located mainly on the "distal(?)" surface of the grain and on the terminal ends of the equatorial protrusions, and (3) as illustrated and described by Mtchedlishvili, has equatorial protrusions tending to be longer in proportion to their width and becoming narrower at their ends. *A. calvus* has (1) equatorial protrusions that either do or do not form angles at contact with the body, (2) verrucae and baculae scattered over all of surface except on major polar dome, (3) equatorial protrusions tending to be shorter in proportion to their width, and usually with broadly rounded ends, and (4) a row of sharp-pointed retroflexed spines along colpal margins at free ends of equatorial protrusions (this feature is not mentioned or shown in illustrations of *"M." anchoriforme*).

A. calvus also resembles *"Mancicorpus" notabile* Mtchedlishvili (*in* Samoilovitch and others, 1961, p. 220-221, Pl. 71, figs. 1a-2b) from the Senonian from the eastern part of the western Siberian lowlands of the U.S.S.R., except that *"M." notabile* does not have verrucae (or) baculae ornamenting the surface of the exine.

A. calvus resembles *"Triceropollis" unicus* Chlonova (1961, p. 86, Pl. XV, fig. 110), and the latter species appears very similar to and may be synonymous with *"Mancicorpus" anchoriforme.*

Srivastava's (1968b. Pl. II, figs. 15, 16) specimen assigned to *Mancicorpus anchoriforme* Mtchedlishvili resembles *A. calvus;* however, Srivastava does not mention the row of large spines along the colpal margins at the free ends of the equatorial protrusions that characterize *A. calvus.*

A. calvus is distinguishable from *"Mancicorpus" albertensis* Srivastava (1968b, p. 1486, Pl. I, figs. 1, 2) by its verrucate and baculate surface ornamentation and by the distinct row of spines along the colpal margins at the free ends of the equatorial protrusions.

Aquilapollenites tripodiformis n. sp.
(Pl. 1, figs. 8A-8B; Pl. 3, figs. 9A-10)

Holotype. USGS paleobotany locality D1330-16, slide 4, at 78.9 × 15.2; Plate 3, figures 9A-9B; Plate 1, figure 8A; NW¼NE¼ sec. 23,

T. 38 N., R. 62 W.; Black Hills area Niobrara County, Wyoming; lower unnamed shale member, Pierre Shale, Upper Campanian.

Paratype. USGS paleobotany locality D1330-16, slide 4, at 101.9 × 2.0; Plate 3, figure 10; same locality data as for holotype.

USGS Records of Occurrence

Formation	*Stage*	*USGS paleobotany locality*	*Locality*
Pierre Shale, lower unnamed shale member	Upper Campanian	D1330-16	Wyoming
Fruitland	Upper Campanian	D3980-A	Colorado
Judith River (middle part)	Upper Campanian	D4069	Montana
Pierre Shale, Gregory Member	Middle Campanian	D1375-14	South Dakota

Description. This description is based on 12 specimens in equatorial view (variously inclined) and 2 specimens in polar view. Trisyncolpate, heteropolar pollen grains with three equatorial protrusions. *In equatorial view*: shape of grains tripodal; major pole flat, slightly rounded, or somewhat pointed; minor pole concave; polar axis short (about 7 to 12 μ); equatorial diameter (including equatorial protrusions) 35 to 45 μ; equatorial protrusions paddle-shaped, slightly curved, flattened in plane parallel to polar axis of grain, dipping toward minor pole at about 45° from polar axis. *In polar view*: grains triangular, with plane sides and extended corners. *Apertures*: colpi extending full length of equatorial protrusions and joining at both major and minor poles (trisyncolpate). *Pollen wall*: exine apparently two-layered; endexinal costae if present not visible in our specimens; wall thick and unstructured at major pole; wall thinning from base of equatorial protrusions toward their tips; ektexine (incorporating sculpture) may be as much as 2 μ thick at equator on body, but endexine not discernible; sculpture of ektexine rugulate on body of grain at concave pole, echinate and striate on equatorial protrusions, striate to granulate on major pole; spinae at free ends of equatorial protrusions retroflexed; from tips of equatorial protrusions inward, spinae and ridges forming more or less continuous parallel striations; striations becoming discontinuous on poles.

Remarks. Included in the group of specimens described under *Aquilapollenites tripodiformis* n. sp. are 13 specimens (from USGS paleobotany locality D1330-16) that appear to be very similar to each other in size, shape, and sculpture; a single specimen from USGS paleobotany locality D1375-14 also included in the description is very similar but has slightly finer striations and spines and more prominent granules. Several additional specimens that were recently found in USGS paleobotany localities D3980-A and D4069 conform to the species as described above.

Three specimens from the Hell Creek Formation of northeastern Montana at localities USGS D3011 and D3012 resemble *A. tripodiformis* yet appear to be distinct from that species. These grains are larger, have longer polar axes, have narrower equatorial protrusions, are reticulate on the body at the minor pole, and have finer striations on the equatorial protrusions. Two of the grains have slightly protruding major and minor poles. From our limited material we are unable to ascertain whether these specimens are colpate or syncolpate. These specimens can probably be assigned to *"Mancicorpus" rostratus* Srivastava (1968b, p. 1487-1488, Pl. II, figs. 11, 12) or to *"Mancicorpus" vancampoi* Srivastava (1968b, p. 1489-1490, Pl. II, figs. 13, 14).

Similar Species. In shape and size, equatorial views of *A. tripodiformis* resemble *"Mancicorpus" anchoriforme* Mtchedlishvili (*in* Samoilovitch and others, 1961, p. 219-220, Pl. 70, figs. 3a-3e); however, in polar views the two species do not resemble each other. *"M." anchoriforme,* as illustrated, is not syncolpate, whereas *A. tripodiformis* is distinctly syncolpate. The striate sculpture is a conspicuous feature of *A. tripodiformis* but is not described for, or shown in the illustrations of, *"M." anchoriforme.*

Our specimens also resemble *"Tricerapollis" unicus* Chlonova, 1961, and the latter species strikingly resembles *"M." anchoriforme* Mtchedlishvili. Chlonova did not mention whether *"T." unicus* is simply colpate or syncolpate; she merely stated that the pore opening was narrow, elongate, slitlike. Drawings of the holotype of *"T." unicus* (Chlonova, 1961, Pl. XV, figs. 110, 110a) do not show striations on the equatorial protrusions. Chlonova (1961, p. 86) stated (translation by I. Mittin), "sculptural ornaments are uniform and of dense dotted appearance and there are sparsely distributed light [colored] spines on the body of pollen and on the membrane of the pore." She further stated (translation by I. Mittin), "sometimes on the projection there is well-observed striated structure of exine." It is possible that Chlonova was dealing with more than one species, because her drawings of the holotype do not show striations and because she stated that "sometimes" striated structure was present. Our specimens always have striations on the equatorial protrusions.

As illustrated by Mtchedlishvili and by Chlonova, *"M." anchoriforme* and *"T." unicus* appear to have endexinal thickenings that end abruptly some distance from the free ends of the equatorial protrusions. In our specimens, the wall thins gradually from the base of the equatorial protrusions to the free ends of the protrusions.

Isopolar (to Subisopolar) Species

Aquilapollenites trialatus Rouse, 1957

Aquilapollenites trialatus Rouse, 1957, p. 371, Pl. 2, figs. 14, 15.

Holotype. Rouse, 1957, Plate II, figure 14, Mc. U.P.C.; slide Leth. No. 8, 1:5(ii); Lethbridge City Collieries, south-central Alberta, Canada; Oldman Formation, Upper Cretaceous.

Description. This description is based on many specimens, including (1) a few specimens from a slide of the type material loaned to us by Dr. Rouse, (2) many specimens from one sample collected along the Colville River, Alaska, and (3) many specimens from several localities in Montana, South Dakota, and Wyoming. The following description of *A. trialatus* is expanded to include two new varieties from the Rocky Mountain localities. Tricolpate, isopolar to subisopolar pollen grains with three equatorial protrusions. Polar axis 33 to 93 μ (60 to 100 μ, according to original description by Rouse, 1957); equatorial diameter of body (measured about halfway between equator and pole) 11 to 38 μ; equatorial protrusions (measured in equatorial view of grain) 14 to 29 μ long, 7 to 14 μ wide. *In equatorial view*: body of grain oval, with broadly rounded poles, usually distended in equatorial area where equatorial protrusions attach; equatorial protrusions gradually tapering from base to tip, or fairly abruptly tapering near free end. *In polar view*: shape of grains triangular, with narrow to rounded corners and plane to irregularly concave-convex sides. *Apertures*: colpi extending entire length of equatorial protrusions, in plane parallel to polar axis of grain, extending onto body for approximately two-thirds the distance to poles. *Pollen wall*: exine two-layered; body endexine (exclusive of costae) as much as 0.5 μ thick; axillary endexinal costae (when present) bow-shaped, as much as 4 μ thick in thickest part, tapering to thin ends, extending nearly to ends of equatorial protrusions and onto body of grain for about two-thirds the distance to poles; body ektexine 0.5 to 2 μ thick (in polar view of grain, ektexine as much as 3 μ thick at equator in interequatorial protrusion areas); ektexinal elements columnar, fused at tips—resulting in fine- to medium-reticulate surface sculpture; lumina of reticulum less than 0.5 or as much as 1.5 μ in diameter, muri as wide as 0.4 μ; reticulum of body uniform in size, or in wide equatorial belt becoming finer and often somewhat striate in fingerprint pattern; endexine of equatorial protrusions thinning toward free ends of protrusions; ektexine of equatorial protrusions thinner than body ektexine, thinnest along colpal margins at free ends of protrusions; equatorial protrusions smooth, finely granulate, or extremely finely reticulate, sometimes having fingerprint pattern adjacent to body of grain.

Aquilapollenites trialatus Rouse var. *trialatus*

Holotype. Same as for *Aquilapollenites trialatus* Rouse, 1957.

USGS Records of Occurrence

Formation	*Series*	*USGS paleobotany locality*	*Localty*
Colville Group	Upper Cretaceous	D3124-H	Alaska

Description. This description is based on specimens from Rouse's type material and from USGS paleobotany locality D3124-H; included were 20 specimens in equatorial view, 3 in oblique views, and 2 in polar view. Description same as for *A. trialatus* (described above, p. 143-144), except as follows: polar axis 52 to 93 μ; diameter of body (measured about halfway between one pole and equator) 21 to 38 μ; equatorial protrusions (measured in equatorial view of grain) 19 to 29 μ long, 10 to 14 μ wide. In equatorial view of grains some equatorial protrusions gradually tapering from base to tip but usually having more or less parallel sides for one-half to two-thirds their length, then abruptly tapering to rounded point at free end. Ektexine 1 to 2 μ thick, as much as 3 μ thick in interequatorial protrusion areas as seen in polar view of grains. In broad equatorial belt, reticulum finer to much finer than on remainder of body, sometimes arranged in fingerprint pattern; lumina of reticulum (except in equatorial belt) 0.5 to 1.5 μ in diameter, muri about 0.4 μ wide.

Remarks. This northern variety of *A. trialatus* in the New World tends to be larger than the two Rocky Mountain varieties, but the size ranges of the three varieties do overlap. *A. trialatus* var. *trialatus* is almost always more finely reticulate in a broad equatorial belt, whereas *A. trialatus* var. *uniformis* has a uniform or almost uniform reticulum over the entire body. *A. trialatus* var. *trialatus* tends to have (1) broader, less taper-pointed equatorial protrusions, and (2) less contrast between size of reticulum at polar and equatorial areas than *A. trialatus* var. *variabilis*.

In all respects the grains from the Alaskan sample (USGS paleobotany locality D3124-H) are similar to those of *A. trialatus* on the slide from Rouse's type material from Leth. No. 8.

Aquilapollenites trialatus Rouse var. *uniformis* n. var.
(Pl. 1, fig. 7; Plate 4, figs. 8A-8B)

Holotype. USGS paleobotany locality D1330-4, slide 1 at 106.1 × 6.7; Plate 4, figures 8A-8B; NE¼NE¼ sec. 23, T. 38 N., R. 62 W.; Niobrara County, Wyoming; Mitten Black Shale Member, Pierre Shale, middle or lower Campanian.

USGS Records of Occurrence

Formation	*Stage*	*USGS paleobotany locality*	*Locality*
Judith River (upper part)	Upper Campanian	D3725-C	Montana
Pierre Shale, Verendrye Member	Upper Campanian	D1375-12	South Dakota
Bearpaw (basal part) or Judith River (uppermost part)	Upper Campanian	D4067	Montana
Pierre Shale, Red Bird Silty Member	Middle Campanian	D1330-13A	Wyoming
Claggett Shale	Middle and lower Campanian	D3724-B, -E	Montana
Pierre Shale, Mitten Black Shale Member	Middle or lower Campanian	D1330-4	Wyoming
Pierre Shale, Sharon Springs Member	Lower Campanian	D1845	Wyoming

Description. This description is based on six specimens in equatorial view and three specimens in oblique and polar views. Grains resembling *A. trialatus* var. *trialatus,* except as follows: polar axis 48 to 62 μ; equatorial diameter of body (measured about halfway between one pole and equator) 14 to 17 μ; equatorial protrusions (measured in equatorial view of grains) 18 to 21 μ long, about 7 μ wide; equatorial protrusions usually tapering gradually from base to a narrow point at free end; body reticulum finer than that of *A. trialatus* var. *trialatus;* lumina of reticulum varying in shape from nearly isodiametric to irregular; body ornamentation at equator not noticeably finer than at

PLATE 4. NEW TAXA OF *AQUILAPOLLENITES*

All magnifications ×750.

Figure

1. *Aquilapollenites rigidus* n. sp. Equatorial view of holotype. USGS paleobotany locality D3718-I, slide 2, at 86.6 × 10.2.
2. *Aquilapollenites rigidus* n. sp. Equatorial view. USGS paleobotany locality D3718-I, slide 1, at 102.8 × 16.4.
3. *Aquilapollenites rigidus* n. sp. Oblique view. USGS paleobotany locality D3718-I, slide 1, at 99.3 × 1.6.

4A, 4B. *Aquilapollenites trialatus* var. *variabilis* n. var. Equatorial view of holotype, two focal levels. USGS paleobotany locality D3718-I, slide 2, at 88.4 × 3.7.

5. *Aquilapollenites trialatus* var. *variabilis* n. var. Oblique view. USGS paleobotany locality D3718-I, slide 1, at 101.8 × 8.8.
6. *Aquilapollenites trialatus* var. *variabilis* n. var. Equatorial view of paratype. USGS paleobotany locality D3718-I, slide 2, at 105.9 × 12.9.
7. *Aquilapollenites trialatus* var. *variabilis* n. var. Oblique equatorial view. USGS paleobotany locality D3718-I, slide 1, at 97.2 × 16.6.

8A, 8B. *Aquilapollenites trialatus* var. *uniformis* n. var. Equatorial view of holotype, two focal levels. USGS paleobotany locality D1330-4, slide 1, at 106.1 × 6.7.

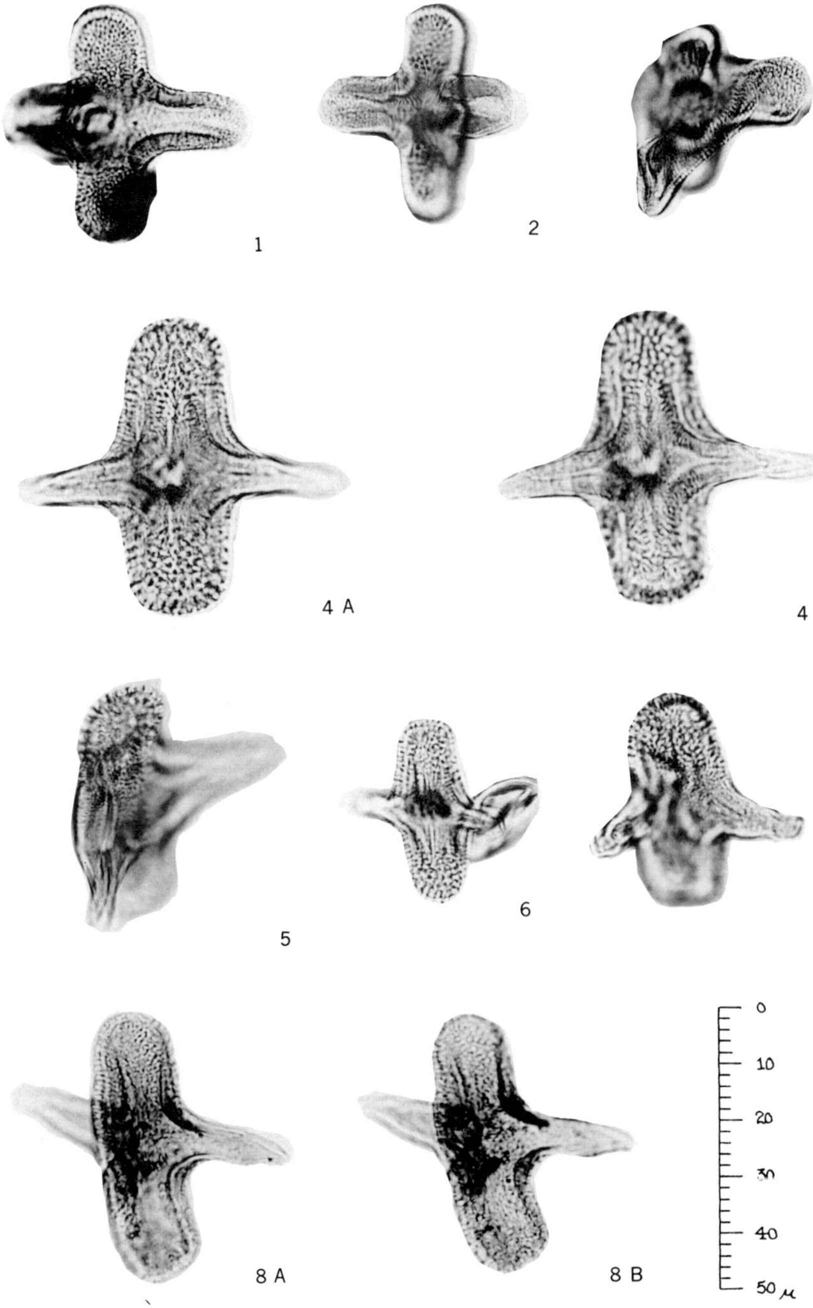

NEW TAXA OF *AQUILAPOLLENITES*

TSCHUDY AND LEOPOLD, PLATE 4
Geological Society of America Special Paper 127

poles; bases of equatorial protrusions sometimes having fingerprint pattern.

Remarks. This variety has been given the name *uniformis* because of its uniform body sculpture.

Similar Varieties. This Rocky Mountain variety is separable from *A. trialatus* var. *trialatus* and from *A. trialatus* var. *variabilis* by its uniform pole-to-equator body ornamentation. It usually has narrower, more gradually tapered equatorial protrusions and usually is smaller than var. *trialatus.*

Aquilapollenites trialatus Rouse var. *variabilis* n. var.
(Pl. 1, fig. 3; Pl. 4, figs. 4A-7)

Holotype. USGS paleobotany locality D3718-I, slide 2 at 88.4 × 3.7; Plate 4, figures 4A-4B; NW¼NW¼ sec. 13, T. 22 N., R. 17 E.; Fergus County, Montana; Eagle Sandstone, lower Campanian.

Paratype. USGS paleobotany locality D3718-I slide 2 at 105.9 × 12.9; Plate 4, figure 6; same locality data as for holotype.

USGS Records of Occurrence

Formation	*Stage*	*USGS paleobotany locality*	*Locality*
Claggett Shale	Middle and lower Campanian	D3724-C, -D	Montana
Eagle Sandstone (upper part)	Lower Campanian	D3718-I	Montana

Description. This description is based on 23 specimens in equatorial view and 5 specimens in oblique and polar views. Grains resembling *A. trialatus* var. *trialatus* except as follows: polar axis 33 to 59 μ; equatorial diameter of body (measured about halfway between one pole and equator) 11 to 31 μ; equatorial protrusions (measured in equatorial view of grains) 14 to 25 μ long, 7 to 10.5 μ wide, (measured at widest part, near body); equatorial protrusions tapering throughout their length to a narrow point at free end; lumina of body reticulum in some specimens commonly 1.5 to 2 μ in diameter, in other specimens commonly 0.5 to 1 μ in diameter, and in others lumina intermediate in size. Reticulum of body much finer in equatorial belt than on remainder of body—this gradation, as well as a tendency toward fingerprint ornamentation in the equatorial belt, is more conspicuous than in *A. trialatus* var. *trialatus.*

Remarks. This Rocky Mountain variety from the middle and lower Campanian has been given the name *variabilis* because of its wide size range, and the variability in coarseness of surface sculpture among different specimens. Polar axes in specimens of this variety commonly range from 48 to 55 μ in length, but the polar axes of many specimens from the same sample range from 47 to 33 μ. Except for size, the smaller specimens are strikingly similar to the larger ones (*see* Pl. 4, figs. 4A-4B, 6).

Similar Varieties and Species. *A. trialatus* var. *variabilis* can usually be distinguished from *A. trialatus* var. *trialatus* by its narrower, gradually tapering, more pointed equatorial protrusions. In both of these varieties, the reticulate sculpture is much finer in equatorial areas than on polar areas; however, this variance, as well as a tendency toward the fingerprint pattern in equatorial areas, is more obvious in var. *variabilis* than in var. *trialatus*. *A. trialatus* var. *variabilis* is usually smaller than var. *trialatus*, but there is a slight overlap in size ranges.

A. trialatus var. *variabilis*, which has much finer equatorial than polar ornamentation, can easily be distinguished from *A. trialatus* var. *uniformis*, which has uniform body ornamentation.

A. trialatus var. *variabilis* resembles *A. trialatus* Rouse *in* Bolkhovitina (1959, p. 127, Pl. VIII, fig. 113c). Bolkhovitina's figures are drawings, but she stated that her specimens (from the Upper Cretaceous, Yakutsk, U.S.S.R.) agreed entirely with the species (*A. trialatus*) described by Rouse. Our specimens also resemble illustrations of a grain named "*Integricorpus*" *bellum* Mtchedlishvili (*in* Samoilovitch and others, 1961, p. 217-218, Pl. 70, figs. 1a, 1b). Mtchedlishvili stated that there was no doubt that her grains were identical to those of Bolkhovitina (1959), but she believed that her grains differed from *A. trialatus* Rouse in the structure of exine and furrows. As illustrated, Mtchedlishvili's species from the western Siberian lowlands has much shorter furrows than does *A. trialatus* var. *variabilis*. Mtchedlishvili stated that her grains varied but little in dimensions, in outline of the body, and in thickness of the exine and that equatorial projections were often broken off. The variety we have described, as was pointed out above, is extremely variable, and broken-off equatorial protrusions are not characteristic of our specimens.

Aquilapollenites reticulatus (Mtchedlishvili, 1961) n. comb.
(Pl. 1, figs. 12A, 12B)

Parviprojectus reticulatus Mtchedlishvili, 1961 (July), *in* Samoilovitch and others, 1961, p. 226-227, Pl. 73, figs. 2a-3.

Holotype. Paleophytological Laboratory of the VNIGRI, preparation no. 6189-X; Mtchedlishvili, *in* Samoilovitch and others, 1961, Plate

73, figures 2a-2c; Western Siberian lowlands (River Sym, Brusov Yar specimen 7201); Upper Symian subsuite, Maestrichtian to Danian.

Aquilapollenites reticulatus Stanley, 1961 (November), p. 348-349, Pl. 8, figs. 1-12; Crow Butte, Harding County, South Dakota; Hell Creek Formation, Maestrichtian,

not *Aquilapollenites reticulatus* Chlonova, 1961 (August), p. 84, Pl. 14, fig. 107.

Aquilapollenites triauritus Takahashi, 1964, p. 250-251, Pl. 37, figs 1a-2c;Hokkaido, Japan; Hakobuchi-Schichtengruppe, Campanian.

Aquilapollenites mtchedlishvilii nom. nov. Srivastava, 1968a, p. 692-693.

USGS Records of Occurrence

Formation	*Stage*	*USGS paleobotany locality*	*Locality*
Hell Creek (upper part)	Upper Maestrichtian	D3730-G	Montana
Laramie	Upper Maestrichtian	D3282	Colorado
Hell Creek (middle part)	Upper and middle Maestrichtian	D3730-B, -E	Montana
Laramie (lower part)	Middle Maestrichtian	D1537	Colorado
Hell Creek (lower part)	Middle Maestrichtian	D3147-B D3690-C, -D, -E, -F, -H, D3472-B, C, -E, D3754-A	Montana
Fox Hills Sandstone, Colgate Member	Middle or lower Maestrichtian	D3690-A, -B	Montana
Fox Hills Sandstone	Middle and (or) lower Maestrichtian	D1970	Wyoming
Bearpaw Shale	Lower Maestrichtian or upper Campanian	D3011 D3012	Montana
Cokedale	Upper Campanian	D1815-1	Montana

Remarks. We believe that *Aquilapollenites reticulatus* Stanley is the same species as *"Parviprojectus" reticulatus* Mtchedlishvili. Mtchedlishvili described *"P." reticulatus* as three-furrowed (tricolpate); we have examined the type specimens of *A. reticulatus* Stanley and have found that they are tricolpate instead of tridemicolpoidate as they were originally described. In a single sample (USGS paleobotany locality D3147-B) from the Hell Creek Formation of Montana, we found otherwise identical specimens with intergrading types of sculpture on the equatorial protrusions: (1) faintly striate as in Stanley's types, (2) faintly striate on equatorial protrusion base but granulate (or pilate) on the end, and (3) granulate (or pilate) all over equatorial protrusion as in Mtchedlishvili's type. The evidence in this population indicates an intergrading series that we consider assignable to one species.

Mtchedlishvili (1965, p. 27) also was of the opinion that the American and Russian species are identical.

"Parviprojectus" reticulatus Mtchedlishvili (1961, July, *in* Samoilovitch and others) has priority over *Aquilapollenites reticulatus* Stanley (1961, November). *"Parviprojectus" reticulatus* has the defining characteristics of the genus *Aquilapollenites* (Rouse) Funkhouser and we hereby transfer that species.

Aquilapollenites reticulatus Chlonova (1961, August) appears from her drawings and description to be a different species, having much coarser reticulate sculpture than does *A. reticulatus* (Mtchedlishvili) n. comb.

Takahashi stated that *A. reticulatus* Stanley possesses thinner exine and higher muri than *A. triauritus*. We have observed considerable variation in wall thickness and muri height in Stanley's type material as well as in our own specimens; we believe that *A. triauritus* is not significantly different and consider it synonymous with *A. reticulatus*.

Aquilapollenites reductus Norton, 1965
(Pl. 1, figs. 11A, 11B)

Aquilapollenites reductus Norton, 1965, p. 140, 142, Pl. 2, figs. 5-7; Pl. 3, figs. 8, 9.

Holotype. Norton, 1965, Plate 2, figures 5-7; slide 6P-13-I; S½ sec. 31, T. 22 N., R. 37 E., Fort Peck Reservoir, Garfield County, Montana; Hell Creek Formation, Maestrichtian.

Aquilapollenites funkhouseri Srivastava, 1966, p. 541, Pl. IX, figs. 5-8, 10; Scollard, Alberta, Canada; Upper Cretaceous, Maestrichtian.

Aquilapollenites ceriocorpus Srivastava, 1968a, p. 674, 676, Pl. II, figs. 1-4; Horseshoe Canyon, Alberta, Canada; Edmonton Formation, Maestrichtian.

Aquilapollenites medeis Srivastava, 1968a, p. 678, 680, Pl. III, fig. 6; Pl. IV, figs. 1, 2; Alberta, Canada; late Campanian?-Maestrichtian.

USGS Records of Occurrence

Formation	*Stage*	*USGS paleobotany locality*	*Locality*
Fox Hills Sandstone	Upper and middle Maestrichtian	D1375-1	South Dakota
Hell Creek (basal part)	Middle Maestrichtian	D3690-C	Montana
Fox Hills Sandstone, Colgate Member	Middle or lower Maestrichtian	D3754-B	Montana
Fox Hills Sandstone	Lower Maestrichtian	D1331-2, -5	Wyoming
Bearpaw Shale	Lower Maestrichtian or upper Campanian	D3011 D3012 D3472-A	Montana
Pierre Shale, lower unnamed shale member	Upper Campanian	D1330-19	Wyoming

Remarks. We have assigned to *A. reductus* more than 30 specimens that we have found in equatorial, polar, and oblique views in samples from the localities listed above. The grains are extremely variable in shape of body and shape of equatorial protrusions and in coarseness and gradations of reticulum on the body and on equatorial protrusions. Equatorial protrusions (in equatorial views of grains) are usually nearly isodiametric and with rounded ends; in polar and oblique views of grains some of the equatorial protrusions are tapered at the outer end (as shown in the equatorial protrusion at the left in Srivastava's illustration of *A. medeis* (Srivastava, 1968a, Pl. IV, figs. 1, 2). In our specimens, polar axes vary from 33 to 63 μ and equatorial diameters (including equatorial protrusions) range from 29 to 58 μ. The group examined includes an intergrading series of specimens that fall within the span of the four species of Norton and Srivastava; inasmuch as these taxa cannot be clearly separated we hereby place them in synonymy.

Aquilapollenites rigidus n. sp.
(Pl. 1, fig. 1; Pl. 4, figs. 1-3)

Holotype. USGS paleobotany locality D3718-I, slide 2, at 86.6 × 10.2; Plate 4, figure 1; NW¼NW¼ sec. 13, T. 22 N., R. 17 E.; Fergus County, Montana; Eagle Sandstone, lower Campanian.

USGS Records of Occurrence

Formation	*Stage*	*USGS paleobotany locality*	*Locality*
Claggett Shale	Middle and lower Campanian	D3724-B, -C, -D -E	Montana
Eagle Sandstone (upper part)	Lower Campanian	D3718-I	Montana

Description. This description is based on the examination of 17 specimens in equatorial view and 8 specimens in oblique polar views. Tricolpate, isopolar to subisopolar pollen grains with three equatorial protrusions. *In equatorial view*: body oval, with nearly parallel sides and broadly rounded poles; polar axis 34 to 50 μ, equatorial diameter (including equatorial protrusions) 33 to 44 μ, diameter of body 10.5 to 14 μ; equatorial protrusions 12 to 17.5 μ long, 8 to 12.5 μ wide, extending at right angles from body of grain, with nearly parallel sides and rounded ends, commonly approximately the same length as polar protrusions. *In polar view*: grains triangular, with concave sides; equatorial protrusions paddlelike, flattened in plane parallel to polar axis of grain. *Apertures*: colpi extending entire length of equatorial protrusions, apparently not extending onto body (if so, for only a very short distance). *Pollen wall*: exine two-layered; endexine of body (exclusive of costae) as much as 0.5 μ thick; axillary endexinal costae as much as 3 μ thick in thickest part, confined mostly to equatorial protrusions, extending two-thirds to three-fourths the distance to free ends of equatorial protrusions, extending only a short distance onto body of grain; quarter-moon-shaped supplementary endexinal thickenings appear as swellings in axillae, tending to merge with main costal thickenings (*see* Pl. 4, fig. 2); ektexine of body about 1 μ thick; ektexinal elements columnar, fused at tips—resulting in a foveolate to reticulate surface sculpture; lumina mostly less than 0.5 μ in diameter; muri about same width as diameter of lumina; ektexinal elements of equatorial protrusions finer and shorter than those of body; surface sculpture on equatorial protrusions reticulate-striate, striations (sometimes indistinct) encircling protrusions (*see* Pl. 4, fig. 2); at junction of equatorial protrusions with body of grain, striations continuing onto body in a more or less distinct fingerprint pattern.

Remarks. This is not a variable species; in the 25 specimens we have examined, variations in shape, size, and sculpture are minor. We have not previously encountered the small quarter-moon-shaped swellings on the endexinal thickenings in any other species of *Aquilapollenites*. The stiff crosslike form in equatorial view and the heavy axillary thickenings make appropriate the name *rigidus* for this species.

Similar Species. In shape and size, *A. rigidus* n. sp. is similar to *A. procerus* Samoilovitch (1965, p. 126); however, *A. procerus* has sharply defined smooth polar domes, whereas the sculpture of the polar domes of *A. rigidus* is undifferentiated.

A. rigidus resembles *A. stelckii* Srivastava (1968a, p. 683, Pl. V, figs., 7-9) but *A. rigidus* is clearly colpate, not demicolpate, and has equatorial protrusions that are typically narrower (in equatorial view of grains) than the polar protrusions. In *A. rigidus* the endexinal thickenings have their greatest thickness in the axillae, whereas in *A. stelckii* the greatest thickness of the costae (in the holotype illustrated) appears to be near the outer ends of the equatorial protrusions. The quarter-moon-shaped supplementary endexinal thickenings typically present in the axillae of *A. rigidus* are not shown or described for *A. stelckii.*

Aquilapollenites catenireticulatus Srivastava, 1968a
(Pl. 1, fig. 9; Pl. 3, figs. 7A-8)

Aquilapollenites catenireticulatus Srivastava, 1968a, p. 674, Pl. I, figs. 7-9.

Holotype. Srivastava, 1968a, Pl. I, figs. 7-9; preparation EC1A/2; East Coulee, Alberta, Canada; Edmonton Formation, late Campanian?-Maestrichtian.

USGS Records of Occurrence

Formation	*Stage*	*USGS paleobotany locality*	*Locality*
Hell Creek (lower part)	Middle Maestrichtian	D3147-A, -B D3690-E, -F	Montana
Fox Hills Sandstone, Colgate Member	Middle or lower Maestrichtian	D3690-A	Montana
Fox Hills Sandstone	Lower Maestrichtian	D1331-5	Wyoming
Bearpaw Shale	Lower Maestrichtian or upper Campanian	D3012	Montana
Pierre Shale, Virgin Creek Member	Lower Maestrichtian or upper Campanian	D1375-9	South Dakota
Judith River (upper part)	Upper Campanian	D3725-A	Montana

Description. This description expands the scope of Srivastava's species. It is based on 37 USGS specimens in equatorial view and 8 specimens in oblique and polar views. Tricolpate, isopolar to subisopolar pollen grains with three equatorial protrusions. *In equatorial view*: body of grain oval, distended in equatorial region near junction with equatorial protrusions; polar axis 27 to 50 μ, equatorial diameter (including equatorial protrusions) 29 to 41 μ; equatorial protrusions, extending at right angles from body of grain, knoblike, 7 to 14 μ long, 8 to 13 μ wide, broadly rounded at free ends; some protrusions slightly constricted at base. *In polar view*: grains triangular, with plane to slightly convex sides, with corners slightly protruding or not protruding. *Apertures*: colpi extending entire length of equatorial protrusions, extending onto body two-thirds to three-fourths the distance to the poles. *Pollen wall*: exine two-layered; endexine of body 0.5 to 1.5 μ thick; axillary endexinal costae when present bow-shaped, as much as 2 μ thick in thickest part, extending two-thirds to three-fourths the distance to poles, reaching a short distance onto base of equatorial protrusions; ektexine 0.5 to 2 μ thick; ektexinal elements on body of grain fused at tips—resulting in a finely reticulate-striate appearance to surface of grain; striations on body ladderlike (with lirae connected by horizontal muri), randomly oriented but striations near colpi tending to be oriented perpendicular to colpi margins; sculpture on equatorial protrusions striate, lirae not usually connected by horizontal muri; surface of equatorial protrusions often appearing hirsute or bristly.

Remarks. Both in size of grain and in grade of coarseness of ornamentation, this is a variable species. Intermediate forms link the group together to such an extent that we cannot reliably distinguish varieties within the species.

Aquilapollenites attenuatus Funkhouser, 1961
(Pl. 1, fig. 10)

Aquilapollenites attenuatus Funkhouser, 1961, p. 194, 196, Pl. 2, figs. 1a-1c.

Holotype. Funkhouser, 1961, Plate 2, figures 1a-1c; USNM 41303 (JPR 214-1); western flank of Rawlins uplift, Rawlins, Wyoming, sec. 11, T. 21 N., R. 89 W.; Lance Formation, Maestrichtian.

USGS Records of Occurrence

Formation	*Stage*	*USGS paleobotany locality*	*Locality*
Hell Creek (upper part)	Upper Maestrichtian	D3472-I, -J D3730-G	Montana
Hell Creek (middle part)	Upper and middle Maestrichtian	D3730-B	Montana

Formation	*Stage*	*USGS paleobotany locality*	*Locality*
Lance (lower part)	Middle Maestrichtian	D1603	Wyoming
Hell Creek (lower part)	Middle Maestrichtian	D3147-B D3472-B, -C, -D, -E, -F D3690-C, -D, -H	Montana
Fox Hills Sandstone, Colgate Member	Middle or lower Maestrichtian	D3690-B D3754-B	Montana
Fox Hills Sandstone	Lower Maestrichtian	D1331-2, -5	Wyoming
Bearpaw Shale	Lower Maestrichtian or upper Campanian	D3011 D3012	Montana
Meeteetse	Lower Maestrichtian or upper Campanian	D1455-A	Wyoming
Pierre Shale, Virgin Creek Member	Lower Maestrichtian to upper Campanian	D1375-9	South Dakota
Miner Creek	Upper Campanian	D1613	Montana
Judith River (upper part)	Upper Campanian	D3725-A, -C	Montana
Pierre Shale, lower unnamed shale member	Upper Campanian	D1330-19	Wyoming

Remarks. *A. attenuatus* is one of the more commonly encountered species of *Aquilapollenites* in Maestrichtian and upper Campanian rocks of the Rocky Mountain area. Wide variations in size and shape of grains were found among the nearly 100 specimens examined from the localities listed below. Because variations were gradational, we believe that these specimens belong to one species. Several of Srivastava's (1969) species—*A. accipiteris, A. ascriptivus, A. aucellatus, A. augustus, A. drumhellerensis, A. hispidus,* and *A. regalis*—appear to merge within this widely variable but not distinctly divisible taxon. This group of closely intergrading forms deserves further evaluation.

Grains resembling *A. attenuatus* but having characteristics that distinguish them from *A. attenuatus* are being described as a new species in another publication. These are abundant in USGS paleobotany locality D1799 from Alaska (*see* Appendix).

Aquilapollenites spinulosus Funkhouser, 1961
(Pl. 1, figs. 17A, 17B)

Aquilapollenites spinulosus Funkhouser, 1961, p. 194, Pl. 1, figs. 4-6.

Holotype. Funkhouser, 1961, Plate 1, figures 4a, 4b; USNM 41310 (JPR 219-3); sec. 11, T. 21 N., R. 89 W., western flank of Rawlins uplift, Rawlins, Wyoming; Fort Union Formation, Paleocene.

USGS Records of Occurrence

Formation	*Stage or series*	*USGS paleobotany locality*	*Locality*
Fort Union, Sentinel Butte Member	Paleocene	D1387-2, -3	North Dakota
Fort Union, Tongue River Member	Paleocene	D3818-A	North Dakota
Coalmont (lower part)	Paleocene	D1823-2	Colorado
Hell Creek (lower part)	Middle Maestrichtian	D3147-B	Montana

Aquilapollenites turbidus n. sp.
(Pl. 1, fig. 6; Pl. 2, figs. 7A-10)

Holotype. USGS paleobotany locality D1330-14, slide 3, at 79.0 × 15.3; Plate 2, figures 7A, 7B; NW¼NE¼ sec. 23, T. 38 N., R. 62 W.; Black Hills area, Niobrara County, Wyoming; Red Bird Silty Member, Pierre Shale, middle Campanian.

in part *Aquilapollenites quadrilobus* Rouse, 1957, p. 371, probably Pl. II, fig. 9 (*see* Remarks, p. 159),

in part *Triporina unica* Chlonova, 1957, p. 44, probably Pl. 1, fig. 3 (not Pl. 1, figs. 1, 2).

Aquilapollenites quadrilobus Rouse, *in* Bolkhovitina, 1959, p. 127-128, Pl. VIII, fig. 114a.

Aquilapollenites quadrilobus Rouse, *in* Mtchedlishvili, 1961, *in* Samoilovitch and others, 1961, p. 210-211, Pl. 67, fig. 3a.

Aquilapollenites quadrilobus Rouse, 1957, *in* Bratzeva, 1965, p. 12-13, Pl. II, figs. 7, 8.

USGS Records of Occurrence

Formation	*Stage*	*USGS paleobotany locality*	*Locality*
Dawson Arkose (lower part)	Upper Maestrichtian	D1078	Colorado
Laramie (lower part)	Middle Maestrichtian	D1537	Colorado
Kirtland Shale	Upper Campanian	D3980-H	Colorado
Pierre Shale, Verendrye Member	Upper Campanian	D1375-12	South Dakota
Pierre Shale, DeGrey Member	Upper Campanian	D1375-13	South Dakota
Pierre Shale, lower unnamed shale member	Upper Campanian	D1330-16	Wyoming
Judith River (middle part)	Upper Campanian	D4069	Montana
Pierre Shale, Red Bird Silty Member	Middle Campanian	D1330-14	Wyoming
Pierre Shale, Mitten Black Shale Member	Middle or lower Campanian	D1330-8, -5	Wyoming
Claggett Shale (lower part)	Lower Campanian	D3724-B	Montana

Description. This description is based on 12 grains in equatorial view and 6 in oblique polar views. Tricolpate, isopolar to subisopolar pollen grains with three equatorial protrusions. *In equatorial view*: polar and equatorial protrusions with broadly rounded ends, equatorial protrusions almost always longer than polar protrusions; polar axis 25 to 35 μ, equatorial diameter (including equatorial protrusions) 31 to 45 μ; diameter of polar protrusions (measured near junction with equatorial protrusions) 11 to 16 μ; width of equatorial protrusions (measured near junction with polar protrusions) 11 to 20 μ; polar axis usually shorter than equatorial diameter; equatorial protrusions somewhat flattened in plane parallel to polar axis of grain. *In oblique polar view*: grains triangular, with rounded corners. *Apertures*: colpi apparently extending full length of equatorial protrusions, in some grains (oblique views) colpi appearing to extend for a short distance onto body; some colpi gaping at free ends of equatorial protrusions. *Pollen wall*: exine two-layered; endexine as much as 0.5 μ thick on polar protrusions; prominent lens-shaped endexinal costae, as much as 5 μ thick in thickest part, occupying axillae and extending a short distance onto bases of polar and equatorial protrusions; in some specimens axillary costae indistinct or lacking (probably having been dissolved by chemical treatment of grains during their removal from rock matrix); ektexine 0.5 to 1.5 μ thick on polar protrusions, very thin at colpal margins at free ends of equatorial protrusions; ektexinal elements columnar, fused at tips—resulting in a foveolate to finely reticulate appearance to surface of grains (in some cases reticulum very fine, visible only at 1000 magnification); conical spinae, 1 to 2 μ long, ornamenting equatorial protrusions and a wide equatorial belt on body of grain; on equatorial protrusions spinae usually oriented toward polar axis of grain; polar domes lacking spinae or some having a few reduced spinae.

Remarks. The species described above has been named *A. turbidus* (=confused) because specimens that are synonymous with it have been confused with and called *A. quadrilobus* Rouse. Grains that we believe are equal to *A. turbidus* have been called *A. quadrilobus* by Bolkhovitina (1959), Mtchedlishvili (*in* Samoilovitch and others, 1961), and Bratzeva (1965), who may have considered their specimens equal to Rouse's (1957, Pl. II, fig. 9) illustration of *A. "quadrilobus."*

It is unfortunate that at the time *A. quadrilobus* was named and described it was identified with three different specimens that are at least two, and probably three, different species of *Aquilapollenites*. These three are: (1) Plate I, figure 14, *in* Radforth and Rouse, 1954, Canadian Jour. Botany, v. 32. This is the holotype of *A. quadrilobus* Rouse. (2) Plate II, figure 8, *in* Rouse, 1957, Canadian Jour. Botany, v. 35. This is clearly an isopolar grain and cannot be a specimen of *A. quadrilobus*. It has been placed in synonymy with *A. procerus* Samoilovitch (1965, p. 126). (3) Plate II, figure 9, *in* Rouse, 1957, Canadian Jour. Botany, v. 35. We believe this specimen is a subisopolar specimen of *A. turbidus* n. sp.; it conforms to that species in that its polar domes are nonornamented, its polar axis is less than its equatorial diameter, and its polar protrusions are smaller than its equatorial protrusions. Because it lacks spines on its polar domes, this grain could not be a specimen of *A. quadrilobus*.

Similar Species. *Aquilapollenites turbidus* n. sp. in shape and size resembles *Aquilapollenites rombicus* Samoilovitch (1965, p. 124-125, Plate I (p. 403), figs. 3a-3d) from Yakutia, Siberia, U.S.S.R.; however *A. turbidus* can be distinguished by its thick axillary endexinal thickenings and by its spinose ornamentation.

Aquilapollenites bertillonites Funkhouser, 1961

(Pl. 1, figs. 16A, 16B)

Aquilapollenites bertillonites Funkhouser, 1961 (April), p. 196, Pl. 2, figs. 5a-5c.

Holotype. Funkhouser, 1961, Plate 2, figures 5a-5c; USNM 41308 (JPR 218-1); sec. 1, T. 21 N., R. 89 W., western flank of Rawlins uplift, Rawlins, Wyoming; Lance Formation, Maestrichtian.

Aquilapollenites murus Stanley, 1961 (November), p. 347, Pl. 5, figs. 1-9; Pl. 6, figs. 1-9; Crow Butte, Harding County, South Dakota; Hell Creek Formation, Maestrichtian.

USGS Records of Occurrence

Formation	*Stage*	*USGS paleobotany locality*	*Locality*
Hell Creek (lower part)	Middle Maestrichtian	D3147-B D3472-C D3690-D, -E, -F, H	Montana
Fox Hills Sandstone, Colgate Member	Middle or lower Maestrichtian	D3690-A, -B	Montana

Remarks. We have examined the type specimens of *A. bertillonites* and *A. murus* and believe that the two species are synonymous. *A. bertillonites* has priority.

Aquilapollenites conatus Norton, 1965
(Pl. 1, fig. 15)

Aquilapollenites conatus Norton, 1965, p. 142, Pl. 3, figs. 10, 11; Pl. 4, figs. 12-16.

Holotype. Norton, 1965, Plate 3, figures 10, 11; Plate 4, figures 12, 13; Slide 6P20-I; S½ sec. 31, T. 22 N., R. 37 E., Fort Peck Reservoir, Garfield County, Montana; Hell Creek Formation, Maestrichtian.

USGS Records of Occurrence

Formation	*Stage*	*USGS paleobotany locality*	*Locality*
Lance (upper part)	Upper Maestrichtian	D1602	Wyoming
Hell Creek (lower part)	Middle Maestrichtian	D3147-B D3472-C D3754-A	Montana

SUMMARY

A short historical account and a new description of the genus *Aquilapollenites* are presented. The restrictions placed on *Aquilapollenites* by Mtchedlishvili and by Srivastava deprive the genus of its type species and therefore are not acceptable under the International Code of Botanical Nomenclature.

The holotype of the type species of *Aquilapollenites* is redescribed and illustrated; four new species and three new varieties are named and described.

From data compiled in this paper, the following ranges of 17 taxa of *Aquilapollenites* are indicated:

(1) Four taxa—*A. trialatus* var. *uniformis, A. tripodiformis, A. rigidus,* and *A. trialatus* var. *variabilis*—are restricted to the Campanian, the latter two being confined principally to the lower half of the Campanian.

(2) Three taxa—*A. delicatus* var. *collaris, A. conatus,* and *A. bertillonites*—are confined to the Maestrichtian.

(3) Four species—*A. senonicus, A. calvus, A. quadrilobus,* and *A. turbidus*—occur intermittently between lower or middle Campanian and upper or uppermost Maestrichtian.

(4) Five taxa—*A. catenireticulatus, A. attenuatus, A. reductus, A. reticulatus,* and *A. delicatus* var. *delicatus*—occur in upper Campanian and Maestrichtian sediments.

(5) One taxon—*A. spinulosus*—commonly found in the Paleocene was also found in rocks as old as middle and lower Maestrichtian.

The genus *Aquilapollenites* in the Rocky Mountain region ranges from lower Campanian through middle and probably upper Paleocene but is apparently absent from the lowermost Paleocene.

ACKNOWLEDGMENTS

We acknowledge the cooperation of Glenn E. Rouse, John W. Funkhouser, and Edward A. Stanley for the loan of type specimens. We wish to thank Wayne W. Brideaux for locating the misplaced slide that contains the holotype of *A. quadrilobus,* the type species of the genus *Aquilapollenites.* We appreciate the help of Robert H. Tschudy for making his collections from many Upper Cretaceous localities available for this work and for bringing to our attention many *Aquilapollenites* specimens. We thank Robert M. Kosanke and Jack A. Wolfe for their technical reviews and helpful suggestions. We are deeply indebted to Ivan J. Mittin for translations from several Russian publications.

REFERENCES CITED

Agranovskaia, I. A., Aliushinskii, IU. A., Asatkina, E. F., Boitsova, E. P., Bocharnikova, A. D., Voevodova, E. M., Gromova, N. S., Zauer, V. V., Martynova, Z. I., Panova, L. A., Pokrovskaia, I. M. Romanovskaia, G. M., Sedova, M. A., Stel'mak, N. K., Kaikina, S. L., and Edel'shtein, L. I., 1960, Atlas of Upper Cretaceous, Paleocene and Eocene spore-pollen complexes from certain regions of the USSR (translation): Leningrad, Vses. Geol. Inst. Trudy, new ser., v. 30, 576 p.

Belsky, C. Y., Boltenhagen, E., and Potonié, R., 1965, Sporae dispersae der Oberen Kreide von Gabun, Äquatoriales Afrika: Paläont. Zeitschr., v. 39, no. 1/2, p. 72-83, 2 pls.

Bolkhovitina, N. A., 1959, Sporovo-pyl'tsevye kompleksy mezozoishikh otlozlenii Viliuiskoi vpadiny i ikh znachenie dlia stratigrafii. (Mesozoic Sporopollen complexes in the Viliuoi basin and their stratigraphic significance): Moskva, Akad. Nauk SSSR, Geol. Inst., Trudy, v. 24, 187 p., 8 pls.

Bratzeva, G. M., 1965, Pyl'tsa i spory Maastrikhtskikh otlozlenii Pal'nego Vostoka (Pollen and spores in Maestrichtian deposits of the far east): Moskva, Akad. Nauk SSSR, Geol. Inst., Trans., v. 129, 43 p., 42 pls.

Chlonova, A. F., 1961, Spory i pyl'tsa verkhnei poloviny verkhnego mela vostochnoi chasti Zapadno-Sibirskoi nizmennosti (Spores and pollen from the upper half of the Upper Cretaceous of the eastern part of the western Siberian depression): Novosibirsk, Akad. Nauk SSSR, Sibirskoye Otdeleniye, Inst. Geologii i Geofiziki Trudy, v. 7, 140 p., 17 pls.

Cobban, W. A., and Reeside, J. B., 1952, Correlation of the Cretaceous formations of the western interior of the United States: Geol. Soc. America Bull., v. 63, no. 10, p. 1011-1044, 1 pl., 2 figs.

Drew, C. M., and Tschudy, B. D., 1968, *Aquilapollenites*: fossil pollen as seen under the scanning electron microscope: Geol. Soc. America Bull., v. 79, p. 1829-1832.

Funkhouser, J. W., 1961, Pollen of the genus *Aquilapollenites*: Micropaleontology, v. 7, no. 2, p. 193-198, 2 pls.

Gill, J. R., and Cobban, W. A., 1966, The Red Bird Section of the Upper Cretaceous Pierre Shale in Wyoming: U.S. Geol. Survey Prof. Paper 393-A, p. A1-A73, 12 pls.

Herrick, S. M., and Tschudy, R. H., 1967, Microfossil evidence for correlation of Paleocene strata in Ballard County, Kentucky, with the lower part of the Porters Creek Clay, p. B40-B44 *in* Geological Survey Research 1967: U.S. Geol. Survey Prof. Paper 575-B.

Jardiné, Serge, and Magloire, Lili, 1965, Palynologie et Stratigraphie du Crétacé des Bassins du Sénégal et de Côte d'Ivoire, p. 187-245, 11 pls. *in* Colloque Internat. de Micropaléontologie, Dakar 1963: Bur. Recherches Géol. et Minières Mem. 32, 369 p.

Jones, E. L., 1962, Palynology of the Midway-Wilcox boundary in south-central Arkansas: Trans. 12th Ann. Meeting Gulf Coast Assoc. Geol. Societies, Oct. 31, Nov. 1, 2: New Orleans, p. 285-294, 3 pls.

Keefer, W. R., 1965, Stratigraphy and geologic history of the uppermost Cretaceous, Paleocene, and lower Eocene rocks in the Wind River Basin, Wyoming: U.S. Geol. Survey Prof. Paper 495-A, p. A1-A77.

Lanjouw, Joseph (Chairman), and others, 1961, International code of botanical nomenclature: Utrecht, Netherlands, Internat. Bur. Plant Taxonomy and Nomenclature, Internat. Assoc. Plant Taxonomy, 372 p.

Martin, A. R. H., 1968, *Aquilapollenites* in the British Isles: Palaeontology, v. 11, pt. 4, p. 549-553.

Mtchedlishvili, N. D., 1965, Znachenie pokrytosemennykh rastenii dlia stratigrafii verkhnemelovykh otlozhenii (Importance of angiosperm plants in stratigraphy of Upper Cretaceous deposits), p. 5-34 *in* Samoilovitch, S. R., *Editor,* Paleofitologicheskii Sbornik: Moskva, Vses. Neft. Nauchno-Issled. Geol.-Ramedochnyĭ Inst. Trudy, v. 239, 506 p.

Norton, N. J., 1965, Three new species of *Aquilapollenites* from the Hell Creek Formation, Garfield County, Montana (1): Mus. Natl. D'Histoire Naturelle, Pollen et Spores, v. 7, no. 1, p. 135-143, 4 pls.

Radforth, N. W., and Rouse, G. E., 1954, The classification of recently discovered Cretaceous plant microfossils of potential importance to the stratigraphy of western Canadian coals: Canadian Jour. Botany, v. 32, p. 187-201, 1 pl.

Roberts, A. E., 1963, The Livingston Group of south-central Montana, p. B86-B92 *in* Short papers in geology and hydrology: U.S. Geol. Survey Prof. Paper 475-B.

——1965, Correlation of Cretaceous and lower Tertiary rocks near Livingston, Montana, with those in other areas of Montana and Wyoming, p. B54-B63 *in* Geological Survey Research 1965: U.S. Geol. Survey Prof. Paper 525-B.

Robinson, F. M., 1958, Test Wells, Gubik area, Alaska, *with* Micropaleontologic study of the Gubik test wells, northern Alaska by H. R. Bergquist: U.S. Geol. Survey Prof. Paper 305-C, p. 207-264.

Rouse, G. E., 1957, The application of a new nomenclatural approach to Upper Cretaceous plant microfossils from western Canada: Canadian Jour. Botany, v. 35, no. 3, p. 349-375, 3 pls.

Samoilovitch, S. R., 1965, Pyl'tsa novykh vidou verkhnemelovykh pokrytosemennykh rastenii Iakutii (Pollen of new species of Upper Cretaceous Angiosperm plants of Yakutia), p. 121-141, 5 pls. *in* Samoilovitch, S. R., *Editor,* Paleofitologicheskii Sbornik: Moskva, Vses. Neft. Nauchno-Issled. Geol.-Ramedochnyi Inst. Trudy, v. 239, 506 p.

Samoilovitch, S. R., Mtchedlishvili, N. D., Rusakova, L. IA., Iashchurzhinskaia, A. B., *Editors,* 1961, Pyl'tsa i spory zapadnoĭ Sibiri, ·Uira-Paleotsen (Pollen and spores of western Siberia, Jurassic to Paleocene): Leningrad, Vses. Neft. Nauchno-Issled. Geol.-Razved. Inst. VNIGRI) Trudy, v. 177, 657 p., 149 pls.

Sato, Seiji, 1961, Pollen analysis of Carbonaceous matter from the Hakobuchi Group in the Enbetsu District, northern Hokkaido, Japan—Palynological Study on Cretaceous sediments (I): Hokkaido Univ., Faculty Sci. Jour., ser. 4, Geology and Mineralogy, v. 11, no. 1, p. 77-93, 2 pls.

Simpson, J. B., 1961, The Tertiary pollen-flora of Mull and Ardnamurchan: Royal Soc. Edinburgh Trans., v. 64, no. 16, p. 421-468, 20 pls.

Srivastava, S. K., 1966, Upper Cretaceous microflora (Maestrichtian) from Scollard, Alberta, Canada: Mus. Natl. D'Histoire Naturelle, Pollen et Spores, v. 8, no. 3, p. 497-552, 11 pls.

——1968a, Reticulate species of *Aquilapollenites* and emendation of genus *Mancicorpus* Mtchedlishvili: Mus. Natl. D'Historie Naturelle, Pollen et Spores, v. 10, no. 3, p. 665-699.

——1968b, Eight species of *Mancicorpus* from the Edmonton Formation (Maestrichtian), Alberta, Canada: Canadian Jour. Botany, v. 46, no. 12, p. 1485-1490, 2 pls.

——1969, New spinulose *Aquilapollenites* spp. from the Edmonton Formation (Maestrichtian), Alberta, Canada: Canadian Jour. Earth Sci., v. 6, no. 1, p. 133-144, 8 pls.

Stanley, E. A., 1961, The fossil pollen genus *Aquilapollenites*: Mus. Natl. D'Histoire Naturelle, Pollen et Spores, v. 3, no. 2, p. 329-352, 8 pls.

Takahashi, K., 1964, Sporen und pollen der oberkretazeischen Hakobuchi-Schichtengruppe, Hokkaido: Kyushu Univ., Faculty Sci. Mem., ser. D, Geology, v. 14, no. 3, p. 159-271, 22 pls.

Tourtelot, H. A., and Schultz, L. G., 1961, Core from the Irish Creek well, Ziebach County, South Dakota: U.S. Geol. Survey Open-File Report, 20 p.

Traverse, Alfred, 1958, Locating plant microfossils on mixed slides: Micropaleontology, v. 4, no. 2, p. 207-208.

Tschudy, R. H., 1966, Associated megaspores and microspores of the Cretaceous genus *Ariadnaesporites* Potonié, 1956, emend.: U.S. Geol. Survey Prof. Paper 550-D, p. D76-D82.

Voitsel', Z. A., Ivanova, E. A., Kruchinina, N. V., Klimko, S. D., Markova, L. G., Asatkina, E. F., and Stel'mak, N. K., 1956, Brief description of spores and pollen from the Oligocene deposits of the central part of the western Siberian lowland (translation), p. 175-216 *in* Pokrovskaia͡, I. M., *Editor*, Atlas of Oligocene spore-pollen complexes of different regions of the USSR (translation): Materialy Vses. Nauchno-Issled. Geol. Inst. Trudy (VSEGEI) Ministerstva Geologii i Okhrany Nedr. SSSR, new ser., v. 16, Paleontologiia i Stratigrafiia, 313 p.

PUBLICATION AUTHORIZED BY THE DIRECTOR, U.S. GEOLOGICAL SURVEY

APPENDIX: LIST OF POLLEN LOCALITIES

D1078: Dawson Arkose (lower part); SE¼NE¼ sec. 18, T. 13 S., R. 61 W., El Paso County, Colorado.

D1225-K: Denver Formation (lower part); center NW¼ sec. 31, T. 3 S., R. 69 W., Jefferson County, Colorado.

D1330-4: Pierre Shale; NE¼NE¼ sec. 23, T. 38 N., R. 62 W., Niobrara County, Wyoming.

D1330-5, -8, -13A, -14, -16: Pierre Shale; NW¼NE¼ sec. 23, T. 38 N., R. 62 W., Niobrara County, Wyoming.

D1330-19: Pierre Shale; NE¼NW¼ sec. 23, T. 38 N., R. 62 W., Niobrara County, Wyoming.

D1331-2, -5: Fox Hills Sandstone; SW¼SW¼ sec. 14, T. 38 N., R. 62 W., Niobara County, Wyoming.

D1375-1: Fox Hills Sandstone; core hole, SE¼ sec. 17, T. 15 N., R. 20 E., Ziebach County, South Dakota.

D1375-9, -10, -12, -13, -14: Pierre Shale; core hole, SE¼ sec. 17, T. 15 N., R. 20 E., Ziebach County, South Dakota.

D1387-2 -3: Fort Union Formation Sentinel Butte Member; SE¼ sec. 7, T. 137 N., R. 102 W., Billings County, North Dakota.

D1455-A: Meeteetse Formation; NW¼NW¼ sec. 20, T. 6 N, R. 1 E., Fremont County, Wyoming.

D1501-C: Lance Formation (lower part); SW¼ sec. 10, T. 34 N., R. 90 W., Fremont County, Wyoming.

D1537: Laramie Formation (lower part); NW¼SW¼ sec. 24, T. 4 S., R. 70 W., Jefferson County, Colorado.

D1602: Lance Formation (upper part); center sec. 3, T. 37 N., R. 64 W., Niobrara County, Wyoming.

D1603: Lance Formation (lower part); NE¼SE¼ sec. 7, T. 37 N., R. 64 W., Niobrara County, Wyoming.

D1612: Miner Creek Formation; SE¼ sec. 19, T. 2 S., R. 9 E., Park County, Montana.

D1613: Miner Creek Formation; NW¼ sec. 20, T. 2 S., R. 9 E., Park County, Montana.

D1750-B: Undated Upper Cretaceous; NW¼ sec. 12, T. 15 N., R. 92 W., Carbon County, Wyoming.

D1799:[3] Undated Upper Cretaceous; Charley River B-1 quadrangle, SE¼ sec. 23, T. 6 N., R. 30 E.; lat. 65° 19.6′ N., long. 141° 29.6′ W.; eastern bank of Nation River, Alaska.

D1815-1: Cokedale Formation; sec. 23, T. 2 S., R. 8 E., Park County, Montana.

D1823-2: Coalmont Formation (lower part); SW¼SE¼ sec. 19, T. 9 N., R. 79 W., Jackson County, Colorado.

D1845: Pierre Shale; NE¼ sec. 22, T. 45 N., R. 62 W., Weston County, Wyoming.

D1970: Fox Hills Sandstone; sec. 27, T. 39 N., R. 62 W., Niobrara County, Wyoming.

D3011, D3012: Bearpaw Shale; sec. 31, T. 22 N., R. 38 E., Garfield County, Montana.

D3014-A: Hell Creek Formation (upper part); sec. 32-33, T. 21 N., R. 37 E., Garfield County, Montana.

D3124-B, -G, -H:[3] Colville Group; 151° 25′ W. long., 69° 40′ N. lat., Colville River, northern Alaska.

D3147-A, -B: Hell Creek Formation (lower part); sec. 1, T. 21 N., R. 37 E., Garfield County, Montana.

D3282: Laramie Formation; SE¼ sec. 34, T. 3 S., R. 70 W., Jefferson County, Colorado.

D3463: Lance Formation (upper part); NE¼SW¼ sec. 9, T. 31 N., R. 82 W., Natrona County, Wyoming.

D3472-A: Bearpaw Shale; sec. 29, T. 21 N., R. 34 E., Garfield County, Montana.

D3472-B, -C, -D, -E, -F, -I, -J: Hell Creek Formation; sec. 29, T. 21 N., R. 34 E., Garfield County, Montana.

D3474-B: Hell Creek Formation (upper part); sec. 33, T. 21 N., R. 34 E., Garfield County, Montana.

D3690-A, -B: Fox Hills Sandstone, Colgate Member; and D3690-C, -D -E, -F, -H: Hell Creek Formation (lower part); sec. 14, T. 141 N., R. 55 E., Dawson County, Montana.

D3718-E, -I: Eagle Sandstone; NW¼NW¼ sec. 13, T. 22 N., R. 17 E., Fergus County, Montana.

---

[3]The localities of these collections are not plotted on the index map (Fig. 1).

D3724-B, -C, -D, -E: Claggett Shale; NE¼SW¼ sec. 12, T. 22 N., R. 17 E., Fergus County, Montana.

D3725-A, -C: Judith River Formation (upper part); NW¼ sec. 26, T. 24 N., R. 17 E., Blaine County, Montana.

D3726-E: Judith River (lower part); center NE¼ sec. 31, T. 23 N., R. 17 E., Fergus County, Montana.

D3730-A, -B, -C, -D, -E, -G, -H, -I: Hell Creek Formation; sec. 2, T. 141 N., R. 55 E., Dawson County, Montana.

D3740-B; Judith River Formation (lower part); NE¼NW¼NW¼ sec. 19, T. 57 N., R. 86 W., Sheridan County, Wyoming.

D3754-B: Fox Hills Sandstone, Colgate Member, and D3754-A: Hell Creek Formation (lower part); sec. 29, T. 21 N., R. 34 E., Garfield County, Montana.

D3818-A: Fort Union Formation; SW corner NW¼NW¼SE¼ sec. 16, T. 138 N., R. 87 W., Morton County, North Dakota.

D3980-A, -E: Fruitland Formation, and D3980-H: Kirtland Shale; core hole, center sec. 36, T. 47 N., R. 7 W., Montrose County, Colorado.

D4067: Bearpaw Shale (basal part) or Judith River Formation (uppermost part); NW¼NE¼ sec. 31, T. 22 N., R. 24 E., Phillips County, Montana.

D4069: Judith River Formation (middle part); SW¼SW¼SW¼ sec. 15, T. 23 N., R. 22 E., Phillips County, Montana.

Palynological Interpretation of the Type Section, Chuckanut Formation, Northwestern Washington

PETER H. GRIGGS
Shell Oil Company, Houston, Texas

ABSTRACT

A palynologic study of the lower part of the Samish Bay section of the Chuckanut Formation in northwestern Washington has been made to determine the age of these rocks and to interpret the nature of the source flora and environment of deposition.

The Chuckanut Formation is a series of strongly folded, terrestrial sandstones and shales. On the basis of plant megafossils, this formation has previously been assigned an age of Late Cretaceous to early Eocene. Only the lower 5100 feet of the approximately 9500-foot section of the Chuckanut Formation along Samish Bay are adequately represented by the 22 samples reported on in this study.

Seventy-nine palynomorphs are illustrated here. Of these, 19 have been used to zone the lower 5100 feet of strata into three parts. This zonation is based on changes in relative frequency and the stratigraphic ranges of these palynomorphs. It is demonstrated that these changes were brought about by changing environmental conditions during the deposition of the sediments.

An age of late Late Cretaceous to early Eocene is indicated for the lower 5100 feet of the Chuckanut section exposed along Samish Bay, on the basis of analysis of the palynomorphs.

CONTENTS

INTRODUCTION

General Statement

This report is concerned with the stratigraphic significance of fossil pollen and spores of the Chuckanut Formation in northwestern Washington. Little previous work has been done on the palynology of the Chuckanut and associated sedimentary rocks in this area. The primary purpose of the study was to determine the age of the Chuckanut Formation and to subdivide it, if possible, by palynologic analysis. Illustrations of the fossils are given to clarify their use for the zonation and interpretation of the Samish Bay section, which is considered here as a standard section, as well as the type section,[1] for the Chuckanut Formation.

---

[1]*Editor's note*: Glover (1935) gives a detailed account of the exposures of the Chuckanut Formation and identified and described *two* "Type sections" (p. 15). These are the "Section along the east shore of Samish Bay" (p. 15-19), largely exposed in road cuts of Chuckanut Drive, and the "Section from South Bay to Lake Louise, along Lake Whatcom" (p. 19-23). Weaver (1937) stated (p. 78) "For the purpose of presenting 2 standard sections of the Chuckanut formation on the north side of the San Juan Island upwarp which would be representative of the stratigraphic sequence, exposures were chosen along Chuckanut Drive on the shore of Samish Bay and along the west side of Lake Whatcom." These sections, remeasured and redescribed by Weaver, are identical to those first measured and described by Glover as "Type sections" (Glover, 1935, p. 15). McLellan (1927), who apparently first used the name "Chuckanut formation" (p. 93) for exposures of Eocene rocks (referred to as "Lower Eocene" in his graphic section, p. 93, but as "Eocene" in his description, p. 136) on the northern end of Lummi Island (p. 136), also stated that these strata were identical to the rocks exposed on Chuckanut Drive along Samish Bay. He did not describe or locate a type section for the Chuckanut Formation. The name "Chuckanut" had been used earlier for one of the members of this thick sequence of rocks. Shedd (1903) had used the name "Chuckanut Sandstone" for sandstone quarried along Chuckanut Drive south of Inspiration Point and had described and located this bed adequately. Jenkins (1923, p. 50-52) repeated Shedd's description for the Chuckanut Sandstone in full and included it as part of the "Description of the Formation" (p. 50). The "formation" to which he apparently referred the Chuckanut Sandstone is the "Coal Measures" (p. 46). A type section for this "Coal Measures Formation" was not defined or described. There is no reference to a "Chuckanut Formation" by either Shedd or Jenkins. Therefore, it seems that Glover first defined and validated the Chuckanut Formation type section with his description of the section along Samish Bay. Weaver's redescription and remeasurement of this section as one of two "standard sections," in 1937, appears to be adequate for this purpose. The author here indicates that the Chuckanut Drive section along Samish Bay, of Glover's two type sections, is to be treated as the type section of the Chuckanut Formation.

A. T. Cross

Previous Work

Plant fossils from the Chuckanut were first studied by Newberry (1863). The fossils, collected by the United States Exploring Expedition, 1838-1842, were given to Newberry in 1842, although he did not publish a description of them until 1863. He interpreted the plants as Late Cretaceous in age. In later work by Lesquereux (1859) and Knowlton (1902), the strata containing these fossil plants were placed in the Eocene Puget Group. Chaney and La Motte (La Motte, 1938) collected and identified some fossil plants from the Chuckanut. Pabst (1962) reported on the Equistales, Filicales, and Coniferales of the Chuckanut Formation. This was the first attempt to systematically collect from different zones in the formation and place the floral assemblages in chronological order and compare them with other Pacific Coast Tertiary floras. Pabst, on the basis of these studies, concluded that the Chuckanut is of Late Cretaceous to early Eocene age. Berry (1926), after examination of the plant fossils from the Burrard Formation, which underlies the city of Vancouver, British Columbia, equated its megaflora with the Eocene Puget Group.

Several recent studies have given some basis for stratigraphic identification of comparable series in the general area on the basis of plant microfossils. Rouse (1962) suggests that both Late Cretaceous and middle Eocene age are represented in the microfossil flora of the Burrard Formation. Crickmay and Pocock (1963) present evidence for probable Late Cretaceous or possible Paleocene age on the basis of palynomorphs identified from a sample collected from the Chuckanut Formation south of Deming, Washington, on the Nooksack River. Miller and Misch (1963) correlate the lower part of the Chuckanut Formation with the upper part of the Nanaimo Group, on the basis of unpublished palynologic studies by John Fisher of Union Oil Company of California. The Huntingdon Formation, which overlies the Chuckanut, is interpreted as being middle Eocene by Miller and Misch (1963). Rouse (1964, personal commun.) is of the opinion, from studies conducted in his laboratory, that the microfossil flora of the Huntingdon may be as young as late Eocene to Oligocene(?).

GEOLOGY OF THE CHUCKANUT FORMATION

Regional Setting

In northwestern Washington and southwestern British Columbia, rocks of two groups of formations are found: a basement complex of pre-Cretaceous metasedimentaries, metamorphics, and igneous intrusives and extrusives and a series of generally unaltered sedimentary rocks of Upper Cretaceous or Tertiary strata. The sedimentary rocks were deposited in a large structural basin, the axis of which was northwest-southeast, essentially parallel to the Strait of Georgia. During the Cascade Range orogeny, which affected all these deposits except the late Eocene to Oligocene, the rocks were uplifted and folded. They now crop out along the mountain flanks on either side of the Strait of Georgia (Fig. 1).

This Cretaceous-Cenozoic basin has become structurally divided to form several basins. Two of these, the Comox and Nanaimo Basins, are on Vancouver Island. The rocks in these two basins consist of partly marine, coal-bearing beds that are dated as Campanian on the basis of molluscan faunas (Crickmay and Pocock, 1963). On the mainland side of the Strait of Georgia there is one large basin. Miller and Misch (1963) have named this the Bellingham Basin.[2] The rocks of the Bellingham Basin are of

---

[2]*Editor's note*: The name "Bellingham Basin," as used here, follows the use of Miller and Misch (1963), which appears to have priority in valid publication. However, the name "Bellingham Basin" appears to have been first applied by Glover (1935, p. 31 and Pl. I) to the small, shallow, synclinal trough plunging southwest from the juncture of King Mt.-Lookout Mt. anticline through the townsite of Bellingham to Bellingham Bay (and probably continuing southeast from there as the Samish Lake-Padden syncline complex). Glover applied this name to the southwestern limb of the Lake Whatcom syncline of Jenkins (1923, Pl. I) after he found it necessary to modify Jenkins' earlier postulation of continuity of this synclinal fold with the western branch of the northwestern bifurcate limb of the Lake Whatcom syncline.

A much larger basin northwest of Bellingham, which underlies the Bellingham-Sumas coastal plain of the lower Nooksack River and the lower portion of the Fraser River drainage area (from American and Canadian Sumas Mountains on the east to the Coast Mountains on the north and west into the Strait of Georgia and the city of Vancouver), has been named and defined as

continental origin. They crop out along the northern side of the Fraser River and in the coastal mountains of northwestern Washington. Their age is in dispute. Small remnants are also found on the mainland side of Malaspina Strait and on the American and Canadian San Juan Islands. The stratigraphic relations of the rocks of the several basins are illustrated in Figure 2.

Chuckanut Formation

Overlying pre-Cretaceous strata in northwestern Washington is a series of continental sandstones, shales, and conglomerates. White (1888) assigned these rocks to the Eocene Puget Group. Later, Willis (1898), Shedd (1903), and Jenkins (1923, 1924) mapped these as part of the Puget Group but referred to them as the "Chuckanut Bay Sandstone." McLellan (1927) named the coal-bearing sandstones and conglomerates exposed on the northern part of Lummi Island and in the vicinity of Bellingham Bay the "Chuckanut Formation" and assigned them to the lower part of the Puget Group. Glover (1935) retained the name "Chuckanut Formation" in his investigation of the coal-bearing beds in Whatcom County.

Both Glover (1935) and Weaver (1937) gave detailed stratigraphic descriptions for standard Chuckanut sections. They reported 9484 and 11,272 feet, respectively, for the section on the eastern shore of Samish Bay along Chuckanut Drive and 8800 feet along the southwestern shoreline of Lake Whatcom. Miller and

---

the "Bellingham Basin" by Miller and Misch (1963, p. 171, 173, and Fig. 1, p. 164-165). This same large basin has been referred to as the "Whatcom Basin" by Moen (1962, p. 30, 64, 65) without definition, apparently following the definition and use by Newcomb and others (1949, open-file water resources report) and also in a recent PhD. dissertation by Hopkins (1966, Fig. D). There seems to be no valid publication of this name, and therefore the name Bellingham Basin as described by Miller and Misch is used here for the "major tectonic downwarp of middle-upper Eocene rocks" (1963, p. 171) of the area of northwestern Whatcom County, Washington, and the Vancouver-lower Fraser River-Sumas Mountain area, British Columbia. It is important to note that the Chuckanut rocks of the upper Upper Cretaceous or lower Tertiary may have been spread over some of the area of the Bellingham Basin prior to the orogeny that resulted in the folding of the Chuckanut Formation south of Bellingham and before the deposition of the now broadly warped and folded Huntingdon Formation (middle Eocene) found in the Bellingham-Sumas coastal plain and areas to the west. This is on the evidence of micro- and macro-paleobotany of boreholes and outcrops (Miller and Misch, 1963, p. 171, 173).
A. T. Cross

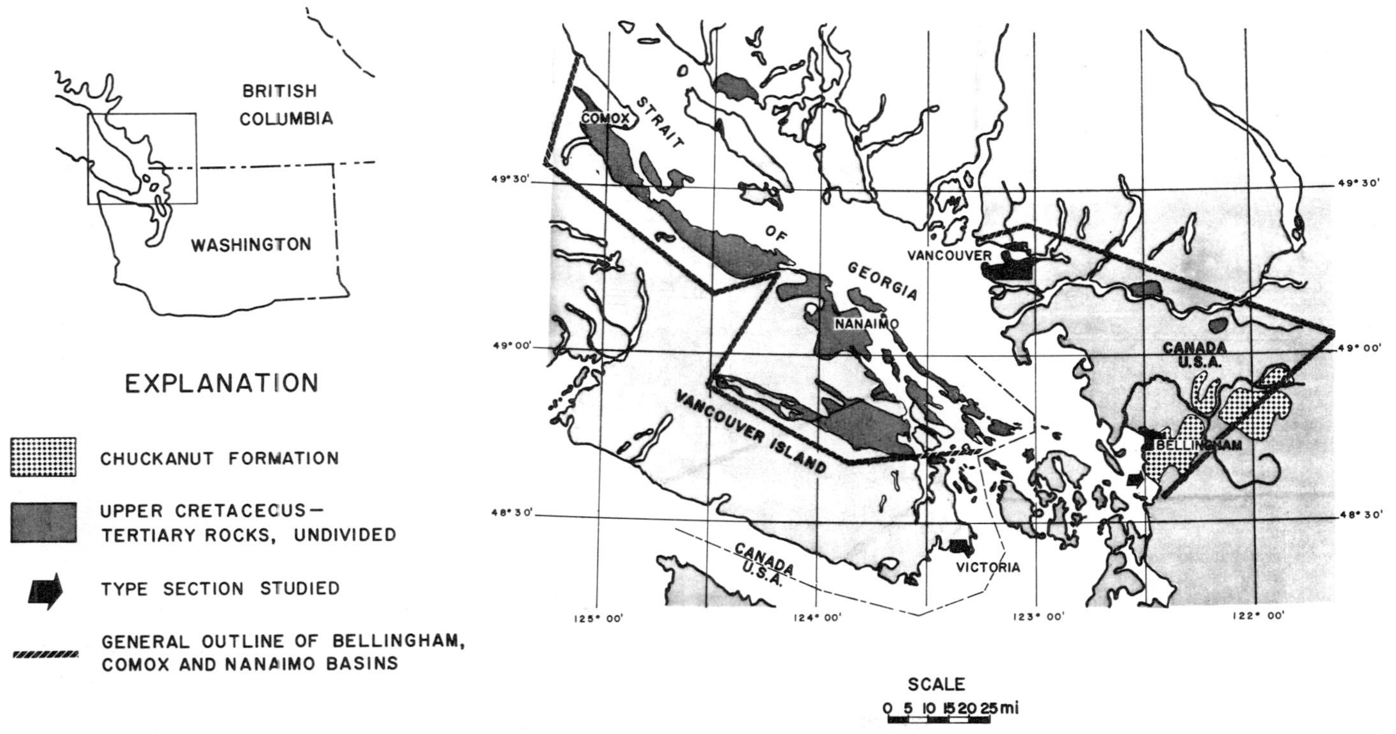

Figure 1. Upper Cretaceous-Tertiary outcrops in northwestern Washington and southwestern British Columbia indicating outlines of principal basins and location of the section studied along Samish Bay (*from* Griggs, 1965; in part *after* Miller and Misch, 1963, and Crickmay and Pocock, 1963).

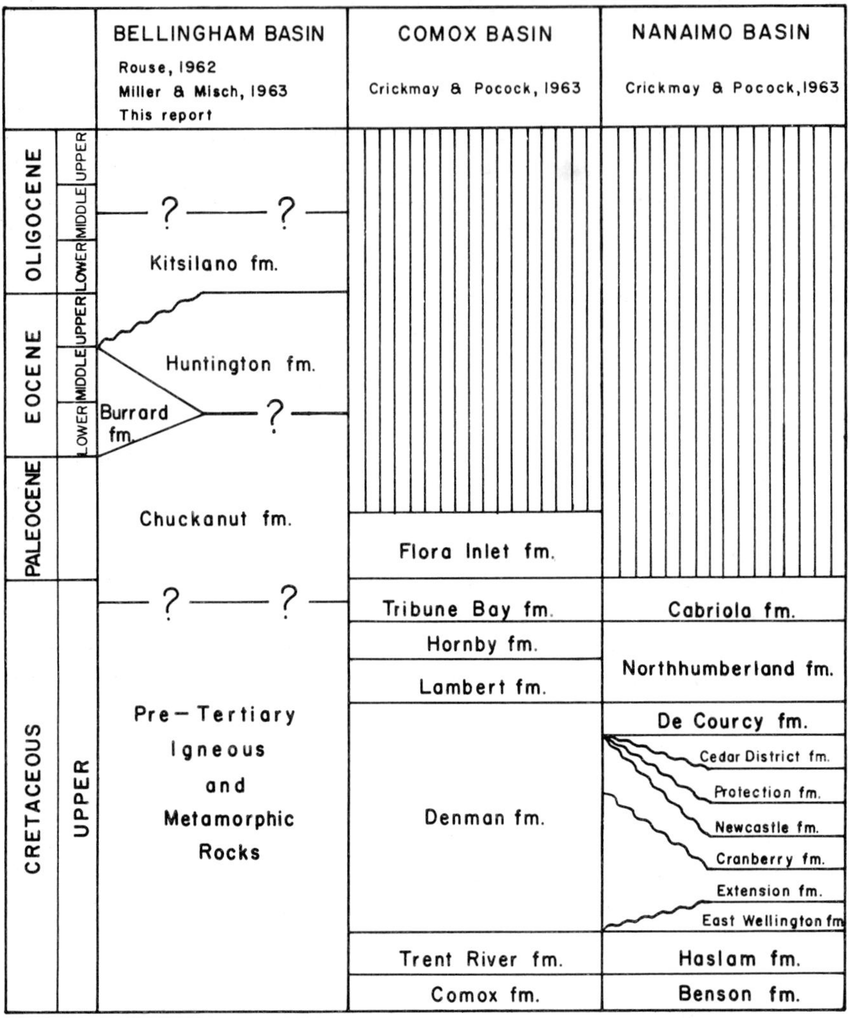

Figure 2. Generalized stratigraphic relations of formations in Bellingham, Comox, and Nanaimo Basins.

Misch (1963) report that 15,000 feet and possibly as much as 20,000 feet of section may be present in the area east of American Sumas Mountain.

The Chuckanut Formation outcrops in an area that trends northeast-southwest from Samish Bay, on the west, to the town of Glacier, on the east. It rests unconformably on pre-Cretaceous rocks to the south, is overlain unconformably by the Huntingdon Formation on the northwest, and is in fault contact with pre-Cretaceous rocks in the north and northeast along the Boulder Creek fault.

The rocks of the Chuckanut Formation are strongly folded into a series of anticlines and synclines, the axes of which trend northwest-southeast in the area south of Bellingham and Lake Whatcom. Northeast of Lake Whatcom, the axes of the folds trend northeast-southwest. The structure of the rocks between the North Fork and the Middle Fork of the Nooksack River is not completely known. Dips within the Chuckanut range from a few degrees to vertical and locally the rocks are overturned. There is some evidence of minor reverse faulting and overthrusting.

The Chuckanut Formation consists, for the most part, of massive cross-bedded to evenly stratified, medium- to coarse-grained sandstones and subordinate amounts of sandy shales. The shales vary from pure claystones to sandy and sometimes carbonaceous varieties. The shales often contain abundant fragments of plant leaves and stems. The coal seams range from a few inches up to 8 feet in thickness. The coal and shale sequences contain abundant pollen and spores, whereas the sandstones and their associated shales were found to be practically barren. Intercalated within the sandstones, and in some places at the base of the formation, are thick lenses of conglomerate.

The section along the eastern shore of Samish Bay, exposed in road cuts along Chuckanut Drive (Alternate U.S. 99) is referred to here as the standard section. Good exposures are also found along the northern shore of Lake Samish (Interstate 5) and in the road cuts along the southern shore of Lake Whatcom. Few other good outcrops are available because of dense forest cover.

DESCRIPTION OF FIELD INVESTIGATION AND LABORATORY METHODS

Field Investigation

Seventy-five of the samples collected in 1963 and 1964 from the type section of the Chuckanut along the eastern shore of Samish Bay, as described by Glover (1935), were prepared for this study. Only 26 of these were satisfactory for analysis, and all of these were from the lower 5100 feet of the 9484-foot section. There are some portions of the lower part of this type section that are concealed for as much as 250 feet. Most of the upper part of the section is concealed, except for an outcrop of conglomerate at 7700 feet above the base and a shale unit at the top of the section. The vertical distribution of the usable samples is indicated on Plates 1 and 2.

The samples were collected from gray shales, carbonaceous shales, and coals. Unweathered samples were collected in deep road cuts where possible and by digging to a depth of several feet elsewhere, to be beyond the more extreme effects of recent weathering. A maximum sample interval of 50 feet was maintained where outcrops permitted.

The base of the section is located along Chuckanut Drive in the SW¼SW¼ sec. 9, T. 36 N., R. 3 E. Due to the forest cover, it was impossible to collect a satisfactory, unweathered sample in the lower 377 feet of the section. Glover (1935) was able to differentiate and describe several beds in this 377-foot interval. From 261 to 200 feet above the schist, which underlies the Chuckanut Formation here, several shale and sandstone beds included one shaly sandstone with fossil leaves and one shale with carbonized wood fragments. Weaver (1937) was able to describe additional strata to within 100 feet of the base of the section.

The top of the section is north of here, near the site of the old Wildcat Cove station, in the NE¼SE¼ sec. 25, T. 37 N., R. 3 E. The upper part of the section is not continuously exposed and the thickness of many of the concealed units has been determined by calculation from the attitude of the beds and the horizontal distance between them. The top of the section is poorly exposed but is identified at the place where the dip is reversed and the section is repeated northward.

Laboratory Methods

Two preparation procedures were used to free the palynomorphs from coal and shale. The coals were crushed to 0.25-inch size and thoroughly mixed. A 5-gram portion from each sample was covered with 52-percent hydrofluoric acid for 24 hours. After being washed several times with distilled water to remove the acid, the residue was mixed with an equal amount (by volume) of dry potassium chlorate and concentrated nitric acid. This mixture (Schulze's solution) was allowed to react for varying periods of time, from 30 minutes to 12 hours. The residues were then washed in distilled water, treated with a 5-percent solution of potassium hydroxide for 5 to 30 minutes, and washed again until neutral. The shales were macerated in more or less the same manner with usually more time in the hydrofluoric acid and less time in Schulze's solution. All residues were then stained with Safranin O. Preparations were stored in an aqueous solution containing a few drops of acetic acid as a preservative.

The residues were mounted in Clearcol on cover slips and allowed to dry. The cover slips were then inverted and mounted on the slide with HSR (Harleco Synthetic Resin). Five slides were made from each residue.

Photographs were taken with a Leitz Ortholux microscope and a Leitz Orthomat camera using a fluorite 95/1 oil, 54/1 oil, or 40/1 dry objective and no filter. The photographic film used was Kodak Panatomic-X.

All slides have been deposited with the paleontology collection of the Geology Department, Michigan State University.

Analytical Procedure

Of the 75 samples prepared from the standard section, only 26 contained pollen and spores adequately preserved for study. Twenty-two samples were selected for examination. These represent only the lower 5100 feet of the total 9500-foot section, as previously mentioned.

Each sample was first examined qualitatively, and from results of this analysis a stratigraphic range chart was constructed for the type section (Pl. 1). A minimum of 500 spores and pollen grains were then counted from each sample (approximately 100 grains per slide and five slides per sample) for analysis of relative frequency of occurrence.

An informal classification system was used in this study, modeled after the "formulae" proposed by Tschudy (1957). Each morphological type was assigned a code designation as illustrated in Table 1.

Further subdivision was effected by a study of the sculpturing elements of the pollen grain or spore. Faegri and Iversen (1964, p. 26) presented a key to the sculpturing elements that was helpful in this study. The code designations for sculpturing features used here (Table 2) follow the usage of Tschudy (1957).

Using these "formulae," a tricolpate pollen grain with reticulate sculpturing would be designated C_3r. Each different taxon in

TABLE 1. POLLEN AND SPORE "CODE" DESIGNATIONS*

Inaperturate	O
Sulcate	S_1
	S_3
Colpate	C_1
	C_2
	C_3
	C_∞
Colporate	CP_2
	CP_3
	CP_∞
Porate	P_1
	P_2
	P_3
	P_∞
Trilete	Tl
Monolete	M
Vesiculate	V

**After* Tschudy (1957).

TABLE 2. SHORTHAND "CODE" DESIGNATIONS FOR SCULPTURE FEATURES*

Psilate (smooth)		sm
Foveolate		f
Scabrate	(projections)	p
Verrucate		
Gemmate		
Baculate		
Clavate		
Echinate (spiny)		sp
Fossulate	(striate)	st
Rugulate		
Striate		
Reticulate		r
Cicatricose		cic

*Slightly modified *from* Tschudy (1957, Table 3).

each such category is given a number (C_3r-1, C_3r-2, and so on). The first taxon of smooth trilete spores would be assigned the "formula" Tlsm-1, the second Tlsm-2, and so on.

RESULTS OF PALYNOLOGICAL INVESTIGATION

Introduction

Plant microfossils are useful for providing detailed stratigraphic and ecologic information in many instances. Their use in stratigraphy and the zonation of sedimentary rocks, especially as applied to oil exploration, is clearly indicated in the Society of Economic Paleontologists and Mineralogists symposium "Palynology in Oil Exploration" (Cross, 1964).

The application of palynological methods to certain correlation problems is particularly effective if the palynomorphs in the rocks through the complete type section or standard section of the formation are studied first. After examination of the residues the differences found in successive floras may be sufficient to allow the formulation of biostratigraphic (palynologic) zones.

Zonation of the Lower Chuckanut Section along the Eastern Shore of Samish Bay

Zonal Divisions. Three zones based on plant microfossils have been established for the lower Chuckanut section. The zonation is based on an analysis of the stratigraphic range and relative frequency of 19 selected forms. The data on which the zonation is based are shown on Plate 2.

Zone A

Triporate pollen are less abundant than *Liliacidites* sp. (S_1r-1) in Zone A (*compare* graphs D and E). *Liliacidites* sp. (S_1r-1) has greater than 10 percent relative frequency in the majority of samples in the section, reaching a high of 62 percent in sample No. 6. Although the ranges of three species of triporate pollen, P_3sm-2, *Carya juxtaporites* (P_3sm-3), and *Corylus tripollenites* (P_3sm-10), extend throughout Zone A, they are rarely abundant enough to be encountered when counting 500 grains (graphs A, B, C). *Corylus tripollenites* (P_3sm-10) attains its maximum abundance at the boundary between Zones A and B. Tricolporate pollen attain a relative frequency of up to 3.5 percent of the total microfossil flora in Zone A (graph F).

Only one significant form has a stratigraphic range that is restricted to Zone A. This is the triporate grain P_3sm-9. Two types first appear near the top of Zone A, S_1r-3 and *Cyathidites*(?) sp. (Tlsm-10). These two were found only at the boundary between Zones A and B.

Zone B

The flora of Zone B is markedly different from that found in Zones A and C. Triporate pollen are more abundant in this zone than in Zone A (graph D); especially the two forms P_3sm-2 and *Carya juxtaporites* (P_3sm-3) (graphs A, B). In the upper part of Zone A and throughout the lower third of Zone B, *Liliacidites* sp. (S_1r-1, graph E) attains high percentages in several samples. The percentages of all tricolporate pollen (graph F) drop to very low levels in the lower part of Zone B. There is a peak in two samples at the upper boundary of Zone B, an increase that is of uncertain significance.

The lower part of Zone B is marked by the first occurrence of a number of new forms. Besides the two previously mentioned, which occur first at the top of Zone A, four distinctive forms first occur in the lower part of Zone B. Two of these, *Carya*(?) sp. (P_4sm-2) and *Quercus*(?) *sp.* (C_3sm-7), occur only in Zone B and appear to be a good marker species for this zone. The other forms that first occur in the lower part of Zone B are *Tricolpites* sp. (C_3p-1) and *Magnosporites staplinii* (Msm-1).

The boundary between Zones B and C is marked, in this section, by the disappearance of 13 taxa. *Corylus tripollenites* (P_3sm-10) and *Tricolpites* sp. (C_3p-1) were not found above sample No. 20. *Proteacidites* sp. (P_3f-1), *Salix discoloripites* (C_3r-4), *Tricolporopollenites* sp. (CP_3sm-1), and P_3sm-12 are absent above sample No. 18. *Triatriopollenites*(?) sp. (P_3sm-20), *Tricolpites anguloluminosus* (C_3r-8), *Quercus*(?) sp. (C_3sm-13), *Carya*(?) sp. (P_4sm-2), *Quercus*(?) sp. (C_3sm-7), and C_3st-1 have their tops in sample No. 17.

Zone C

The relative frequencies of spores and pollen in Zone C is more nearly similar to Zone A than to Zone B. The upper limit of the zone has not been established because fossiliferous sample material was not available from the upper part of the section. The triporate pollen are less abundant than in Zone B, although they are more abundant than in Zone A (graph D). The triporate and tricolporate pollen and *Liliacidites* sp. (S_1r-1) all show increasing abundance in the top two samples (graphs D, E, F). There are no good stratigraphic forms that are restricted to Zone C.

DISCUSSION

Depositional Environment

The major characteristic changes in the microfossil flora that differentiate Zone B from Zones A and C are interpreted as being due to environmental, particularly climatic, conditions controlling the flora rather than to evolutionary changes, although serious consideration must be given the possibility that unconformities or diastems of some time significance do exist here and thus some of the changes may have been evolutionary.

The sediments of the Chuckanut Formation were deposited subaerially on a wide alluvial plain close to sea level. The nearest highlands were to the north and northeast, beyond the present Cascades. Large streams flowing from the highlands to the sea deposited sediment in the Bellingham Basin from perhaps as early as Late Cretaceous time, continuing on into the late Eocene or early Oligocene.

The sequence is characterized by thick sandstone and thinner shale units that show great lateral variation. Most of the gray shales, carbonaceous shales, and minor coal seams are richly fossiliferous. In some places the fossil leaves are so densely matted on the bedding surfaces that it is very difficult to distinguish the outlines of single leaves in the mass. Casts of large logs are often found embedded in many of the sandstones. Some of these stand vertically, *in situ,* with their roots still discernible in position in the thin coal seams or shales in which they grew. Marine fossils have not been reported from the Chuckanut Formation. I have found freshwater molluscs in some localities, but information on them has not been published.

The sediments exhibit a great variety of sedimentary structures. Cross-bedding, ripple marks, and mud cracks are common. Armored mudballs and flute casts can also be found in many outcrops.

On the basis of my studies of the Chuckanut sections, the evidence from sedimentary features, fossils, general patterns of distribution, and stratigraphy indicate that the sediments were deposited in forested, coastal swamps and backwater areas near river mouths where rates of sedimentation were quite varied. Both brackish and fresh-water swamps were intermittently buried by irregular sheets and great lenses of sand during high-water inundations. Continuous subsidence in the depositional basins was balanced by deposition. The surface of the area was thus maintained at or near sea level.

Thick sandstone units are characteristic of Zones A and C. These sandstone units are usually associated with thinly bedded shales and, in some instances, conglomerates. In Zone B the massive sandstones are not as prominent and are split by interbedded carbonaceous shales and thin coal seams. The shales and coals in Zone B contain abundant pollen and spores throughout the interval (Pl. 2).

Zone B represents a time when the eastern shore of Samish Bay was occupied by freshwater swamps and brackish lagoons. The swamps and backwater areas were at some distance from the large streams that carried coarse sediment into the area during deposition of Zone A sediments and those that followed in Zone C.

Composition of the Flora

The microfossil flora of the Chuckanut Formation indicates that a warmer and more humid climate than that of today existed in this area of the Pacific coast during early Tertiary time. This is in line with the evidence presented by Pabst (1962) from the megaflora of the Chuckanut and by Rouse (1962) from the microfossil flora of the Burrard.

The 79 palynomorphs illustrated in this paper are listed by coded designation in Table 3, along with their corresponding generic names. Genera and species that also occur in the Burrard or other pertinent strata are indicated.

Although Pabst limited herself to the study of nonflowering plants in her dissertation on the Chuckanut megaflora, she was able to present considerable evidence for reconstruction of the climatic and environmental conditions. Her paper includes 22 species: 1 scouring rush, 14 genera of ferns in 7 families and 2 orders, and 5 genera of conifers in 2 families. The megaflora and possible spore and pollen affiliations are listed in Table 4.

Of the 20 types of spores illustrated in this paper, 8 are assigned possible affiliations with the Chuckanut megaflora. The monolete spore Mp-2 (Pl. 3, fig. 5) may belong to the genus *Danaea*. Pabst described a new species, *Danaea borealis,* and noted that it was the first time *Danaea* had been described from the Tertiary. The living genus *Danaea* is restricted to the Neotropics.

Four spores are assigned to the family Schizaeaceae. Two of these, Tlcic-1 and Tlcic-3 are placed in the form genus *Cicatricosisporites* (Pl. 4, figs. 1, 2). The other two, Tlst-1 and Tlst-2, are possibly affiliated with the genus *Anemia* (Pl. 5, figs. 3, 4). Pabst described two genera from this family, *Lygodium kaulfussii* and *Anemia elongata.* Both of these are represented by modern species in tropical and subtropical regions in both hemispheres.

The three spores, Msm-2, Msm-3, Msm-4, are conspecific with *Laevigatosporites albertensis, L. discordatus,* and *L. ovatus,* respectively (Pl. 3, figs. 1-3). These three species also occur in the Burrard. Rouse stated that they are probably related to *Dryopteris* and *Asplenites.* Three species of *Dryopteris* are described from the Chuckanut megaflora. *Dryopteris* is probably the most frequently recognized genus of ferns in the North American Tertiary. The present distribution of the modern genus is temperate in both the New World and the Old. Pabst stated that "the Chuckanut species find their closest modern relatives in Mexican and Central American species."

Three spores, Osp-1 (Pl. 3, fig. 8), and Tlp-1 (Pl. 5, fig. 8), and Tlp-3 (Pl. 5, fig. 7) appear to be related to the genus *Osmunda* and are placed in the fossil genus *Osmundacidites.* Although *Osmunda* spores occur in the Burrard, Osp-1 is the only Chuckanut spore that resembles Rouse's *Osmundacidites* (Designate T_1). Tlp-1 and Tlp-3 appear to have no counterpart in the Burrard microfossil flora. *Osmunda* is a temperate and tropical fern found in moist woodlands and swamps in the northern hemisphere.

The large, nondescript spore Msm-1 appears to be identical to *Magnosporites staplinii* (Pl. 3, fig. 4), a new genus and species described by Rouse from the Burrard. Although Rouse did not indicate its botanical affinity, he suggested that it is probably associated with fresh-water plants, as these spores are found only in fresh-water deposits.

Two spores, Tlsm-6 and Tlsm-8, apparently belong to the genus *Sphagnum* (Pl. 3, figs. 9, 11). No affiliation below the generic level is offered for Tlsm-6. Tlsm-8 appears to be conspecific with *Sphagnum antiquasporites* Wilson and Webster, which Rouse reported from the Burrard. *Sphagnum* is often indicative of acid peat bogs. Another spore present in the Chuckanut, which indicates moist woods or peat bog environments, is *Lycopodium fastigioides* Couper (Tlr-1) (Pl. 5, fig. 6). This genus has not yet been reported in the Burrard flora.

The monolete spore Mp-1 appears to be similar to *Verrucatosporites* (*Polypodiisporites*) *favus* Pflug and Thompson (Pl. 3, fig. 6). Rouse believes that this spore, which is present in the Burrard, represents the perispore from *Laevigatosporites albertensis* or *L. ovatus.*

The trilete spore Tlsm-7 has been assigned to *Gleichenia concavisporites* Rouse (Pl. 3, fig. 12). This spore occurs in the Burrard and is also reported from the Upper Cretaceous Oldman Formation (Rouse, 1957). In his discussion of *G. concavisporites* from the Oldman, Rouse noted the similarity to *G. emarginata* and *G. linearis,* described by Selling (1946) from the Hawaiian Islands, to *G. concavisporites.*

Generic names are assigned tentatively to three other Chuckanut spores. Specimens of Tlsm-1 appear to be similar to *Microlepia* (Pl. 3, fig. 10). Rouse illustrated a spore from the Burrard that he tentatively assigned to this genus. Two other spores, Tlsm-9 and Tlsm-10, are assigned, with reservations, to the genus *Cyathidites* (Pl. 5, figs. 2, 5).

TABLE 3. CHUCKANUT PALYNOMORPHS AND CORRESPONDING BURRARD SPECIES

	Chuckanut	Burrard	Other pertinent occurrences
Op-1	*Inaperturopollenites juniperoides*	*I. juniperoides*	
Osp-1	*Osmundacidites* sp.		
Msm-1	*Magnosporites staplinii*	*M. staplinii*	
Msm-2	*Laevigatosporites albertensis*	*L. albertensis*	
Msm-3	*Laevigatosporites discordatus*	*L. discordatus*	
Msm-4	*Laevigatosporites ovatus*	*L. ovatus*	Comox Formation, Rouse (1957)
Mp-1	*Verrucatosporites (Polypodiisporites) favus*	*V. (Polypodiisporites) favus*	
Mp-2	*Danaea* (?) sp.		
Tlsm-1	*Microlepia* (?) sp.	*Microlepia* (?) sp.	
Tlsm-4	Affiliation unknown		
Tlsm-6	*Sphagnum* sp.		
Tlsm-7	*Gleichenia concavisporites*	*G. concavisporites*	
Tlsm-8	*Sphagnum antiquasporites*	*S. antiquasporites*	Comox Formation, Rouse (1957)
Tlsm-9	*Cyathidites* (?) sp.		
Tlsm-10	*Cyathidites* (?) sp.		
Tlp-1	*Osmundacidites* sp.		
Tlp-3	*Osmundacidites* sp.		
Tlr-1	*Lycopodium fastigioides*		Upper Cretaceous, New Zealand, Couper (1953)
Tlst-1	*Anemia* (?) sp.		
Tlst-2	*Anemia* (?) sp.		
Tlcic-1	*Cicatricosisporites* sp.		
Tlcic-3	*Cicatricosisporites intersectus*	*C. intersectus*	
S_1sm-1	*Metasequoia* (?) sp. *Taxodium* (?) sp. *Glyptostrobus* (?) sp.	*Metasequoia* sp. *Taxodium* sp.	
S_1sm-3	Affiliation unknown		
S_1sm-4	*Ginkgo* (?) sp.		

TABLE 3. CHUCKANUT PALYNOMORPHS AND CORRESPONDING BURRARD SPECIES

	Chuckanut	Burrard	Other pertinent occurrences
S_1r-1	*Liliacidites* sp.		Upper Cretaceous-Paleocene, New Mexico, Anderson (1960)
S_1r-2	*Sabal granopollenites*	*S. granopollenites*	
S_1r-3	Affiliation unknown		
C_3sm-1	*Nyssa*(?) sp.		
C_3sm-4	*Quercus*(?) sp.		
C_3sm-5	*Tricolpopollenites granulosus*	*T. granulosus*	
C_3sm-7	*Quercus*(?) sp.		
C_3sm-10	*Nymphoides tripollenites*	*N. tripollenites*	
C_3sm-13	*Quercus*(?) sp.		
C_3p-1	*Tricolpites* sp.		
C_3p-3	Affiliation unknown		
C_3p-5	Affiliation unknown		
C_3f-2	*Quercus*(?) sp.		
C_3r-2	*Salix discoloripites*(?)	*S. discoloripites*	
C_3r-3	*Tricolpites* sp.		
C_3r-4	*Salix discoloripites*	*S. discoloripites*	
C_3r-5	*Quercus*(?) sp.		
C_3r-6	*Tricolpites interangulus*(?)		Upper Cretaceous, Colorado, Newman (1965)
C_3r-8	*Tricolpites anguloluminosus*		Paleocene, New Mexico, Anderson (1960)
C_3st-1	Affiliation unknown		
Eph-1	*Ephedra notensis*		Eocene, Australia, Cookson (1957)
Eph-2	*Ephedra* sp.		
CP_3sm-1	*Tricolporopollenites* sp.		
CP_3sm-3	*Ulmus* sp.		
CP_3sm-5	*Nyssa* sp.		
CP_3sm-6	Affiliation unknown		

TABLE 3. CHUCKANUT PALYNOMORPHS AND CORRESPONDING BURRARD SPECIES

Chuckanut		Burrard	Other pertinent occurrences
CP_3f-1	*Tricolporopollenites confossus* (?)		Upper Cretaceous, Colorado, Newman (1965)
CP_3f-2	Affiliation unknown		
CP_3r-1	*Cupanieidites reticularis*		Eocene, Australia, Cookson and Pike (1954)
CP_3r-2	*Virburnam* (?) sp.		
CP_3r-3	Affiliation unknown		
P_3sm-1	Affiliation unknown		
P_3sm-2	Betulaceae-Myricaceae		
P_3sm-3	*Carya juxtaporites*	*C. juxtaporites*	
P_3sm-7	*Carya* (?) sp.		
P_3sm-9	Betulaceae-Myricaceae		
P_3sm-10	*Corylus tripollenites*	*C. tripollenites*	
P_3sm-11	Affiliation unknown		
P_3sm-12	Affiliation unknown		
P_3sm-13	Affiliation unknown		
P_3sm-14	*Triorites minor*		Upper Cretaceous, New Zealand, Couper (1953)
P_3sm-15	Affiliation unknown		
P_3sm-16	Affiliation unknown		
P_3sm-20	*Triatriopollenites* (?) sp.		
P_3p-2	*Proteacidites* cf. *marginus*	*P. marginus*	
P_3f-1	*Proteacidites* sp.		
P_4sm-1	*Alnus quadrapollenites*	*A. quadrapollenites*	
P_4sm-2	*Carya* (?) sp.		
P_5sm-1	*Alnus quinquepollenites*	*A. quinquepollenites*	
P_6sm-1	Affiliation unknown		
V-1	*Tsuga* sp.		
V-3	*Podocarpus* sp.		
V-5	*Pinus* sp.		
V-8	*Picea-Abies* spp.		

Table 4. Chuckanut Megaflora (Pabst, 1962) and Possible Spore and Pollen Affiliations

Tracheophyta	
Sphenopsida	
Equisetales	
Equisetaceae	
Equisetum newberryi Knowlton & Cockerell	
Pteropsida	
Filicineae	
Marattiaceae	
Danaea borealis Pabst	Mp-2(?)
Filicales	
Schizaeaceae	Tlcic-1, -3
Lygodium kaulfussii Heer	
Anemia elongata (Newberry) Knowlton	Tlst-1, -2(?)
Pteridaceae	
Pteris whatcomensis Pabst	
Dennstaedtia delicata Pabst	
Cyatheaceae	
Cyathea pinnata (MacGinitie) La Motte	
Aspidiaceae	Msm-2, -3, -4
Rumohra bartona Pabst	
Dryopteris chuckanutensis Pabst	
Dryopteris gibbsi Pabst	
Dryopteris whatcomensis Pabst	
Athyrium gracillium Pabst	
Allantodiopsis erosa Knowlton & Maxon	
Blechnaceae	
Salpichlaena serrata Pabst	
Woodwardia clarus Pabst	
Lorinseria aurora Pabst	
Polypodiaceae	
Polypodium alternatum Pabst	
Gymnospermae	
Coniferales	
Taxodiaceae	
Metasequoia occidentalis (Newberry) Chaney	S_1sm-1(?)
Taxodium dubium (Sternberg) Heer	S_1sm-1(?)
Glyptostrobus dakotensis (Heer) Brown	S_1sm-1(?)
Cupressaceae	
Libocedrus interrupta (Newberry) Pabst	
Juniperus washingtonensis Pabst	Op-1(?)

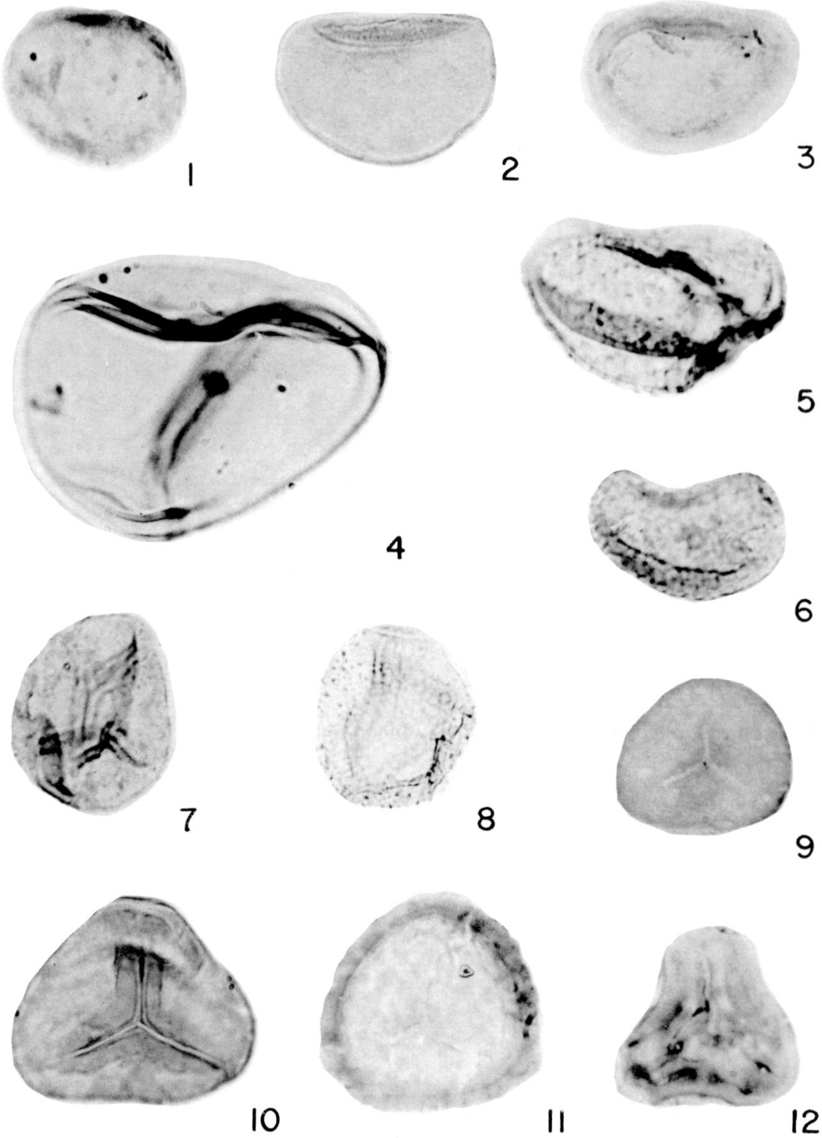

CHUCKANUT PALYNOMORPHS

PLATE 3. CHUCKANUT PALYNOMORPHS

Figures ×1000 unless otherwise indicated.

Figure
1. *Laevigatosporites albertensis* Rouse (Msm-2).
2. *L. discordatus* Thompson and Pflug (Msm-3), ×500.
3. *L. ovatus* Wilson and Webster (Msm-4).
4. *Magnosporites staplinii* Rouse (Msm-1), ×500.
5. *Danaea*(?) sp. (Mp-2), ×750.
6. *Verrucatosporites* (*Polypodiisporites*) *favus* (R. Potonié) Pflug and Thompson (Mp-1), ×500.
7. *Inaperturopollenites juniperoides* Rouse (Op-1).
8. *Osmundacites* sp. (Osp-1), ×500.
9. *Sphagnum antiquasporites* Wilson and Webster (Tlsm-8).
10. *Microlepia*(?) sp. (Tlsm-1), ×750.
11. *Sphagnum* sp. (Tlsm-6).
12. *Gleichenia concavisporites* Rouse (Tlsm-7).

PLATE 4. CHUCKANUT PALYNOMORPHS

Figures ×1000 unless otherwise indicated.

Figure
1. *Cicatricosisporites* sp. (Tlcic-1), ×750.
2. *C. intersectus* Rouse (Tlcic-3), ×750.
3. S_1sm-3, affiliation unknown.
4. *Metasequoia-Taxodium-Glyptostrobus* group (S_1sm-1).
5. *Liliacidites* sp. (S_1r-1), ×750.
6. *Ginkgo*(?) sp. (S_1sm-4).
7. *Sabal granopollenites* Rouse (S_1r-2).
8. *Nyssa*(?) sp. (C_3sm-1), ×750.
9. S_1r-3, affiliation unknown.
10. *Tricolpopollenites granulosus* Rouse (C_3sm-5).
11. *Quercus*(?) sp. (C_3sm-4).
12. *Quercus*(?) sp. (C_3sm-7).
13. *Nymphoides tripollenites* Rouse (C_3sm-10).

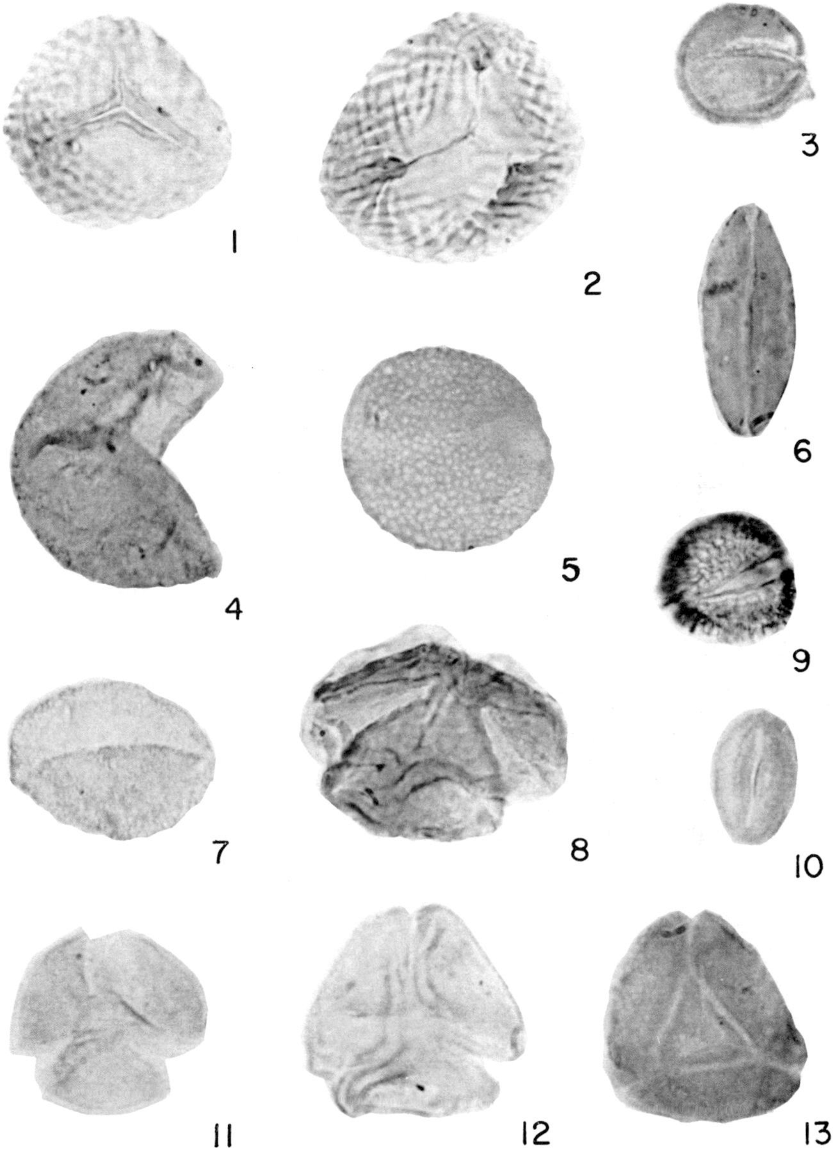

CHUCKANUT PALYNOMORPHS

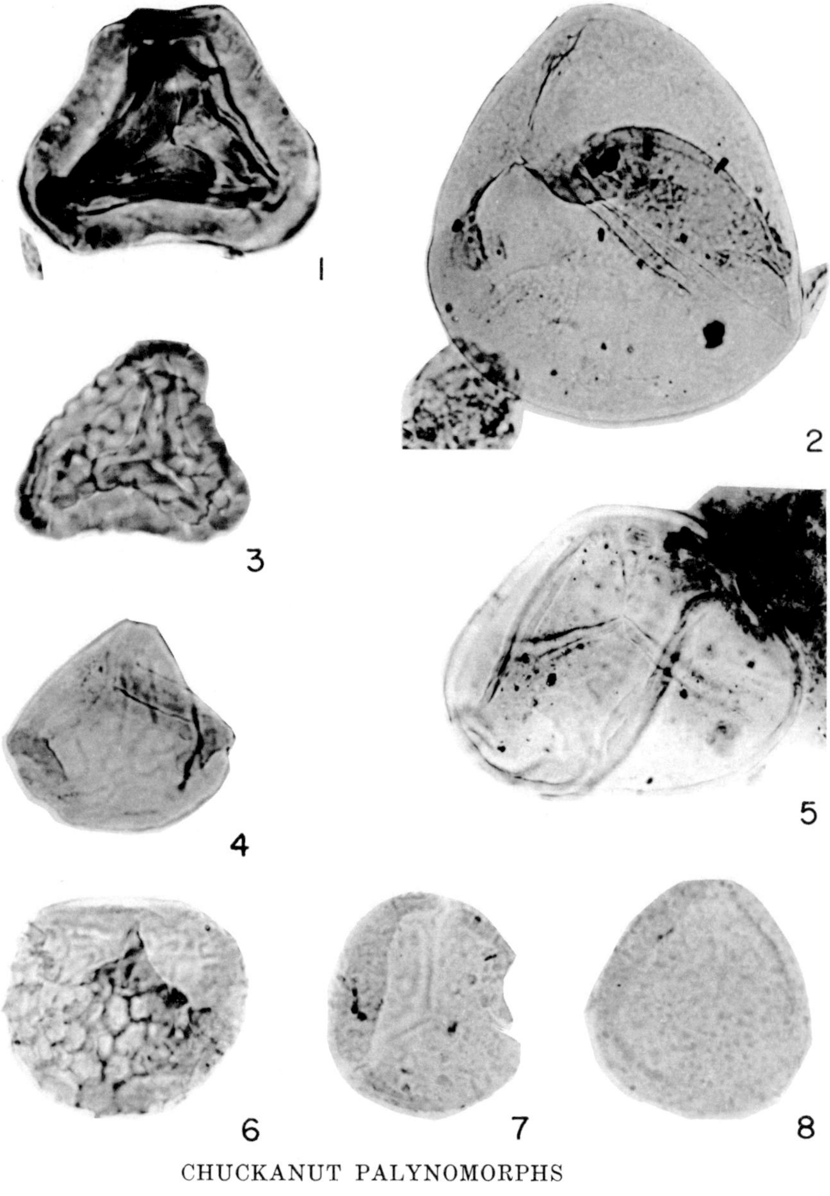

CHUCKANUT PALYNOMORPHS

PLATE 5. CHUCKANUT PALYNOMORPHS

Figure

1. Tlsm-4, affiliation unknown, ×500.
2. *Cyathidites*(?) sp. (Tlsm-9), ×500.
3. *Anemia*(?) sp. (Tlst-1), ×750.
4. *Anemia*(?) sp. (Tlst-2), ×500.
5. *Cyathidites*(?) sp. (Tlsm-10), ×500.
6. *Lycopodium fastigioides* Couper (Tlr-1), ×750.
7. *Osmundacidites* sp. (Tlp-3), ×750.
8. *Osmundacidites* sp. (Tlp-1), ×750.

Of the five genera of conifers that Pabst described from the Chuckanut, four may possibly be represented in the microfossil flora. The form S_1sm-1 has been broadly defined to include the pollen of *Metasequoia, Taxodium,* and *Glyptostrobus* (Pl. 4, fig. 4.) The pollen grains of these three genera are so similar that, for practical purposes, fossil forms are not separated. *Taxodium* and *Glyptostrobus* are the most abundant conifers in the megafossil flora, according to Pabst. Both grow in moist areas in warm-temperate and subtropical regions. Rouse reported *Taxodium hiatipites* Wodehouse and described a new species of *Metasequoia, M. papillapollenites,* from the Burrard. The *Taxodium* specimens from the Burrard are identical with those from the Green River Formation (Wodehouse, 1933). No macrofossils of *Metasequoia* have been reported from the Burrard and, because of this, Rouse indicated that the trees probably grew at some distance from the swamps. In the Chuckanut megaflora, *Metasequoia* is the least abundant of the conifers.

The inaperturate grain Op-1 appears to be identical to *Inaperturopollenites juniperoides* that Rouse described from the Burrard (Pl. 3, fig. 7). Although he had little doubt that it is a species of *Juniperus* pollen, he placed it in the form genus *Inaperturopollenites* because of the difficulty in separating *Juniperus* pollen from other genera of the Cupressaceae. Two genera are reported from the Cupressaceae in the Chuckanut megaflora, *Libocedrus* and *Juniperus.* Fossil pollen Op-1 may be representative of both these genera. *Juniperus* was probably closely associated with the cypress swamps, whereas *Libocedrus* may have had a greater range, extending into the highlands to the north and northeast. Both of these conifers are found today in habitats where there is adequate soil moisture and rainfall.

In addition to the conifers mentioned above, the pollen flora includes at least five, and perhaps six, other conifers. Five of these are *Tsuga* (V-1), *Podocarpus* (V-3), *Pinus* (V-5), and *Picea-Abies* (V-8) (Pl. 6, figs. 8-11). All of these grains are rare and are usually badly corroded. This condition of preservation, along with their absence from the megafossil flora, might indicate that these trees grew in the highlands to the north and northeast and their pollen was carried to the site of deposition, particularly by the streams flowing down from the highlands. Rouse reported two species of *Tsuga,* one of *Picea,* two of *Pinus,* and one of *Podocarpus* from the Burrard. No attempt is made to relate the Chuckanut forms at the species level to those of the Burrard. Another possible conifer is S_1sm-4, which is tentatively assigned to *Ginkgo* (Pl. 4, fig. 6). Grains of this taxon occur throughout the section.

The order Gnetales is represented by two palynomorphs. Both are believed to be pollen of the genus *Ephedra*. *Ephedripites*-1 is very similar to *Ephedra notensis* described by Cookson (1957) from the Eocene of Australia (Pl. 7, fig. 11). No specific epithet is offered for *Ephedripites*-2 (Pl. 7, fig. 8). There are 15 extant species of *Ephedra* in North America, most of which are abundant in the American Southwest and Mexico. Although *Ephedra* pollen have been found in the Chuckanut, none has yet been identified from the Burrard.

Of the 79 palynomorphs described from the Chuckanut microfossil flora, 49 are believed to be from angiosperms. Of these, 23 have been assigned generic names based upon their morphologic affinities to pollen of extant genera, 10 have been placed in form genera, 2 are equated at the family level, and 14 are listed as "affiliation unknown."

The family Palmaceae is represented by two genera in the Chuckanut, *Liliacidites* (S_1r-1) and *Sabal* (S_1r-2) (Pl. 4, figs. 5, 7). Although *Liliacidites* is a form genus, created by Couper (1953) for pollen of apparent liliaceous affinity, the general morphology of the grain and its high abundance in samples known to come from zones containing numerous palm fronds warrant placing it in this family. *Sabal granopollenites* (S_1r-2), described by Rouse from the Burrard, was considered by him to be similar to modern pollen of *Sabal palmetto*. Today the palms have their northern limits between 30° and 40° N. in the northern hemisphere and are considered to be principally tropical. The present distribution of *Sabal* is along the coast of North Carolina into Florida and west along the coastal plain of the Gulf states into southeastern Texas.

The pollen of *Corylus, Quercus,* and *Carya* indicate upland habitats. *Corylus tripollenites* Rouse (P_3sm-10) (Pl. 8, fig. 1) also occurs in the Burrard. Rouse suggested that this species may be conspecific with *Momipites coryloides* described by Wodehouse (1933) from the Green River Formation. *Corylus* is found in woodland thickets in the temperate regions of the northern hemisphere.

Five palynomorphs are tentatively assigned to the genus *Quercus*: C_3sm-4, C_3sm-7, C_3sm-13, C_3f-2, and C_3r-5 (Pl. 4, figs. 11, 12; Pl. 7, figs. 1, 6, 7). Although Rouse has described three new species of *Quercus* from the Burrard, no attempt is made to equate the Chuckanut forms with those in the Burrard, since in most cases the identification, even at the generic level, is in question. *Quercus* is found in a great variety of habitats in the northern hemisphere.

Three palynomorphs are assigned to the genus *Carya*. *Carya juxtaporites* (Wodehouse) Rouse (P_3sm-3) (Pl. 8, fig. 4) was reported from the Burrard. The other two forms, P_3sm-7 and P_4sm-2 (Pl. 8, fig. 12; Pl. 6, fig. 5), are tentatively assigned to *Carya*. P_3sm-7 differs from P_3sm-3 in the shape of the pores. P_4sm-2 is a rare four-pored *Carya* that was found only in Zone B. Today *Carya* is found in temperate climates in eastern Asia and eastern North America in various habitats.

The genus *Nyssa* is represented by CP_3sm-5 (Pl. 9, fig. 11). C_3sm-1 is questionably a *Nyssa* (Pl. 4, fig. 8). Rouse has one form from the Burrard, *Tricolporopollenites* sp. (U_{27}), based on a single specimen that resembles *Nyssa*. *Nyssa* is indigenous to eastern North America and Asia. Its habitats range from dry or moist woods to wet coastal swamps.

Nymphoides tripollenites Rouse (C_3sm-10) (Pl. 4, fig. 13) is present in both the Chuckanut and the Burrard. Rouse stated that this species is almost identical with the modern *Nymphoides glabra;* the latter has a slightly more conspicuous granulation. There are 20 species of *Nymphaea* in the United States, and their greatest distribution is in the temperate regions.

Two palynomorphs, C_3r-2 and C_3r-4 (Pl. 7, figs 5, 9), are referable to the genus *Salix*. Although they have been separated here on the basis of difference in their reticulation, possibly they should be placed together as *Salix discoloripites* Wodehouse. This species is present in the Burrard. *Salix* is almost world-wide in distribution, with present centers of distribution in the north-temperate and subarctic regions.

The form CP_3r-1 appears to be referable to the genus *Cupanieidites*. Cookson and Pike (1954) illustrated a species of *Cupanieidites, C. reticularis,* that appears to be very similar to CP_3r-1 (Pl. 9, fig. 7). They referred this species to the pollen grains of the genus *Lepiderema* and of the genus *Sarcopteryx*. The family Sapindaceae, of which these are members, is primarily pantropical in distribution and abundant in Asia and America.

The family Proteaceae is represented by two palynomorphs, P_3p-2 and P_3f-1 (Pl. 6, figs. 2, 3). The form designated as P_3p-2 appears to be the same as *Proteacidites marginus,* a new species described by Rouse from the Burrard. Although there appears to be no question that the other form, P_3f-1, is affiliated with the

Proteaceae, no attempt is made to classify it beyond the generic level. This family is found almost exclusively in the southern hemisphere today. Plants of the Proteaceae are typically found in warmer and drier sites, both inland and coastal, but they are particularly important and conspicuous in the coastal chaparral. If similar distribution patterns existed in the Upper Cretaceous and lower Tertiary, the presence of proteaceous pollen in the Chuckanut sediments may indicate a climate more nearly comparable to southern California today.

Two species of *Alnus* are represented in the Chuckanut, *Alnus quadrapollenites* (P_4sm-1) and *A. quinquepollenites* (P_5sm-1) (Pl. 6, figs. 4, 7). These two species were first described by Rouse from the Burrard. He suggested that both might fall within *Alnus speciipites* Wodehouse (1933). Wodehouse gave an all-encompassing diagnosis for his species and Rouse feels that he is justified in separating these as two distinct types. *Alnus* is found mostly in wet habitats in north-temperate regions of North America and in the Andes Mountains of South America.

Questionable assignment is given to CP_3sm-3 (*Ulmus*?), CP_3r-2 (*Viburnum*?), and P_3sm-2 and P_3sm-9 (Betulaceae-Myricaceae) (Pl. 9, figs. 4, 5; Pl. 8, figs. 3, 7).

The following palynomorphs are assigned to form genera; *Tricolpopollenites granulosus* Rouse (C_3sm-5) (Pl. 4, fig. 10), described from the Burrard, *Tricolpites* sp. Cookson ex Couper (C_3p-1 and C_3r-3) (Pl. 7, figs. 2, 4), *Tricolpites interangulus* (?) Newman (C_3r-6) (Pl. 7, fig. 10), *Tricolpites anguloluminosus* Anderson (C_3r-8) (Pl. 7, fig. 12), *Tricolporopollenites* Thompson and Pflug (CP_3sm-1) (Pl. 9, fig. 2), *Tricolporopollenites confossus* (?) Newman (CP_3f-1) (Pl. 9, fig. 10), *Triorites minor* Couper (P_3sm-14) (Pl. 8, fig. 9), and *Triatriopollenites* Thompson and Pflug (P_3sm-20) (Pl. 8, fig. 11). The only botanical affiliation offered for this group is for *Triorites minor*, which resembles pollen of the Betulaceae.

The presence of a rich fern flora with good development of angiosperms plus warm-temperate and subtropical conifers indicates a flora that developed on a large coastal plain that was exposed to moist winds and received a large amount of precipitation. The development of such a flora as this would have been possible here only if the boundary between subtropical and temperate zones was much farther north than it is today. There is also evidence of more mixing of species from these two zones than is found today.

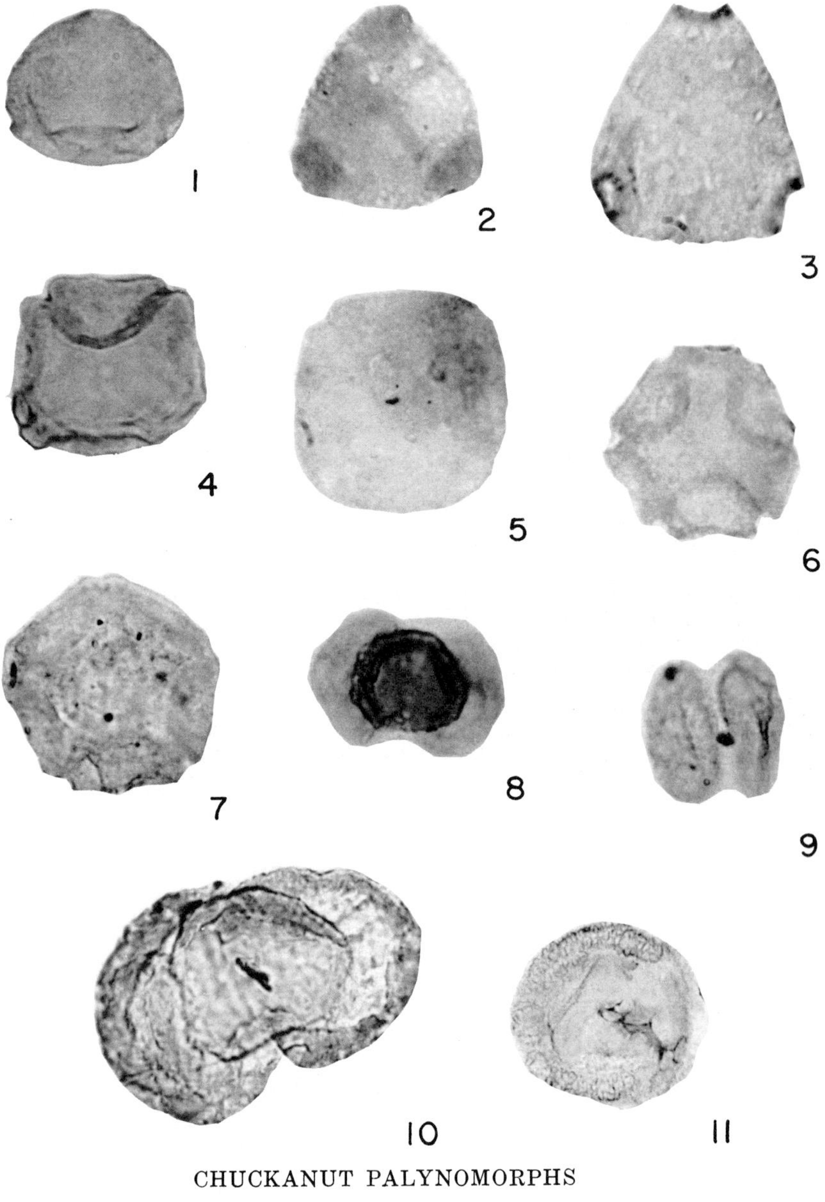

CHUCKANUT PALYNOMORPHS

PLATE 6. CHUCKANUT PALYNOMORPHS

Figures ×1000 unless otherwise indicated.

Figure

1. P_3sm-16, affiliation unknown.
2. *Proteacidites* cf. *marginus* Rouse (P_3p-2).
3. *Proteacidites* sp. (P_3f-1).
4. *Alnus quadrapollenites* Rouse (P_4sm-1).
5. *Carya*(?) sp. (P_4sm-2).
6. P_6sm-1, affiliation unknown.
7. *Alnus quinquepollenites* Rouse (P_5sm-1).
8. *Podocarpus* sp. (V-3), ×500.
9. *Pinus* sp. (V-5), ×500.
10. *Picea-Abies* spp. (V-8), ×500.
11. *Tsuga* sp. (V-1), ×500.

PLATE 7. CHUCKANUT PALYNOMORPHS

Figures $\times 1000$ unless otherwise indicated.

Figure
1. *Quercus*(?) sp. (C_3f-2).
2. *Tricolpites* sp. (C_3p-1), $\times 750$.
3. C_3p-3, affiliation unknown.
4. *Tricolpites* sp. (C_3r-3).
5. *Salix discoloripites*(?) Wodehouse (C_3r-2).
6. *Quercus*(?) sp. (C_3r-5), $\times 750$.
7. *Quercus*(?) sp. (C_3sm-13).
8. *Ephedra* sp. (Eph-2).
9. *Salix discoloripites* Wodehouse (C_3r-4).
10. *Tricolpites interangulus*(?) Newman (C_3r-6).
11. *Ephedra notensis* Cookson (Eph-1).
12. *Tricolpites anguloluminosus* Anderson (C_3r-8).

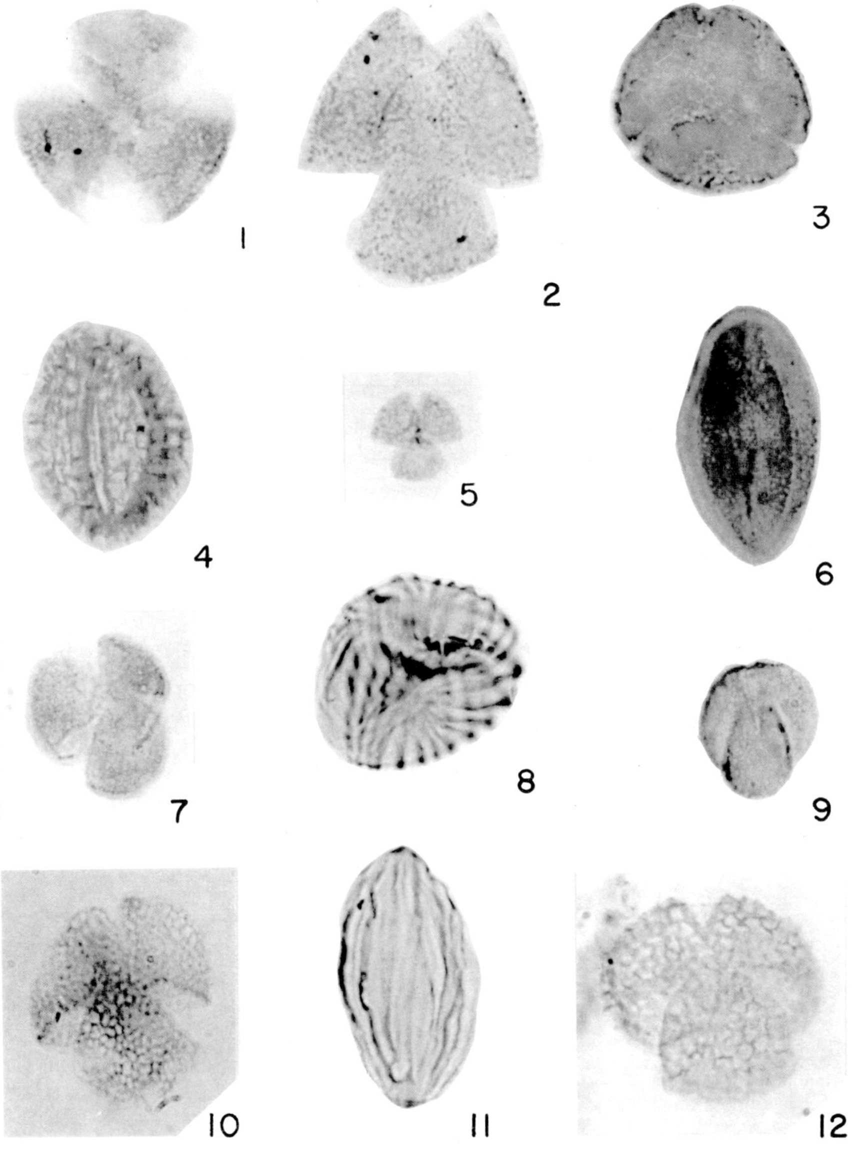

CHUCKANUT PALYNOMORPHS

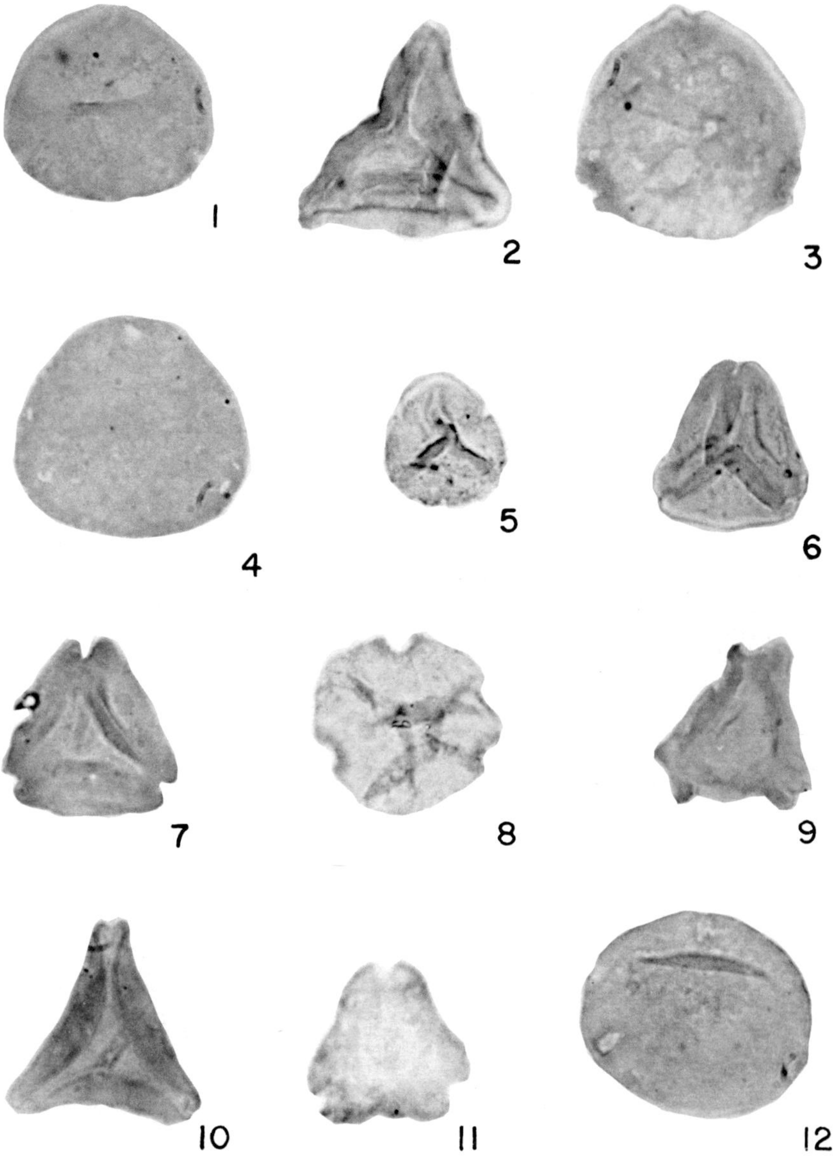

CHUCKANUT PALYNOMORPHS

PLATE 8. CHUCKANUT PALYNOMORPHS

All figures $\times 1000$.

Figure
1. *Corylus tripollenites* Rouse (P_3sm-10).
2. P_3sm-1, affiliation unknown.
3. Betulaceae-Myricaceae (P_3sm-2).
4. *Carya juxtaporites* (Wodehouse) Rouse (P_3sm-3).
5. P_3sm-15, affiliation unknown.
6. P_3sm-12, affiliation unknown.
7. Betulaceae-Myricaceae (P_3sm-9).
8. P_3sm-13, affiliation unknown.
9. *Triorites minor* Couper (P_3sm-14).
10. P_3sm-11, affiliation unknown.
11. *Triatriopollenites*(?) sp. (P_3sm-20).
12. *Carya*(?) sp. (P_3sm-7).

PLATE 9. CHUCKANUT PALYNOMORPHS

Figures ×1000 unless otherwise indicated.

Figure
1. C_3st-1, affiliation unknown.
2. *Tricolporopollenites* sp. (CP_3sm-1).
3. CP_3sm-6, affiliation unknown, ×750.
4. *Ulmus* sp. (CP_3sm-3).
5. *Virburnum*(?) sp. (CP_3r-2).
6. CP_3f-2, affiliation unknown.
7. *Cupanieidites reticularis* Cookson and Pike (CP_3r-1).
8. CP_3r-3, affiliation unknown.
9. C_3p-5, affiliation unknown.
10. *Tricolporopollenites confossus*(?) Newman (CP_3f-1), ×750.
11. *Nyssa* sp. (CP_3sm-5).

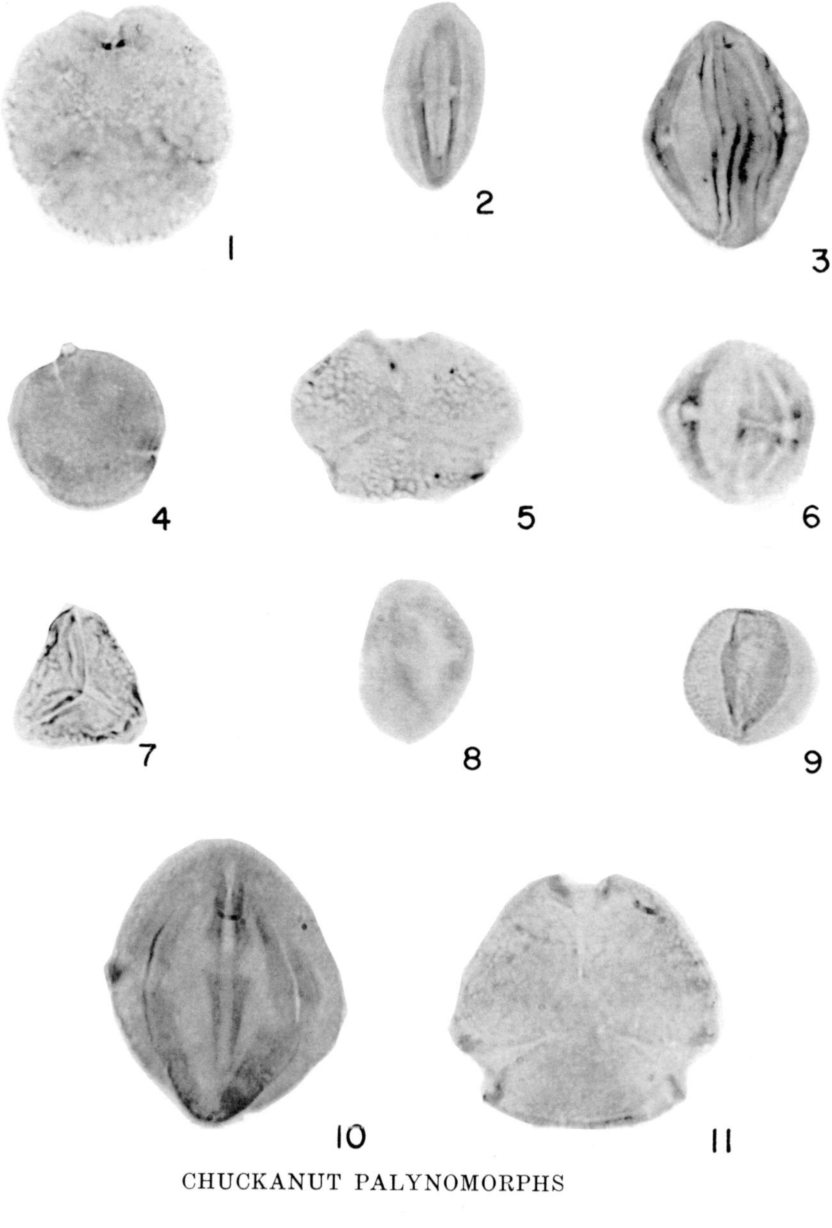

CHUCKANUT PALYNOMORPHS

GRIGGS, PLATE 9
Geological Society of America Special Paper 127

Age and Correlation

The microfossil flora of the Chuckanut Formation includes 22 taxa that are also present in the Burrard (Table 3). However, certain elements of the micro- and macro-fossil flora indicate that rocks of upper Upper Cretaceous to Paleocene and lower Eocene are included in the section studied.

Pabst presented tentative age determinations for the Chuckanut megaflora based upon comparisons with described Tertiary ferns and conifers from other formations. She noted that plants collected from Oyster Creek (base of Samish Bay section) have their closest relationship with species from Upper Cretaceous and lower Paleocene formations such as the Denver, Vermejo, Mesa Verde, and Nanaimo, whereas localities along the middle and upper part of the Samish Bay section (Chuckanut Drive and Pleasant Bay) contain floras that show their closest affinities with plants from the Fort Union, Paskapoo, Raton, and Wilcox Formations. These formations are principally of Paleocene age, although it is difficult to correlate them for such great distances with any degree of precision.

The age of the Samish Bay section, based upon the spore and pollen flora, generally complements that given by Pabst. Two palynological reports by Rouse offer comparison for the Chuckanut microfossil flora. One is the Burrard paper, which has already been referred to extensively, and the other is the Comox microfossil flora (Rouse, 1957), Upper Cretaceous of Vancouver Island. Both the Burrard and the Comox lie within the framework of the Cretaceous-Cenozoic basin of which the Chuckanut is a part.

The Chuckanut microfossil flora has only two genera in common with the Comox, whereas there are about 30 genera in the Burrard that are also found in the Chuckanut. Many of the species are also conspecific. The similarity of the Chuckanut with the Burrard microfossils heavily favors placing part of the Samish Bay section at or near equivalence with the Burrard.

In comparing the Chuckanut microfossil flora with these floras reported by other authors, it is noted that both *Liliacidites* sp. (S_1r-1) and *Tricolpites anguloluminosus* (C_3r-8), described by Anderson (1960), occur in rocks of Upper Cretaceous to Paleocene age in New Mexico. Two other forms that have been tentatively identified here were described by Newman from the Rocky Mountains of Colorado. *Tricolpites interangulus* in Colorado is confined to the Upper Cretaceous and *Tricolporopollenites confossus* is a guide to Upper Cretaceous and Paleocene strata of northwestern Colorado reported on by Newman (1965).

Although the evidence indicates that the age of the lower Chuckanut is late Late Cretaceous to early Eocene, it should be noted that the section studied may represent only the lower one-half to one-third of the total thickness of the formation. If the Chuckanut is 15,000 to 20,000 feet thick, as suggested by Miller and Misch (1963), then the upper part may extend into the middle Eocene and represent sediments that are equivalent to the Burrard.

CONCLUSION

The Chuckanut Formation was deposited subaerially on a wide alluvial plain close to sea level. The source of the sediments was to the north and northeast. Streams flowing into the Bellingham Basin deposited thick sands in shallow fresh-water lakes and swamps. The resulting sequence is characterized by thick sandstone and thinner shale and coal units that show great lateral variation.

Both the megafossil flora of Pabst and the microfossil flora of this study indicate that this area was covered with a rich temperate to subtropical vegetation. The climate was warmer and more humid than that existing in the area today. The flora indicates that there was considerably more mixing of subtropical and temperate elements in the early Tertiary. The subtropical elements in the Chuckanut species have their closest modern relatives among species living in Mexico and Central America.

A comparison of the Chuckanut palynomorphs from the Samish Bay section with those described from the Burrard and Comox Formations of British Columbia indicates that the age of this section is younger than the Comox, perhaps very late Late Cretaceous to Paleocene or early Eocene and nearly the same as the Burrard. Pabst gives an age of Late Cretaceous to Paleocene for this section, based on the megafossil flora. The similarity of the microfossil flora to the Burrard and its almost complete dissimilarity with the Comox is offered as evidence for assignment of a late Late Cretaceous to early Eocene age for this section. It is possible, however, that Upper Cretaceous rocks exist between the lowest sample studied and the base of the section (377 feet) or that they are present elsewhere in the formation but have been removed by erosion from this particular section.

ACKNOWLEDGMENTS

This paper embodies a portion of a Master of Science thesis submitted to the Department of Geology, Michigan State University (Griggs, 1965). My gratitude is expressed to Dr. Aureal T. Cross, who directed the thesis, and to Drs. C. E. Prouty and James H. Fisher for their constructive criticism.

Some portions of this paper, together with some of the information in the Ph.D. dissertation by Dr. Dennis M. Sparks (1966), were presented by Sparks in the symposium, "Palynology of the Late Cretaceous and Early Tertiary," at the 1966 Annual Meeting of The Geological Society of America in San Francisco. The title of that joint paper was "Palynological Investigations of Lower Tertiary Rocks in Western Washington." I wish to thank Dr. Sparks for extending to me joint authorship on that paper and for his advice on this report.

REFERENCES CITED

Anderson, R. Y., 1960, Cretaceous-Tertiary palynology, eastern side of the San Juan Basin, New Mexico: New Mexico Bur. Mines and Mineral Resources Mem. 6, 59 p.

Berry, E. W., 1926, Tertiary floras from British Columbia: Canada Dept. Mines, Geol. Survey Bull. 42 (Geol. ser. no. 45), p. 91-116, pl. X-XIX.

Crickmay, C. H., and Pocock, S. A. J., 1963, Cretaceous of Vancouver, British Columbia, Canada: Am. Assoc. Petroleum Geologists Bull. 47, p. 1925-1942.

Cookson, I. C., 1957, On some Australian Tertiary spores and pollen grains that extend the geological and geographical distribution of living genera: Royal Soc. Victoria Proc., v. 69, p. 41-53.

Cookson, I. C., and Pike, K. M., 1954, Some dicotyledonous pollen types from Cainozoic deposits in the Australian region: Australian Jour. Botany, v. 2, p. 197-219.

Couper, R. A., 1953, Upper Mesozoic and Cainozoic spores and pollen grains from New Zealand: New Zealand Geol. Survey Paleont. Bull. 22, 77 p.

Cross, Aureal T., Editor, 1964, Palynology in oil exploration: A symposium: Soc. Econ. Paleontologists and Mineralogists Spec. Pub. no. 11, 200 p.

Faegri, K., and Iversen, J., 1964, Textbook of pollen analysis: New York, Hafner Publishing Co., 237 p.

Glover, S. L., 1935, Oil and gas possibilities of western Whatcom County, Washington: Washington Div. Mines and Geology Rept. Inv. no. 2, p. 1-69.

Griggs, Peter H., 1965, Stratigraphic significance of fossil pollen and spores of the Chuckanut Formation, northwest Washington: East Lansing, Michigan State Univ., M.S. thesis.

Hopkins, W. S., Jr., 1966, Palynology of Tertiary rocks of the Whatcom Basin, southwestern British Columbia and northwestern Washington: Vancouver, Univ. British Columbia, Ph.D. dissert.

Jenkins, O. P., 1923, Geologic investigation of the coal fields of western Whatcom County, Washington: Washington Div. Mines and Geology Bull. 28, 135 p.

——1924, Geological investigation of the coal fields of Skagit County, Washington: Washington Div. Mines and Geology Bull. 29, 63 p.

Knowlton, F. H., 1902, Preliminary report of fossil plants from the state of Washington, collected by Henry Landes, 1901: Washington Geol. Survey Ann. Rept., 1901, v. 1, p. 32-33.

La Motte, R. S., 1938, An Upper Cretaceous florule from northwestern Washington (abs.): Geol. Soc. America Proc. 1937, p. 283.

Lesquereux, L., 1859, On fossil plants collected by Dr. John Evans at Vancouver Island and Bellingham Bay, Washington Territory: Am. Jour. Sci., ser. 2, v. 27, p. 360-363.

McLellan, R. D., 1927, The geology of the San Juan Islands: Washington Univ. Pub. Geology, v. 2, 185 p.

Miller, G. M., and Misch, P., 1963, Early Eocene angular unconformity at western front of northern Cascades, Whatcom County, Washington: Am. Assoc. Petroleum Geologists Bull., v. 47, p. 163-174.

Moen, W. S., 1962, Geology and mineral deposits of the north half of the Van Zandt quadrangle, Whatcom County, Washington: Washington Div. Mines and Geology Bull. 50, 129 p.

Newberry, J. S., 1863, Descriptions of the fossil plants collected by Mr. George Gibbs . . . (Vancouver Island and Washington Terr.): Boston Jour. Nat. History, v. 7, p. 506-525.

Newman, K. R., 1965, Upper Cretaceous-Paleocene guide palynomorphs from northwestern Colorado: Colorado Univ. Studies Ser. Earth Sci. no. 2, 21 p.

Pabst, M. B., 1962, The flora of the Chuckanut Formation of northwestern Washington. The Equistales, Filicales, and Coniferales: Berkeley, Univ. California, Ph.D. dissert.

Rouse, G.E., 1957, The application of a new nomenclatural approach to Upper Cretaceous plant microfossils from western Canada: Canadian Jour. Botany, v. 35, p. 349-375.

——1962, Plant microfossils from the Burrard Formation of western British Columbia: Micropaleontology, v. 8, p. 187-218.

Selling, O. H., 1946, Studies in Hawaiian pollen statistics; Pt. 1 — The spores of the Hawaiian pteridophytes: Bernice P. Bishop Mus. Spec. Pub. no. 37, 87 p.

Shedd, S., 1903, Building and ornamental stones of Washington: Washington Geol. Survey Ann. Rept., v. 2, p. 62-65.

Sparks, Dennis M., 1966, Microfloral zonation and correlation of some lower Tertiary rocks in southwest Washington and some conclusions concerning the paleoecology of the flora: East Lansing, Michigan State Univ., Ph.D. dissert.

Tschudy, R. H., 1957, Pollen and spore formulae—A suggestion: Micropaleontology, v. 3, p. 277-280.

Weaver, C. E., 1937, Tertiary stratigraphy of western Washington and northwestern Oregon: Washington Univ. Pub. Geol., v. 4, 266 p.

White, C. A., 1888, On the Puget Group of Washington Territory: Am. Jour. Sci., 3rd ser., v. 36, p. 443-450.

Willis, B., 1898, Some coal fields of Puget Sound: U.S. Geol. Survey 18th Ann. Rept., pt. 3, p. 393-436.

Wodehouse, R. P., 1933, Tertiary pollen; II—The oil shales of the Eocene Green River Formation: Torrey Bot. Club Bull., v. 60, p. 497-524.

Palynology of Some Late Cretaceous and Early Tertiary Deposits in British Columbia and Adjacent Alberta

GLENN E. ROUSE
University of British Columbia, Vancouver, British Columbia

W. S. HOPKINS, JR.
Geological Survey of Canada, Calgary, Alberta

K. M. PIEL
Union Oil Company of California, Brea, California

ABSTRACT

Palynological analyses of microfossil assemblages from Upper Cretaceous and lower Tertiary rocks in British Columbia and adjacent Alberta indicate major differences in the evolution of floras between coastal and interior locales. The Santonian-Campanian assemblage on the coast is dominated by an assemblage of ferns, herbaceous angiosperms, and *Proteacidites,* whereas the assemblage from beds of similar age in western Alberta is typified by different species of ferns, probable herbaceous angiosperms, and *Aquilapollenites.* These differences suggest the existence of two floral provinces during Santonian-Campanian time, possibly separated by the western cordillera.

A late Maestrichtian-Danian assemblage from the Rocky Mountain trench contains a taxodiaceous-angiosperm complex that differs from an assemblage of essentially the same age in the western interior region (eastern Montana-western South Dakota) of the United States, suggesting again the presence of two different floral provinces at this time.

A comparison of middle Eocene assemblages on the coast with those of the interior plateau regions of southern and central British Columbia shows a coastal assemblage composed of ferns and mainly herbaceous angiosperms, whereas the comparable assemblage east of the Coast and Cascade Mountains contains a predominantly coniferous-hardwood flora.

CONTENTS

INTRODUCTION

Upper Cretaceous and Tertiary plant-bearing deposits occur in various coastal and interior locales in British Columbia, particularly in western and central regions of the province. Most deposits are relatively small, reflecting sedimentation in localized basins, lake depressions, or former river channels. Preservation of leaves, spores, pollen, and phytoplankton ranges from poor to good and is apparently related to the amount of local metamorphism or physical/chemical degradation. Fortunately, enough well-preserved assemblages have been recovered to provide a preliminary assessment of the floral-evolutionary history along the western coast of Canada. Additional research will certainly enhance the over-all picture of Cretaceous-Tertiary vegetational history and palaeoecology of this region and eventually provide a link in the over-all pattern of floristic history of continental and offshore regions bordering the northern Pacific Ocean.

The purpose of this paper is to outline our present knowledge of microfossil assemblages from selected Upper Cretaceous and lower Tertiary horizons in British Columbia and adjacent Alberta and to suggest the apparent patterns of vegetational history, palaeoecology, and palaeophysiography. Both coastal and interior microfossil assemblages are presented because it has become apparent that some significant differences existed between coastal and interior floral assemblages in Late Cretaceous and early Tertiary times.

The earliest information on floral assemblages came from investigations of leaves, and it was not until 1957 that plant microfossil assemblages were reported from British Columbia. Most of the leaf assemblages are Late Cretaceous or Eocene in age and are included in the reports by Dawson (1883, 1895), Penhallow (1903, 1907, 1908), Johnston (1923), Berry (1926), Arnold (1955a, 1955b), Bell (1957), and Rouse (1967). Studies of plant microfossil assemblages from British Columbia include those from the Upper Cretaceous and Eocene reported by Rouse (1957, 1962), Rouse and Mathews (1961), Crickmay and Pocock (1963), Hills (1965), and Hopkins (1966, 1968) and those from the upper Miocene or lower Pliocene published by Mathews and Rouse (1963) from interior sites and by Martin and Rouse (1966) from the Queen Charlotte Islands off the western coast. An additional study is currently in progress by one of us (K. M. Piel) on assemblages ranging from early Oligocene to late Miocene or early Pliocene in the central part of the province.

ROCK UNITS AND ASSEMBLAGES

The rock units and microfossil assemblages reported here range from Santonian-Campanian to middle Eocene in the coastal region and from Maestrichtian to middle Eocene in the interior. In both cases the succession is broken, and only those assemblages that are considered significant are presented here. The locations of rock units containing microfossil assemblages are shown in the outline map of British Columbia (Fig. 1) along with the main river systems of today. The stratigraphic and probable time relationships of the rock units and assemblages are outlined in Table 1, with the same reference numbers for assemblages as used on the map in Figure 1.

Late Cretaceous Assemblages

The two Late Cretaceous assemblages include one from the Protection Formation of the coastal Nanaimo Group (C-1) and equivalent strata on the coast of Santonian-Campanian age, and one from beds in east-central British Columbia (I-1), of late Maestrichtian-Danian age.

TABLE 1. STRATIGRAPHIC-TIME CHART OF ROCK UNITS AND MICROFOSSIL ASSEMBLAGES

Age		Coastal	Map designates of microfossil assemblages	Interior	Map designates of microfossil assemblages	
Eocene	Upper	Kitsilano Formation and equivalents in Richfield-Pure wells	C-3	Middle Eocene	Princeton Droppingwater	I-2A I-2B
	Middle	Burrard Formation and equivalents in Richfield-Pure wells	C-2		Horsefly(?): Francois Lake:	I-2C I-2D
	Lower				Driftwood Creek:	I-2E
Paleocene		Chuckanut Formation				
Maestrichtian		Gabriola Formation		Reynolds Creek		I-1
Campanian Santonian		Comox Formation Protection Formation Brothers Creek Richfield-Pure wells	C-1A C-1B C-1C C-1-4			

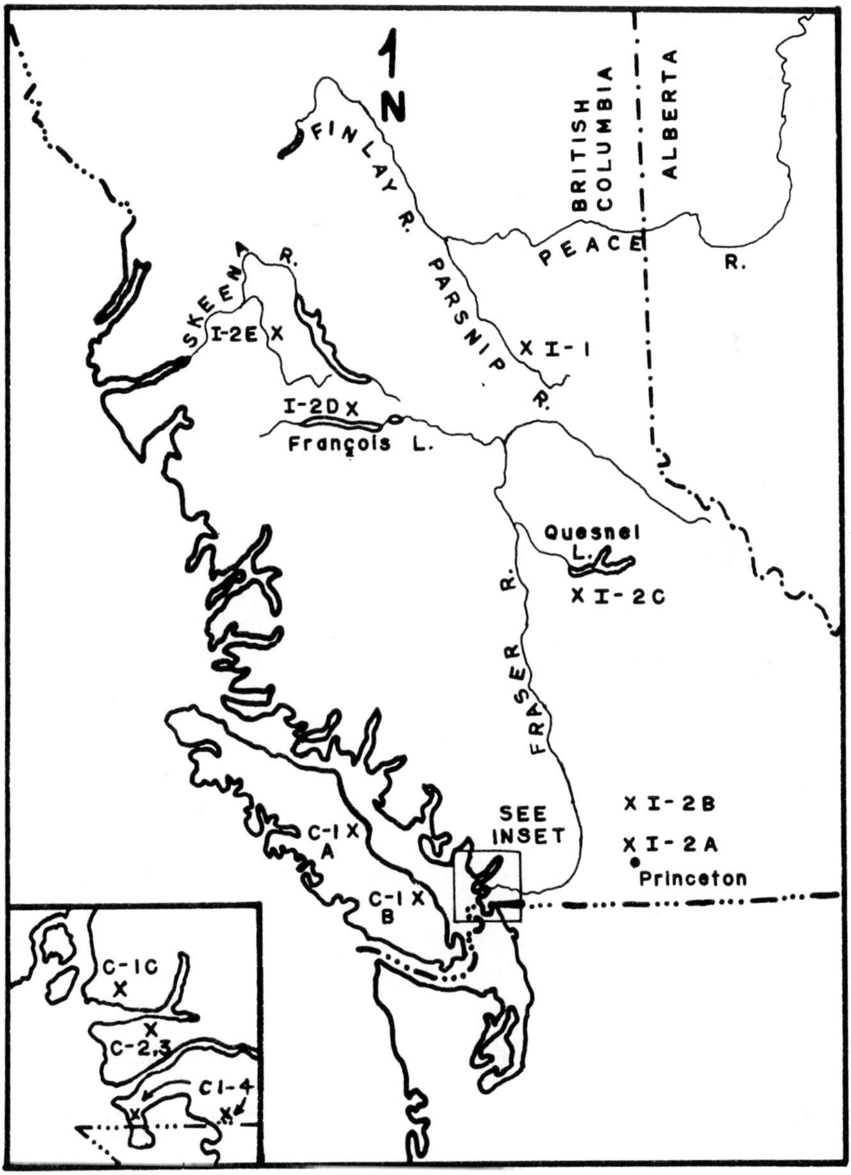

Figure 1. Map of southern and central British Columbia showing locations of coastal and interior rock units and assemblages. C-1A, Comox Formation, Santonian-Campanian. C-1B, Protection Formation, Santonian-Campanian. C-1C, Brothers Creek exposures, Santonian-Campanian. C-1-4, Richfield-Pure wells, Santonian-Campanian-Eocene. C-2, Burrard Formation, middle Eocene. C-3, Kitsilano Formation, mid-upper Eocene. I-1, Reynolds Creek locale, late Maestrichtian-Danian. I-2A, Allenby Formation, Princeton coal basin, middle Eocene. I-2B, Droppingwater Creek locale, middle Eocene. I-2C, Horsefly River beds, middle Eocene. (?) I-2D, François Lake locale, middle Eocene. I-2E, Driftwood Creek locale, middle Eocene.

The Santonian-Campanian coastal assemblage (C-1) includes the following palynomorphs, of which some of the most characteristic are figured on Plate 1, figure 1 through Plate 2, figure 12:

Gleicheniidites senonicus Ross (Pl. 1, figs. 1, 2)
Intrapunctatosporis ellipsoideus (Pflug) Krutzsch (Pl. 1, fig. 4)
Reticulatisporites agathoecus (Potonié) Potonié and Kremp (Pl. 1, fig. 3)
Stereisporites cf. *S. cingulatus* subsp. *cingulatus* Krutzsch (Pl. 1, fig. 5)
Anemia cristata var. *fenestrata* Markova (Pl. 1, figs. 6, 7)
Anemia striosporites Rouse (Pl. 1, fig. 8; Pl. 2, fig. 1)
Deltoidospora microforma Rouse (1962, p. 199, Pl. 3, figs, 6-8)
Hymenophyllumsporites papillosus Rouse (1962, p. 198, Pl. 4, figs. 11-12)
Intrapunctisporis plicatus Rouse (1962, p. 199, Pl. 3, figs. 20, 21)
Proteacidites thalmanni Anderson (Pl. 2, figs. 8, 9)
Proteacidites marginus Rouse (1962, Pl. 2, fig. 26)
Beaupreadites elegansiformis Cookson (Rouse, 1962, Pl. 1, fig. 10)
Symplococites sibericus N. Mtchedlishvili (Pl. 2, figs. 2, 3)
Tricolporopollenites sp. (Pl. 2, fig. 4)
Tricolpites sp. (Pl. 2, fig. 5)
Triporites rhamnoides Bolchovitina (Pl. 2, figs. 6, 7)
Platanus sp. (Pl. 2, fig. 10)
Quercus fusiformipollenites Rouse (Pl. 2, fig. 11)
Aquilapollenites sp. (Pl. 2, fig. 12)
Sparganium globipites Wilson & Webster (Rouse, 1962, p. 360, Pl. 1, figs. 6, 7)
Tricolpopollenites divergens Rouse (1962, p. 360, Pl. 1, figs. 16, 19)

Some of the characteristic palynomorphs of the late Maestrichtian-Danian assemblage from the Rocky Mountain trench in east-central British Columbia (I-1) are illustrated on Plate 2, figure 13, through Plate 4, figure 17. The assemblage includes:

Osmunda regalites Martin and Rouse (Pl. 3, fig. 5)
Sciadopitys sp. (Pl. 2, fig. 13; Pl. 3, fig. 4)
Taxodium sp. (Pl. 3, fig. 1)
Glyptostrobus sp., Type 1 (Pl. 3, fig. 2)
Glyptostrobus sp., Type 2 (Pl. 3, fig. 3)
Tilia sp. (Pl. 3, fig. 11; Pl. 4, figs. 1-3)
Quercus fusiformipollenites Rouse (Pl. 3, figs. 6-8)
Monoporate reticulate (Pl. 3, figs. 9, 10)
Aquilapollenites sp. (Pl. 3, fig. 12)

Pistillipollenites mcgregorii Rouse (Pl. 3, figs. 13, 14)
Tricolpopollenites sp., Type A (Pl. 3, figs. 15, 16)
Tricolpopollenites sp., Type B (Pl. 3, figs. 17, 18)
Pterocarya sp. (Pl. 4, fig. 4)
Potamogeton sp. (Pl. 4, figs. 5, 6)
Alnus sp., Type 1 (Pl. 4, figs. 7, 8)
Alnus sp., Type 2 (Pl. 4, fig. 9)
Tricolpopollenites sp., Type C (Pl. 4, figs. 10, 13)
Tricolpopollenites sp., Type D (Pl. 4, figs. 11, 12)
Tricolpopollenites sp., Type E (Pl. 4, figs. 14, 15)
Polycolpites sp. (Pl. 4, figs. 16, 17)
Acer sp. (Pl. 3, fig. 19)

Eocene Assemblages

Coastal Eocene. The coastal Eocene assemblage is a composite of two recognizable subassemblages. The earler (C-2) comprises the spores and pollen reported by Rouse (1962) from the Burrard Formation and more recently by Hopkins (1966) from well sections near the International Boundary (Fig. 1; C, 1-3) that appear generally equivalent to the Burrard. The latter assemblage (C-3) is from the Kitsilano Formation and Kitsilano equivalents in both the Richfield-Pure wells and outcrop localities in the Whatcom (Bellingham) Basin of southwestern British Columbia and northwestern Washington State. This C-3 assemblage is considered to represent a stage in the Eocene slightly later than the Burrard, most probably middle to late Eocene.

The two Eocene subassemblages contain very similar suites except for several distinguishing features. Pollen of *Sabal* and *Pistillipollenites* appear to be limited essentially to the older C-2 assemblage, whereas *Ilex* and *Liquidambar* appear to be restricted to the younger C-3. In spite of this, the presence of such microfossils as *Cicatricosisporites intersectus* and *Platycarya* sp. indicates a close time interval between the two and that the younger C-3 assemblage was deposited not much later than the middle Eocene C-2 assemblage.

The species of the C-2 and C-3 assemblages are listed below, with some of the more characteristic members illustrated on Plate 4, figure 18, through Plate 6, figure 12.

Anemia poolensis Chandler (Pl. 4, figs. 18, 19)
Cicatricosisporites intersectus Rouse (Pl. 4, figs. 20, 21; Pl. 5, fig. 2)
Laevigatosporites discordatus Thomson & Pflug (Pl. 5, fig. 1)

PLATE 1. COASTAL SANTONIAN-CAMPANIAN ASSEMBLAGE (C-1)

All magnifications ×1000.

Figure

1. *Gleicheniidites senonicus* Ross. Brothers Creek, C-1, C.
2. *Gleicheniidites senonicus* Ross. Comox Formation, C-1, A.
3. *Reticulatisporites agathoecus* (Potonié) Potonié and Kremp. Richfield-Pure wells, C, 1-4.
4. *Intrapunctatosporis ellipsoideus* (Pflug) Krutzsch. Brothers Creek, C-1, C.
5. *Stereisporites* cf. *S. cingulatus* subsp. *cingulatus* Krutzsch, Comox Formation, C-1, A.
6. *Anemia cristata* var. *fenestrata* Markova. Brothers Creek, C-1, C.
7. *Anemia cristata* var. *fenestrata* Markova. Brothers Creek, C-1, C.
8. *Anemia striosporites* Rouse. Brothers Creek, C-1, C.

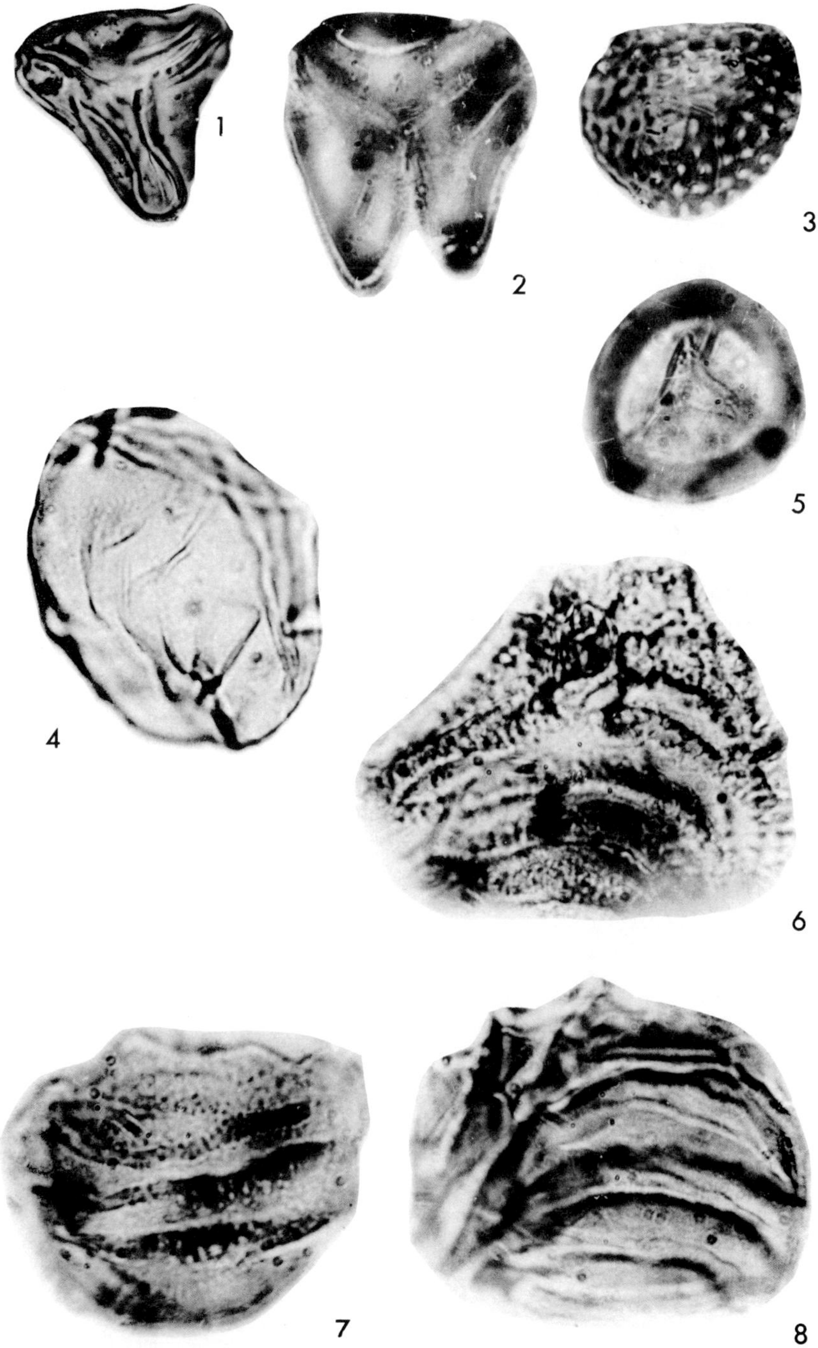

COASTAL SANTONIAN-CAMPANIAN ASSEMBLAGE

ROUSE AND OTHERS, PLATE 1
Geological Society of America Special Paper 127

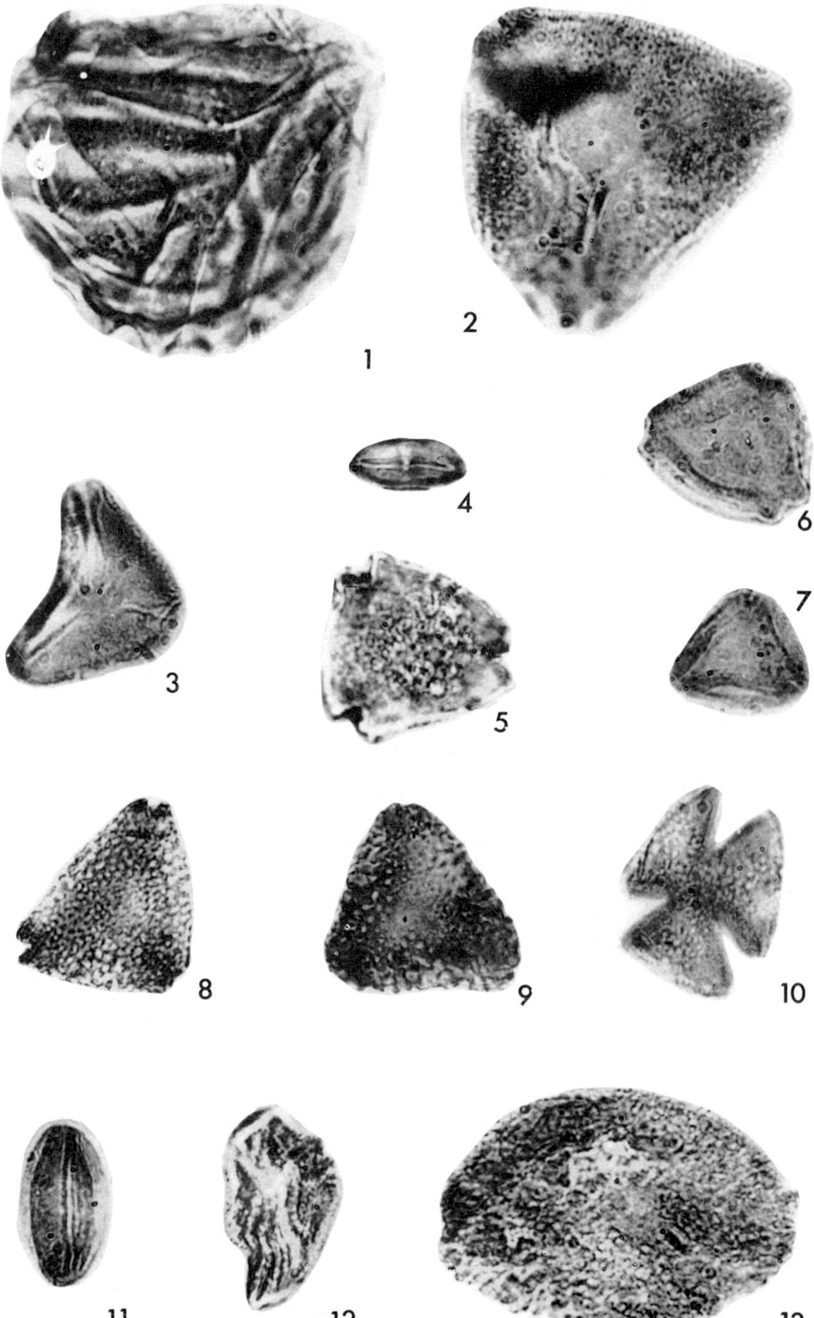

COASTAL SANTONIAN-CAMPANIAN (FIGS. 1-12) AND INTERIOR MAESTRICHTIAN-DANIAN (FIG. 13) ASSEMBLAGES

PLATE 2. COASTAL SANTONIAN-CAMPANIAN (FIGS. 1-12) AND INTERIOR MAESTRICHTIAN-DANIAN (FIG. 13) ASSEMBLAGES

All magnifications ×1000 unless otherwise specified.

Figure
1. *Anemia striosporites* Rouse. Brothers Creek, C-1, C.
2. *Symplococites sibericus* N. Mtchedlishvili. Brothers Creek, C-1, C.
3. *Symplococites sibericus* N. Mtchedlishvili. (×660) Brothers Creek, C-1, C.
4. *Tricolporopollenites* sp. Richfield-Pure wells, C, 1-4.
5. *Tricolpites* sp. Comox Formation, C-1, A.
6. *Triporites rhamnoides* Bolchovitina. Richfield-Pure wells, C, 1-4.
7. *Triporites rhamnoides* Bolchovitina. Richfield-Pure wells, C, 1-4.
8. *Proteacidites thalmanni* Anderson. Brothers Creek, C-1, C.
9. *Proteacidites thalmanni* Anderson. Brothers Creek, C-1, C.
10. *Platanus* sp. Brothers Creek, C-1, C.
11. *Quercus fusiformipollenites* Rouse. Richfield-Pure wells, C, 1-4.
12. *Aquilapollenites* sp. Richfield-Pure wells, C, 1-4.
13. *Sciadopitys sp.* Reynolds Creek, I-1.

PLATE 3. INTERIOR LATE MAESTRICHTIAN-DANIAN ASSEMBLAGE

All specimens from Reynolds Creek, I-1.
Magnifications ×1000 unless otherwise specified.

Figure
1. *Taxodium* sp.
2. *Glyptostrobus* sp., Type 1.
3. *Glyptostrobus* sp., Type 2.
4. *Sciadopitys* sp.
5. *Osmunda regalites* Martin & Rouse.
6. *Quercus fusiformipollenites* Rouse.
7. *Quercus fusiformipollenites* Rouse.
8. *Quercus fusiformipollenites* Rouse.
9. Monoporate reticulate (×660).
10. Monoporate reticulate (×600).
11. *Tilia* sp.
12. *Aquilapollenites* sp. (sp. nov.).
13. *Pistillipollenites mcgregorii* Rouse.
14. *Pistillipollenites mcgregorii* Rouse (×660).
15. *Tricolpopollenites* sp., Type A.
16. *Tricolpopollenites* sp., Type A.
17. *Tricolpopollenites* sp., Type B (×660).
18. *Tricolpopollenites* sp., Type B (×660).
19. *Acer* sp. (×660).

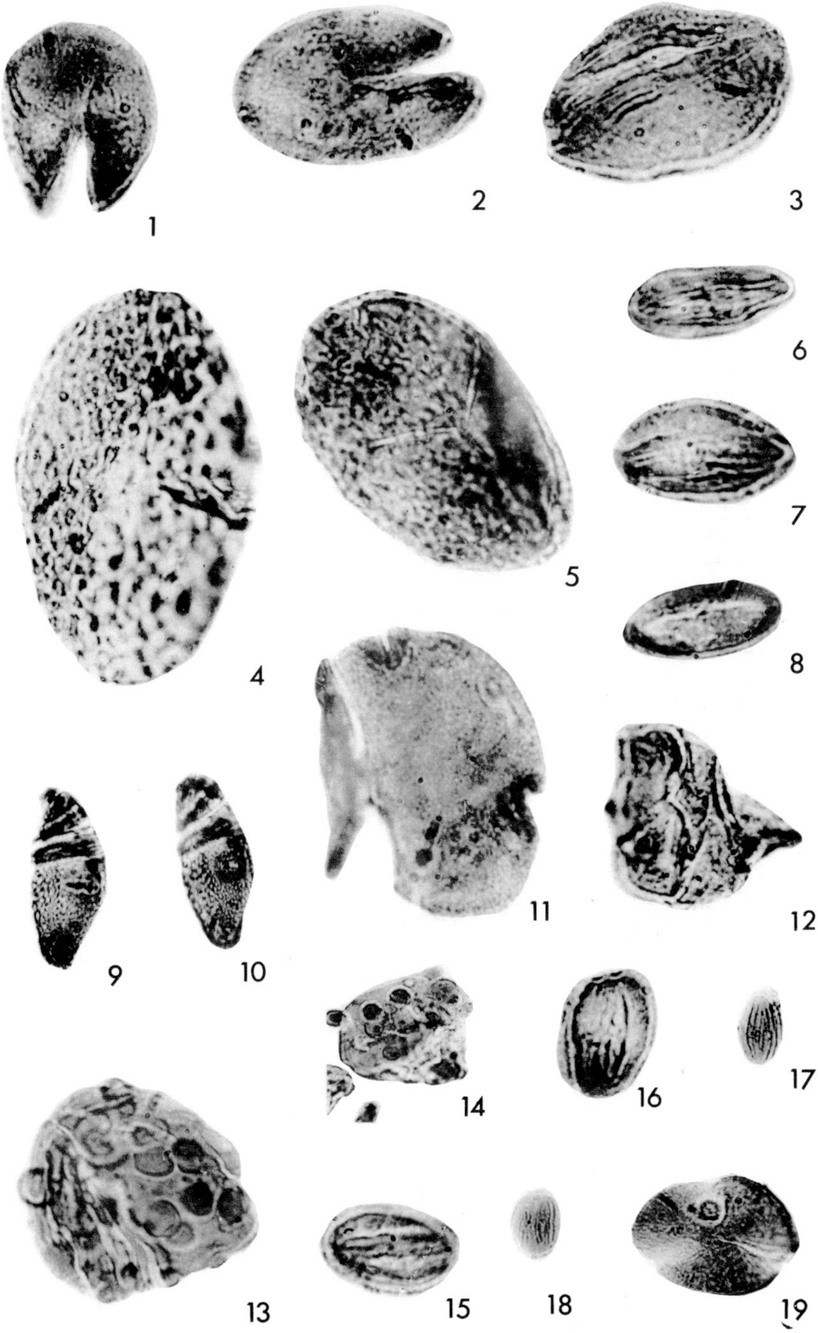

INTERIOR LATE MAESTRICHTIAN-DANIAN ASSEMBLAGE

ROUSE AND OTHERS, PLATE 3
Geological Society of America Special Paper 127

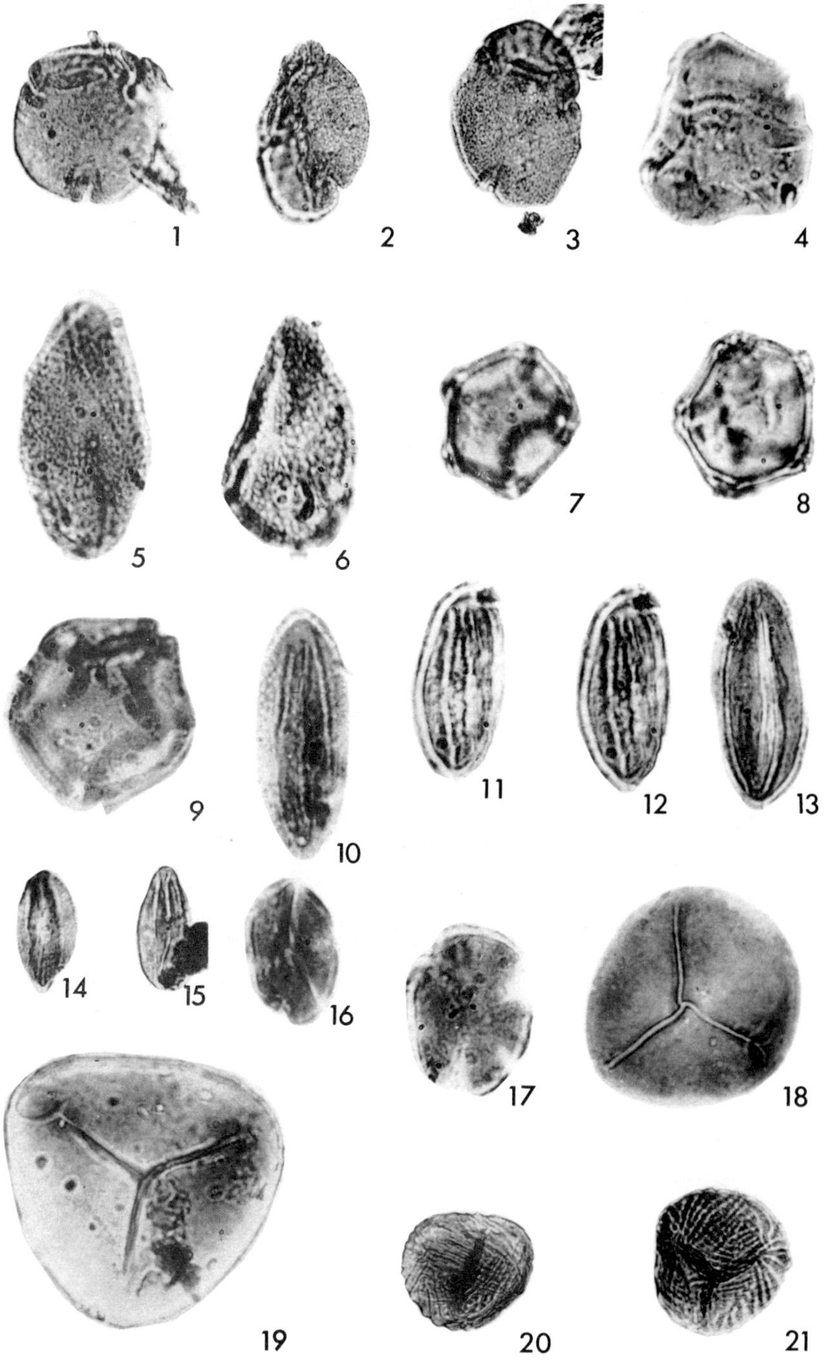

INTERIOR LATE MAESTRICHTIAN-DANIAN (FIGS. 1-17) AND COASTAL EOCENE (FIGS. 18-21) ASSEMBLAGES

PLATE 4. INTERIOR LATE MAESTRICHTIAN-DANIAN (FIGS. 1-17) AND COASTAL EOCENE (FIGS. 18-21) ASSEMBLAGES

Magnifications ×1000 unless otherwise specified.

Figure

1. *Tilia* sp. (×660).
2. *Tilia* sp. (×660).
3. *Tilia* sp. (×660).
4. *Pterocarya* sp.
5. *Potamogeton* sp.
6. *Potamogeton* sp.
7. *Alnus* sp., Type 1.
8. *Alnus* sp., Type 1.
9. *Alnus* sp., Type 2.
10. *Tricolpopollenites* sp., Type C.
11. *Tricolpopollenites* sp., Type D.
12. *Tricolpopollenites* sp., Type D.
13. *Tricolpopollenites* sp., Type C.
14. *Tricolpopollenites* sp., Type E (×660).
15. *Tricolpopollenites* sp., Type E (×660).
16. *Polycolpites* sp.
17. *Polycolpites* sp.
18. *Anemia poolensis* Chandler. (×660) Kitsilano Formation, C-3.
19. *Anemia poolensis* Chandler. Kitsilano Formation, C-3.
20. *Cicatricosisporites intersectus* Rouse (×330). Kitsilano Formation, C-3.
21. *Cicatricosisporites intersectus* Rouse (×330). Kitsilano Formation, C-3.

PLATE 5. COASTAL EOCENE ASSEMBLAGE

Magnifications ×1000 unless otherwise specified.

Figure

1. *Laevigatosporites discordatus* Thomson & Pflug. Kitsilano Formation, C-3.
2. *Cicatricosisporites intersectus* Rouse. Kitsilano Formation, C-3.
3. *Lygodium reticulosporites* Rouse (×330). Kitsilano Formation, C-3.
4. *Laevigatosporites ovatus* Wilson & Webster (×660). Kitsilano Formation, C-3.
5. *Trilites solidus* (Potonié) Krutzsch. Kitsilano Formation, C-3.
6. *Pistillipollenites mcgregorii* Rouse. Burrard Formation, C-2.
7. *Taxodium hiatipites* Wodehouse. Kitsilano Formation, C-3.
8. *Metasequoia papillapollenites* Rouse. Kitsilano Formation, C-3.
9. *Ulmoideipites krempi* Anderson (×660). Kitsilano Formation, C-3.
10. *Platycarya* sp. Richfield-Pure wells, C, 1-4.
11. *Momipites tenuipolus* Anderson. Richfield-Pure wells, C, 1-4.
12. *Myrica annulites* Martin & Rouse (×660). Kitsilano Formation, C-3.
13. *Myrica annulites* Martin & Rouse. Kitsilano Formation, C-3.
14. *Carya juxtaporipites* (Wodehouse) Rouse. Kitsilano Formation, C-3.

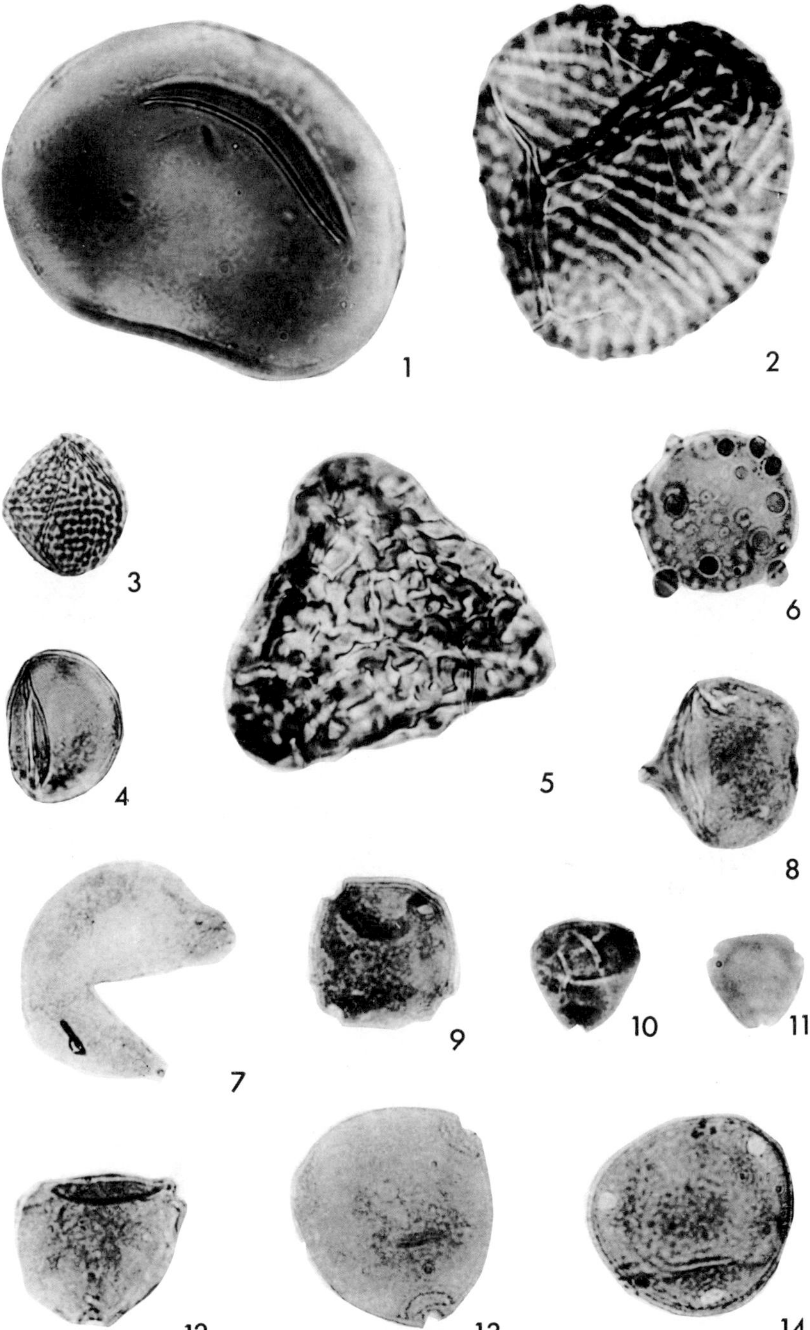

COASTAL EOCENE ASSEMBLAGE

ROUSE AND OTHERS, PLATE 5
Geological Society of America Special Paper 127

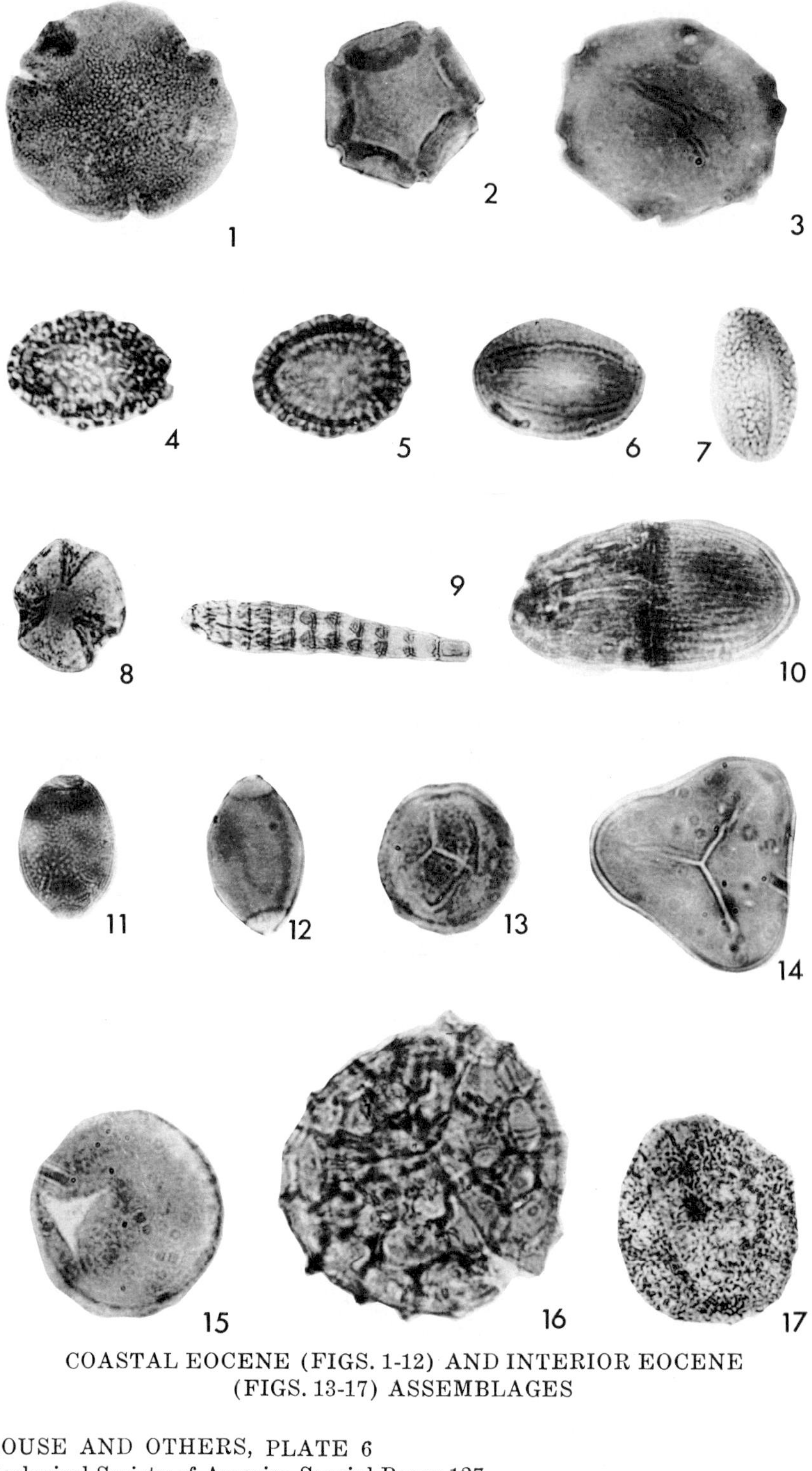

COASTAL EOCENE (FIGS. 1-12) AND INTERIOR EOCENE (FIGS. 13-17) ASSEMBLAGES

ROUSE AND OTHERS, PLATE 6

Geological Society of America Special Paper 127

PLATE 6. COASTAL EOCENE (FIGS. 1-12) AND INTERIOR EOCENE (FIGS. 13-17) ASSEMBLAGES

Magnifications ×1000 unless otherwise specified.

Figure

1. *Tilia vescipites* Wodehouse. Kitsilano Formation, C-3.
2. *Alnus verus* (R. Potonié) Martin & Rouse. Kitsilano Formation, C-3.
3. *Pterocarya stellatus* Martin & Rouse. Richfield-Pure wells, C, 1-4.
4. *Ilex* sp. Kitsilano Formation, C-3.
5. *Ilex* sp. Kitsilano Formation, C-3.
6. *Acer* sp. Kitsilano Formation, C-3.
7. *Liliacidites leei* Anderson. Kitsilano Formation, C-3.
8. ?*Nyssa* sp. (×330) Richfield-Pure wells, C, 1-4.
9. *Pluricellaesporites psilatus* Clarke. Burrard Formation, C-2.
10. *Fusiformisporites* sp. Whatcom Quarry (cf. Kitsilano Formation, C-3).
11. *Diporites granulatus* Rouse (×660). Kitsilano Formation, C-3.
12. *Diporites* sp. (×660). Kitsilano Formation, C-3.
13. Microspore of *Azolla* sp. Princeton, I-2A.
14. *Anemia poolensis* Chandler (×600). Horsefly River, I-2C.
15. Microspore of *Azolla* sp. Princeton, I-2A.
16. *Lycopodium annotinioides* (Krutsch) Martin & Rouse. François Lake, I-2D.
17. *Osmunda* sp: (×660). Francois Lake, I-2D.

Lygodium reticulosporites Rouse (Pl. 5, fig. 3)
Laevigatosporites ovatus Wilson & Webster (Pl. 5, fig. 4)
Trilites solidus (Potonié) Krutzsch (Pl. 5, fig. 5)
Taxodium hiatipites Wodehouse (Pl. 5, fig. 7)
Metasequoia papillapollenites Rouse (Pl. 5, fig. 8)
Pistillipollenites mcgregorii Rouse (Pl. 5, fig. 6)
Ulmoideipites krempi Anderson (Pl. 5, fig. 9)
Platycarya sp. (Pl. 5, fig. 10)
Momipites tenuipolus Anderson (Pl. 5, fig. 11)
Myrica annulites Martin & Rouse (Pl. 5, figs. 12, 13)
Carya juxtaporipites (Wodehouse) Rouse (Pl. 5, fig. 14)
Tilia vescipites Wodehouse (Pl. 6, fig. 1)
Alnus verus (R. Potonié) Martin & Rouse (Pl. 6, fig. 2)
Pterocarya stellatus Martin & Rouse (Pl. 6, fig. 3)
Ilex sp. (Pl. 6, figs. 4, 5)
Acer sp. (Pl. 6, fig. 6)
Liliacidites leei Anderson (Pl. 6, fig. 7)
?*Nyssa* sp. (Pl. 6, fig. 8)
Diporites granulatus Rouse (Pl. 6, fig. 11)
Diporites sp. (Pl. 6, fig. 12)
Fusiformisporites sp. (Pl. 6, fig. 10)
Pluricellaesporites psilatus Clarke, (Pl. 6, fig. 9)

Interior Eocene. This assemblage, obtained from five localities of middle Eocene deposits in the interior regions (I-2), differs mainly from the coastal assemblage by: (1) greater amounts of pinaceous and taxodiaceous pollen, (2) larger numbers of pollen of dicotyledonous trees that are extant in southeastern North America and east-central China, particularly *Carya, Pterocarya, Platycarya, Juglans, Alnus, Betula, Carpinus, Corylus,* and *Tilia,* (3) fewer fern spores, with the exception of microspores of *Azolla primaeva.*

The main microfossils from the interior middle Eocene (I-2) are listed below with examples shown on Plate 6, figure 13, through Plate 9.

Microspores of *Azolla* sp. (Pl. 6, figs. 13, 15)
Glochidia of *Azolla* sp. (Pl. 7, fig. 3)
Anemia poolensis Chandler (Pl. 6, fig. 14)
Lycopodium annotinioides (Krutzsch) Martin & Rouse (Pl. 6, fig. 16)
Osmunda sp. (Pl. 6, fig. 17)
cf. *Keteleeria* sp. (Pl. 7, figs. 1, 2)
Tsuga heterophyllites Martin & Rouse (Pl. 7, fig. 10)
Podocarpus sp. (Pl. 7, fig. 5)
Pinus sp. (*P. sylvestris* type) (Pl. 7, fig. 6)

Pinus strobipites Wodehouse (Pl. 7, figs. 4, 9)
Picea grandivescipites Wodehouse (Pl. 7, figs. 7, 8, 11)
Metasequoia papillapollenites Rouse (Pl. 7, fig. 12; Pl. 8, fig. 2)
Taxodium hiatipites Wodehouse (Pl. 8, figs. 1, 4)
Pistillipollenites mcgregorii Rouse (Pl. 8, figs. 3, 5, 6)
Potamogeton hollickipites Wodehouse (Pl. 8, fig. 7)
Carya viridi-fluminipites (Wodehouse) Wilson & Webster (Pl. 8, fig. 8)
Myrica annulites Martin & Rouse (Pl. 8, figs. 9, 13)
Betula claripites Wodehouse (Pl. 8, figs. 10, 11)
Carpinus ancipites Wodehouse (Pl. 8, figs. 12, 15)
Corylus tripollenites Rouse (Pl. 8, fig 14; Pl. 9, fig. 1)
Alnus verus (R. Potonié) Martin & Rouse (Pl. 9, fig. 3)
Ulmus sp. (Pl. 9, fig. 4)
Juglans sp. (Pl. 9, fig. 5)
Pterocarya stellatus (R. Potonié) Martin & Rouse (Pl. 9, fig. 6)
Pterocarya sp. (Pl. 9, fig. 7)
Quercus sp. (Pl. 9, fig 8)

DISCUSSION

An examination of the assemblages presented above suggests general trends in the evolution of floras in both coastal and interior regions from the Late Cretaceous to Eocene. The main changes in composition can be best appreciated by an examination of Table 2, which lists the most diagnostic genera and species of microfossils for the several horizons.

Coastal Assemblages

The Santonian-Campanian assemblage from the Protection Formation and equivalents on the coast consists mainly of fern spores and angiosperm pollen, with a notable lack of coniferous pollen. The most characteristic fern spores are *Stereisporites* cf. *S. cingulatus* subsp. *cingulatus, Anemia cristata,* and *Anemia striosporites,* although large numbers of *Gleicheniidites senonicus* occur in almost all samples. The most common and characteristic angiosperm pollen are *Proteacidites thalmanni, Proteacidites marginus, Symplococites sibericus,* and *Tricolpopollenites divergens.* Some samples also contain *Beaupreadites elegansiformis, Quercus fusiformipollenites,* and *Sparganium globipites.* Pollen of at least one species of *Aquilapollenites* also occurs in the Protection Formation, but the genus is less strikingly represented in both species

PLATE 7. INTERIOR EOCENE ASSEMBLAGE

Magnifications ×1000 unless otherwise specified.

Figure

1. cf. *Keteleeria* sp. (×660). François Lake, I-2D.
2. cf. *Keteleeria* sp. François Lake, I-2D.
3. *Glochidia* of *Azolla* sp. Princeton, I-2D.
4. *Pinus strobipites* Wodehouse (×660). Horsefly, I-2C.
5. *Podocarpus* sp. (×660). Horsefly, I-2C.
6. *Pinus* sp. (*P. sylvestris* type). (×660). François Lake, I-2D.
7. *Picea grandivescipites* Wodehouse (×660). Horsefly, I-2C.
8. *Picea grandivescipites* Wodehouse (×660). Horsefly, I-2C.
9. *Pinus strobipites* Wodehouse (×660). Horsefly, I-2C.
10. *Tsuga heterophyllites* Martin & Rouse (×660). François Lake, I-2D.
11. *Picea grandivescipites* Wodehouse (×660). Horsefly, I-2C.
12. *Metasequoia papillapollenites* Rouse. Horsefly, I-2C.

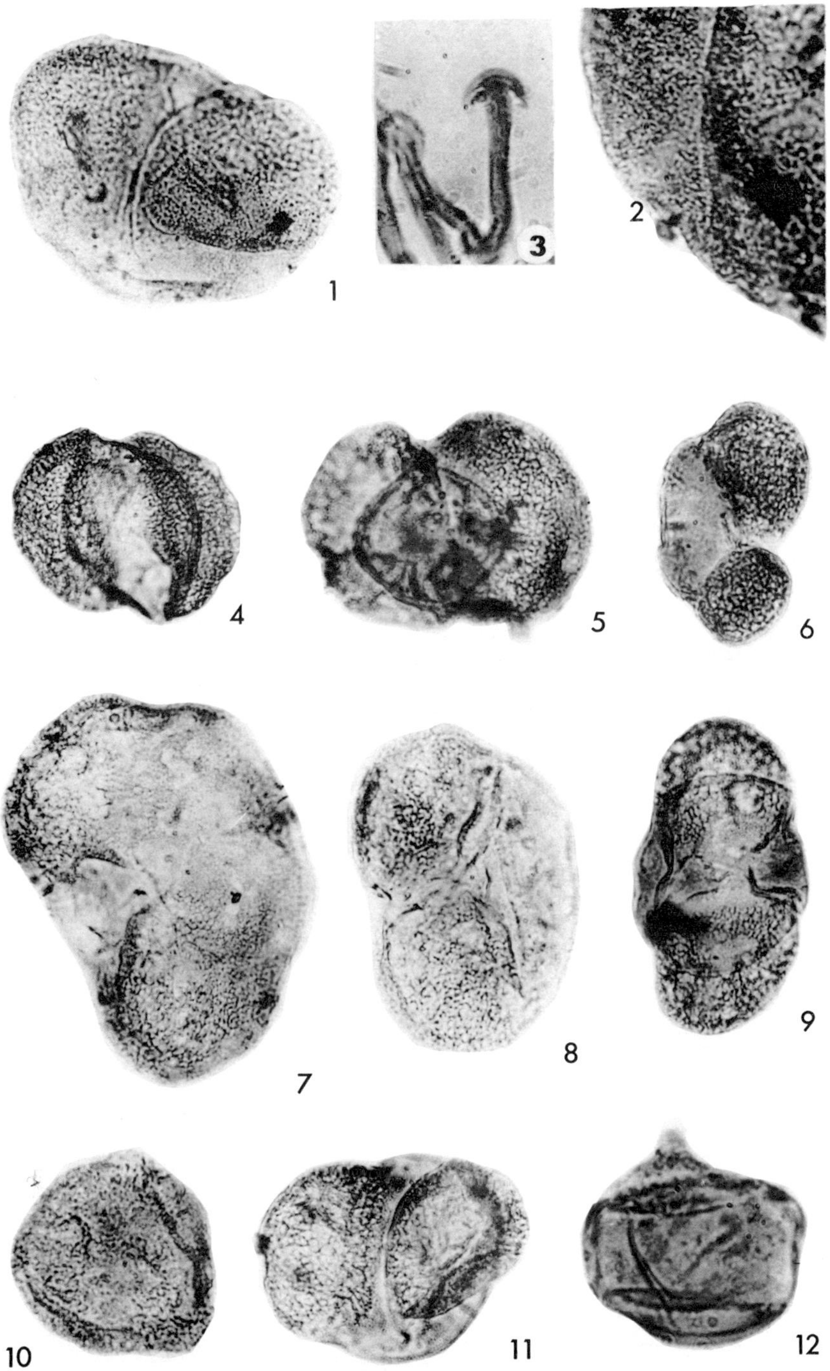

INTERIOR EOCENE ASSEMBLAGE

ROUSE AND OTHERS, PLATE 7
Geological Society of America Special Paper 127

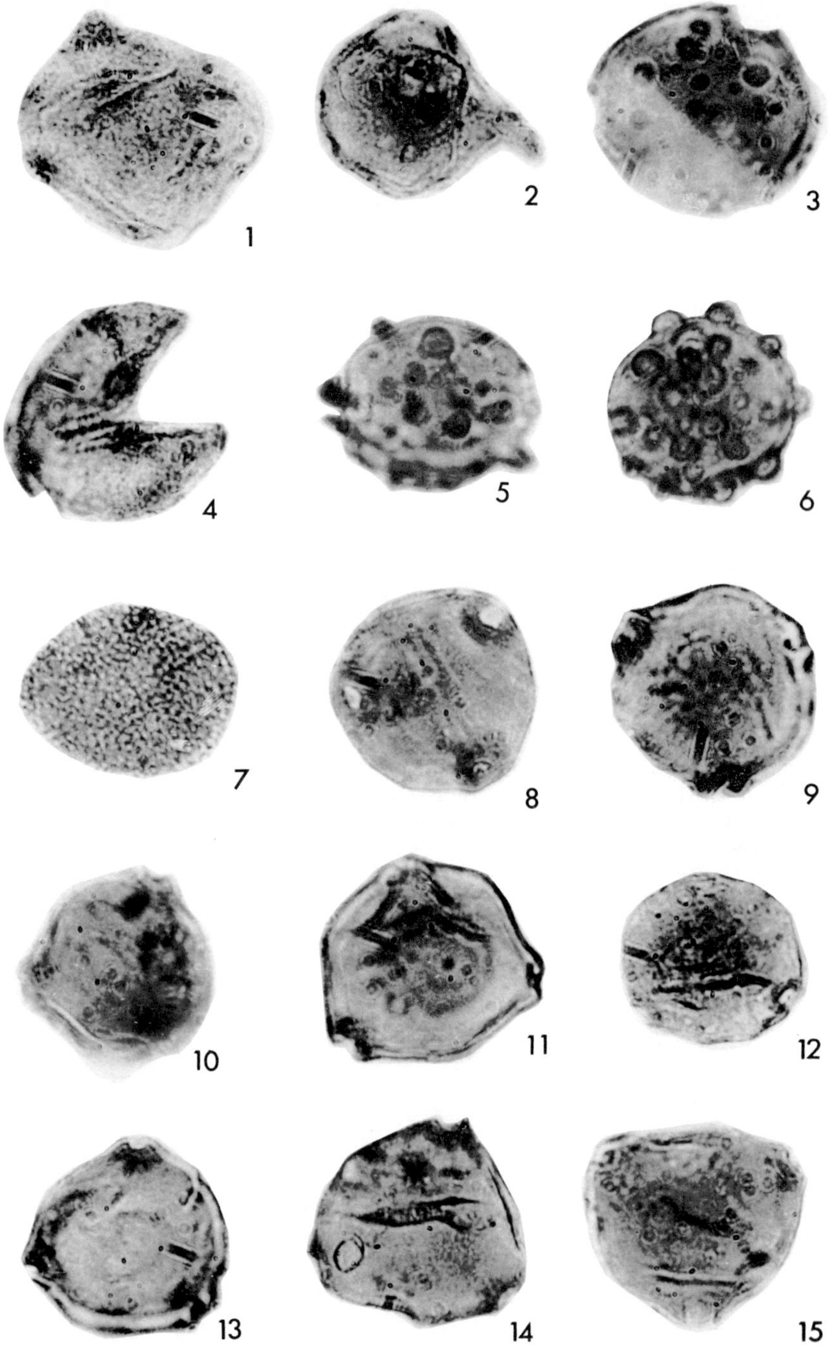

INTERIOR EOCENE ASSEMBLAGE

ROUSE AND OTHERS, PLATE 8
Geological Society of America Special Paper 127

PLATE 8. INTERIOR EOCENE ASSEMBLAGE

All magnifications $\times 1000$.

Figure
1. *Taxodium hiatipites* Wodehouse. Princeton, I-2A.
2. *Metasequoia papillapollenites* Rouse. François Lake, I-2D.
3. *Pistillipollenites mcgregorii* Rouse. Princeton, I-2A.
4. *Taxodium hiatipites* Wodehouse. Horsefly, I-2C.
5. *Pistillipollenites mcgregorii* Rouse. Princeton, I-2A.
6. *Pistillipollenites mcgregorii* Rouse. Princeton, I-2A.
7. *Potamogeton hollickipites* Wodehouse. Princeton, I-2A.
8. *Carya viridi-fluminipites* (Wodehouse) Wilson & Webster. François Lake, I-2D.
9. *Myrica annulites* Martin & Rouse. François Lake, I-2D.
10. *Betula claripites* Wodehouse. Princeton, I-2A.
11. *Betula claripites* Wodehouse. Horsefly, I-2C.
12. *Carpinus ancipites* Wodehouse. François Lake, I-2D.
13. *Myrica annulites* Martin & Rouse. François Lake, I-2D.
14. *Corylus tripollenites* Rouse. François Lake, I-2D.
15. *Carpinus ancipites* Wodehouse. Horsefly, I-2C.

PLATE 9. INTERIOR EOCENE ASSEMBLAGE

All magnifications ×1000.

Figure

1. *Corylus tripollenites* Rouse. François Lake, I-2D.
2. *Myrica annulites* Martin & Rouse. Horsefly, I-2C.
3. *Alnus verus* (R. Potonié) Martin & Rouse. François Lake, I-2D.
4. *Ulmus* sp. François Lake, I-2D.
5. *Juglans* sp. Droppingwater Creek, I-2B.
6. *Pterocarya stellatus* (R. Potonié) Martin & Rouse. François Lake, I-2D.
7. *Pterocarya* sp. Horsefly, I-2C.
8. *Quercus* sp. Droppingwater Creek, I-2B.

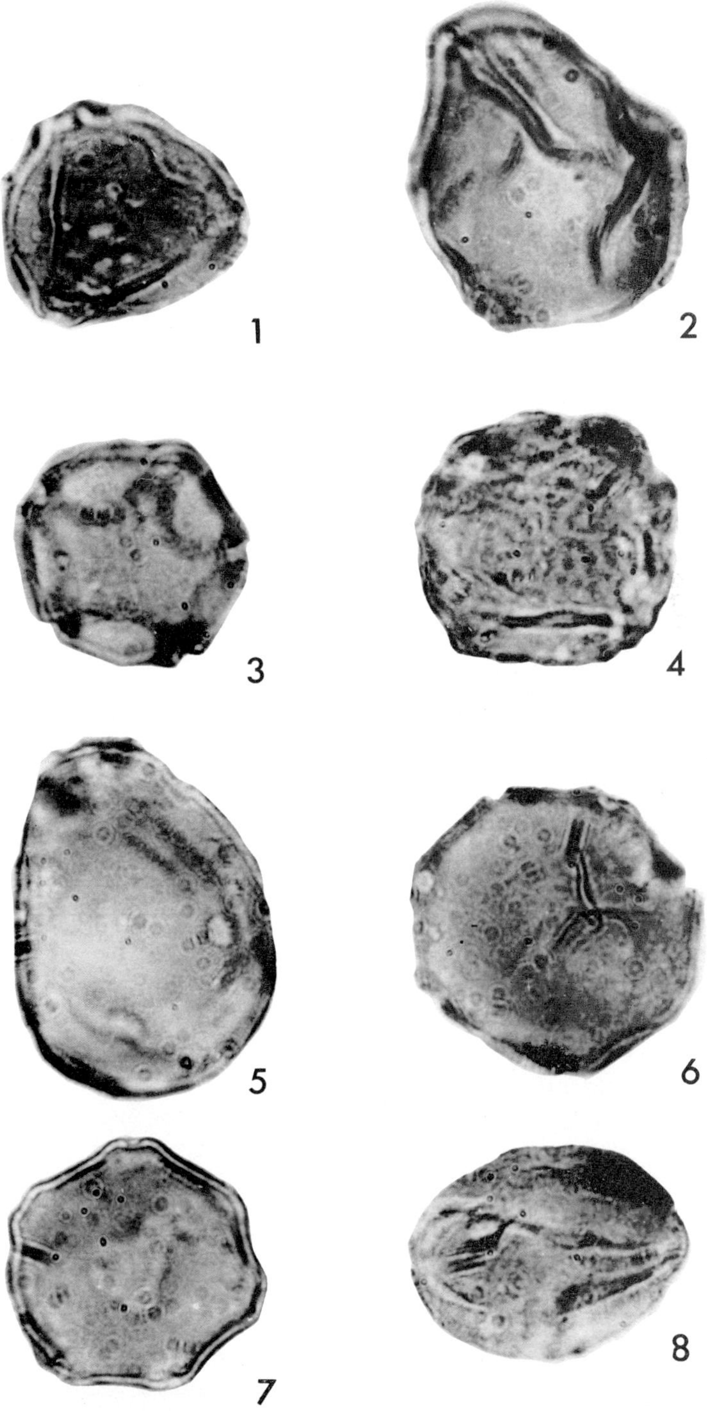

INTERIOR EOCENE ASSEMBLAGE

ROUSE AND OTHERS, PLATE 9
Geological Society of America Special Paper 127

Table 2. A Comparison of Coastal and Interior Microfossil Assemblages Indicating the Most Diagnostic Genera and Species

Coastal	Interior
	Middle Eocene
Anemia poolensis	*Anemia poolensis*
Cicatricosisporites intersectus	*Azolla primaeva*
Lygodium reticulosporites	*Osmunda* spp.
Trilites solidus	*Pinus strobipites*
Laevigatosporites discordatus	*P. sylvestris* type
	Picea grandivescipites
Sabal granopollenites	*Tsuga heterophyllites*
Pistillipollenites mcgregorii	cf. *Keteleeria* sp.
Platycarya sp.	*Metasequoia papillapollenites*
Engelhardtia sp.	*Taxodium hiatipites*
cf. *E. granulata*	*Pistillipollenites mcgregorii*
Quercus spp.	*Betula* sp.
Momipites tenuipolus	*Carpinus* sp.
Liliacidites leei	*Corylus* sp.
? *Nyssa* sp.	*Myrica* sp.
Diporites granulatus	*Alnus* sp.
Fusiformisporites sp.	*Carya* sp.
Pluricellaesporites psilatus	*Pterocarya* sp.
	Platycarya sp.
	Juglans sp.
	Acer sp.
	Tilia sp.
Maestrichtian-Danian	*Osmundacidites primarius*
Unknown	*Glyptostrobus* spp. (2 spp.)
	Taxodium hiatipites
	Sciadopitys sp.
	Sciadopitys sp.
	Aquilapollenites sp.
	Pistillipollenites mcgregorii
	Quercus fusiformipollenites
	Alnus sp.
	Pterocarya sp.
	Tilia sp.
	Tricolpopollenites spp. (5 spp.)
	Santonian-Campanian
Stereisporites cf. *S. cingulatus* subsp. *cingulatus*	*Sphagnum antiquasporites*
Gleicheniidites senonicus	*Gleicheniidites senonicus*
Anemia cristata	*Cicatricosisporites dorogensis*
A. striosporites	*Hymenophyllumsporites deltoida*
Intrapunctisporis plicatus	*Osmundacidites primarius*
Proteacidites thalmanni	*Polypodiisporites favus*
P. marginus	*Aquilapollenites quadrilobus*
Beaupreadites elegansiformis	*A. trialatus*
Symplococites sibericus	*Liliacidites variegatus*
Tricolpopollenites divergens	*Elytranthe striatus*
Quercus fusiformipollenites	*Eucommiidites troedssonii*
Sparganium globipites	*Momipites tenuipolus*

Data taken from Radforth and Rouse (1954); Rouse (1957; 1962); Hills (1965); Hopkins (1966); Srivastava (1966); as well as from the present paper.

and numbers than in assemblages of comparable age in the western interior of Canada and the United States. From the evidence to date, it appears that *Proteacidites* is much more abundant in the later Cretaceous on the coast and *Aquilapollenites* more numerous in the interior regions.

Microfossils from the interval between Campanian and middle Eocene on the coast have not been reported in the published literature to date. A probable Paleocene assemblage from the Chuckanut of northwestern Washington has been investigated by Griggs (1965), however, and we have examined probable Maestrichtian microfossils from the Gabriola Formation of the upper Nanaimo Group. The evidence suggests a gradual change in genera and species throughout this interval. Noteworthy in both the Chuckanut and Gabriola assemblages are *Gleicheniidites senonicus, Cicatricosisporites intersectus, Proteacidites marginus, Liliacidites* sp., *Momipites tenuipolus,* and *Sabal granopollenites.*

Pollen resembling arboreal angiospermous genera that are presently extant in southeastern North America and China, for example *Quercus, Alnus, Carya, Corylus,* and *Ulmus,* are reported by Griggs (1965) from the Chuckanut but are not presently known from the Gabriola or older beds. This suggests that these genera (and possibly others) which are common in Eocene and later Tertiary deposits, made their first appearance on the coast in the interval between deposition of the Gabriola (Maestrichtian) and deposition of the Chuckanut (Paleocene).

The coastal middle Eocene assemblages, represented by the Burrard and Kitsilano suites, show some significant differences from the Late Cretaceous and Paleocene. *Anemia poolensis, Lygodium reticulosporites, Trilites solidus, Pistillipollenites mcgregorii, Platycarya* sp., *Engelhardtia* cf. *granulata,* and *Diporites granulatus* are present and presumably limited to the middle Eocene beds. Also present but not particularly diagnostic are pollen of the dicotyledonous genera *Myrica, Carya, Pterocarya, Tilia, Corylus,* and *Alnus.* Notably absent from the middle Eocene assemblages are characteristic Late Cretaceous species such as *Gleicheniidites senonicus, Anemia cristata, Anemia striosporites, Stereisporites* cf. *S. cingulatus* subsp. *cingulatus, Tricolpopollenites divergens, Symplococites sibericus,* and *Proteacidites* spp.

Interior Assemblages

The Santonian assemblage reported from the Oldman Formation of Southwestern Alberta by Rouse (1957) is most closely allied in age to the Protection assemblage on the coast. As noted in Table 2, the most characteristic microfossils are those of ferns and angiosperms. Most of the species range through at least several epochs of the Late Cretaceous, with the exception of *Aquilapollenites trialatus,* which appears restricted to the Santonian. The characteristic species are notably different from those of the coastal Protection, with the exception of the ubiquitous *Gleicheniidites senonicus.* The differences suggest that coastal and interior assemblages represent two distinct floral provinces during the Santonian, probably the result of significant climatic differences, and possibly caused by physiographic factors such as mountain chains. It would be interesting to see whether or not a microfossil assemblage from a more westerly locale in interior British Columbia would relate closely to the Santonian assemblage of the coast.

The Maestrichtian-Danian microfossil assemblage reported here from Reynolds Creek (*see* I-1) in the Rocky Mountain trench is somewhat younger (Rouse, 1967) than Maestrichtian assemblages previously published by Radforth and Rouse (1954), Rouse (1957), Anderson (1960), Newman (1965), Stanley (1965), and Srivastava (1966) from the western interior of Canada and the United States. The younger age is reflected mainly in the presence of *Pistillipollenites mcgregorii,* which with one exception (Newman, 1965) previously was known to range only from Paleocene to Eocene. There is also an apparently undescribed species of *Aquilapollenites.* The five distinct species of *Tricolpopollenites* and the two species of *Glyptostrobus* dominate the assemblage. If the assemblage is even partly representative of the flora extant during Maestrichtian time, there appear to be significant differences in microfossil composition between the Reynolds Creek assemblage west of the Rocky Mountains and both the Maestrichtian of Alberta (Srivastava, 1966) and the Maestrichtian and Paleocene assemblages reported by Stanley (1965) from South Dakota. Again, it is suggested that the Reynolds Creek assemblage may represent a western floral province essentially separated from a more easterly interior province by a physiographic factor such as the Rocky Mountains.

The middle Eocene assemblage from interior locales consists of two main elements: a relatively high percentage of coniferous pollen and a deciduous arboreal element composed mainly of pollen of the Betulaceae and Juglandaceae. *Pistillipollenites mcgregorii* and microspores, megaspores, and glochidia of *Azolla primaeva* are both characteristic. This assemblage differs from that of the coastal middle Eocene in: (1) having many spores of *Azolla* and *Osmunda,* (2) lacking *Cicatricosisporites intersectus, Lygodium reticulosporites,* and *Trilites solidus,* (3) having much coniferous pollen, particularly *Taxodium, Metasequoia, Pinus, Picea,* and *Tsuga,* (4) having relatively larger numbers of deciduous arboreal dicotyledonous pollen, particularly *Carpinus, Corylus, Betula, Alnus, Carya, Juglans, Pterocarya, Acer,* and *Tilia.* These differences appear to reflect both the inland and upland location of the assemblage, relative to that of the coast, with a more continental climate. The over-all inland assemblage indicates a warm mesothermal climate somewhat cooler than the climate suggested for the coastal middle Eocene assemblage (Rouse, 1962; Hopkins, 1966, 1968). It also appears that during the middle Eocene large mountain ranges such as the present Coast and Cascade Mountains were either very low or nonexistent.

SUMMARY AND CONCLUSIONS

Evidence based on the microfossil assemblages reported here indicates that significant differences developed during the evolution of the floras of coastal and interior regions from Santonian to middle Eocene time. Coastal assemblages consist of two main elements, fern spores and angiosperm pollen. The complement of coniferous pollen, particularly the bladdered forms, is characteristically impoverished in both number of species and number of specimens. The pattern of floral evolution on the coast indicates a transition from a Santonian-Campanian assemblage characterized by spores of *Stereisporites* cf. *S. cingulatus* subsp. *cingulatus, Anemia, Gleicheniidites,* and *Intrapunctisporis plicatus* and pollen assignable to *Proteacidites, Symplococites sibericus,* and *Tricolpopollenites divergens* to a middle Eocene assemblage characterized by such taxa as *Anemia poolensis, Lygodium reticulosporites, Cicatricosisporites intersectus, Trilites solidus, Platycarya* sp., *Engelhardtia* cf. *E. granulata,* and *Pistillipollenites mcgregorii.* Limited evidence from the Maestrichtian and Paleocene on the coast suggests that the above-mentioned characteristic Late Cretaceous microfossils, with the exception of *Gleicheniidites,* had disappeared by the early Paleocene. At the same time, some spores and pollen characteristic of the coastal Eocene had appeared, for example *Cicatricosisporites intersectus, Trilites solidus, Alnus* spp., *Corylus tripollenites, Carya juxtaporipites,* and *Ulmus* sp.

Interior assemblages contain different characteristic microfossils from comparable coastal assemblages and show a different evolutionary pattern. The Santonian assemblage contains a mixture of ferns and angiosperms, but the species differ from those characterizing the comparable coastal assemblage (*see* Rouse, 1957). Distinguishing taxa include *Cicatricosisporites dorogensis, Hymenophyllumsporites deltoida, Osmundacidites primarius, Aquilapollenites quadrilobus, A. trialatus, Liliacidites variegatus,* and *Elytranthe striatus.*

The late Maestrichtian-Danian assemblage from Reynolds Creek contains few fern spores but has taxodiaceous and angiosperm pollen as its main element (Rouse, 1967). This assemblage is interpreted as being transitional between older Maestrichtian (Stanley, 1965; Srivastava, 1966) and Paleocene (Stanley, 1965) assemblages. The Reynolds Creek assemblage is characterized by two species of *Glyptostrobus, Sciadopitys,* a new species of *Aquilapollenites, Pistillipollenites mcgregorii, Quercus fusiformipollenites,* and five species of *Tricolpopollenites.* It also contains pollen of *Alnus, Pterocarya, Tilia, Acer,* and *Potamogeton.* The over-all assemblage appears somewhat younger than the Hell Creek assemblage described by Stanley (1965) and may represent a western floral province of the late Maestrichtian-Danian distinct from that in western interior regions east of the Rocky Mountains.

The interior middle Eocene assemblage contains two major elements: a relatively large number of taxodiaceous and pinaceous pollen and a deciduous arboreal element largely composed of pollen of the Betulaceae and Juglandaceae. Also characteristic are spores of *Azolla* and pollen of *Pistillipollenites mcgregorii.* The interior assemblage appears to reflect the flora of a slightly cooler, mesothermal climate than the comparable assemblage at the coast. The two are not sufficiently different, however, to suggest the presence of high mountain ranges between the coast and interior such as the existing Coast and Cascade Mountains.

The most likely evolutionary patterns may be summed up as follows. Both coastal and interior Late Cretaceous assemblages appear to have been characterized by a fern-angiosperm element containing characteristic species that changed significantly from the Santonian to the middle Eocene, with very few coniferous pollen in any of the assemblages. *Proteacidites* appears to characterize the coastal Cretaceous, whereas *Aquilapollenites* is more representative of interior Cretaceous assemblages. The coastal middle Eocene shows the same fern-angiosperm assemblage with few conifers, as do earlier coastal Cretaceous assemblages, but with different species of ferns and angiosperms. The interior middle Eocene contains a coniferous-deciduous dicotyledon arboreal element reflecting a more continental environment.

The picture of microfossil distribution and evolution presented here is admittedly based on limited information and will necessarily be greatly enhanced by additional research in the future. The general differences in evolutionary patterns between coastal and interior locales is beginning to emerge, however, with this representing but a preliminary appraisal.

REFERENCES CITED

Anderson, R. Y., 1960, Cretaceous-Tertiary palynology, eastern side of the San Juan Basin, New Mexico: New Mexico Inst. Min. Technol. Mem., v. 6, 58 p.

Arnold, C. A., 1955a, A Tertiary *Azolla* from British Columbia: Michigan Univ. Mus. Paleontology Contr., v. 12, no. 4, p. 37-45.

——1955b, Tertiary conifers from the Princeton Coalfield of British Columbia: Michigan Univ. Mus. Paleontology Contr., v. 12, no. 12, p. 245-258.

Bell, W. A., 1957, Flora of the Upper Cretaceous Nanaimo Group of Vancouver Island, British Columbia: Canada Geol. Survey Mem 293, 84 p.

Berry, E. W., 1926, Tertiary floras from British Columbia: Canada Dept. Mines Geol. Survey Bull. 42, p. 91-116.

Crickmay, C. H., and Pocock, S. A. J., 1963, Cretaceous of Vancouver, British Columbia, Canada: Am. Assoc. Petroleum Geologists Bull. 47, no. 11, p. 1928-1942.

Dawson, J. W., 1883, On the Cretaceous and Tertiary floras of British Columbia and the Northwest Territory: Royal Soc. Canada Trans., 1882-1883, v. 1, sec. 4, p. 15-34.

——1895, On collections of Tertiary plants from the city of Vancouver, British Columbia: Royal Soc. Canada Trans., 2nd ser., v. 1, sec. 4, p. 131 167.

Griggs, P. H., 1965, Stratigraphic significance of fossil pollen and spores of the Chuckanut Formation, Northwest Washington: East Lansing, Michigan State Univ., M. S. thesis, 116 p.

Hills, L. V., 1965, Palynology and age of early Tertiary basins, interior British Columbia: Edmonton, Alberta Univ., Ph.D. dissert., 188 p.

Hopkins, W. S., Jr., 1966, Palynology of Tertiary rocks of the Whatcom Basin, southwestern British Columbia and northwestern Washington: Vancouver, British Columbia Univ., Ph.D. dissert., 184 p.

Hopkins, W. S., Jr., 1968, Subsurface Miocene rocks, British Columbia-Washington, a palynological investigation: Geol. Soc. America Bull., v. 79, p. 763-768.

Johnston, W. A., 1923, Geology of the Fraser River delta map-area: Canada Dept. Mines Geol. Survey Mem. 135, 87 p.

Martin, H. A., and Rouse, G. E., 1966, Palynology of late Tertiary sediments from Queen Charlotte Islands, British Columbia: Canadian Jour. Botany, v. 44, p. 171-208.

Mathews, W. H., and Rouse, G. E., 1963, Late Tertiary volcanic rocks and plant-bearing deposits in British Columbia: Geol. Soc. America Bull., v. 74, p. 55-60.

Newman, K. R., 1965, Upper Cretaceous-Paleocene guide palynomorphs from northwestern Colorado: Colorado Univ. Studies Ser. Earth Sci., no. 2, p. 1-21.

Penhallow, D. P., 1903, Notes on Tertiary plants: Royal Soc. Canada Trans., 2nd ser., v. 9, sec. 4, p. 33-95.

——1907, A report on fossil plants from the International Boundary Survey for 1903-1905 collected by Dr. R. A. Daly: Royal Soc. Canada Trans., 3rd ser., v. 1, sec. 4, p. 287-351.

——1908, A report on Tertiary plants of British Columbia, collected by L. M. Lambe in 1906, together with a discussion of previously recorded Tertiary floras: Canada Dept. Mines Geol. Survey, no. 1013, 167 p.

Radforth, N. W., and Rouse, G. E., 1954, The classification of recently discovered Cretaceous plant microfossils of potential importance to the stratigraphy of western Canadian coals: Canadian Jour. Botany, v. 32, p. 187-201.

Rouse, G. E., 1957, The application of a new nomenclatural approach to Upper Cretaceous plant microfossils from western Canada: Candian Jour. Botany, v. 35, p. 349-375.

——1962, Plant microfossils from the Burrard Formation of western British Columbia: Micropaleontology, v. 8, p. 187-218.

——1967, A late Cretaceous plant assemblage from east-central British Columbia: Pt. 1, Fossil leaves: Canadian Jour. Earth Sci., v. 4, p. 1185-1212.

Rouse, G. E., and Mathews, W. H., 1961, Radioactive dating of Tertiary plant-bearing deposits: Science, v. 133, p. 1079-1080.

Srivastava, S. K., 1966, Upper Cretaceous microflora (Maestrichtian) from Scollard, Alberta, Canada: Pollen et Spores, v. 8, no. 3, p. 497-552.

Stanley, E. A., 1965, Upper Cretaceous and Paleocene plant microfossils and Paleocene dinoflagellates and hystrichosphaerids from northwestern South Dakota: Am. Paleontology Bull., v. 49, no. 222, p. 179-384.

Geobiologic Implications of Fossil Phytoplankton Evolution and Time-Space Distribution

HELEN TAPPAN
University of California at Los Angeles, Los Angeles, California

ALFRED R. LOEBLICH, JR.
Chevron Oil Field Research Company, La Habra, California

ABSTRACT

Evolution of the protists is discussed on the basis of their known fossil record and the morphological, biochemical, metabolic, physiological, and distributional data concerning modern representatives. Periods of explosive evolution result from the chance appearance of a major innovation and are followed by extensive exploitation of the possibilities of each innovation. A billion years or more following the origination of life were characterized by a biochemical and metabolic stage of differentiation, procaryote structure, rapid biotic turnover, and simple carbon cycle. This was succeeded by a period of intercellular and physiologic differentiation, about 1.5 b.y. in duration, with continued rapid carbon cycling, eucaryotic protist asexual clonal reproduction, and evolution based on mutation-induced variations.

The last 0.6 b.y. have been characterized by intercellular differentiation (metaphytes and metazoans), with increased size of the individual, dominance of the diploid phase, and hence increased relative importance of genetic recombination in evolution, with smaller populations, and development of an extremely complex carbon cycle that is more susceptible to periodic disruption.

Evolutionary rates in protists are comparable to those of multicellular organisms, although attainment of variability, isolation mechanisms, and selection processes differ. The apparent exponential increase of phytoplankton taxa at successive periods in their geologic history is related to their mutation-based evolution. A hypothetical phylogenetic tree is plotted against the geologic time scale, showing the known extent of the protist fossil record and that of other major groups of organisms, grades of structural and metabolic differentiation, relative age of important Precambrian fossil occurrences, and probable time of appearance of an aerobic environment.

The Phanerozoic fossil record of phytoplankton shows periods of maximum productivity interspersed with times of greatly reduced microfloras. Modern distributions and productivity, together with experimental culture data, are related to the geologic past to indicate both possible causes and effects of these fluctuations, in elaboration of a previously proposed Phytoplankton Periodicity Model.

The photosynthetic activity of phytoplankton, by which CO_2 and H_2O are utilized to produce organic compounds and free oxygen, forms the basis of the oceanic carbon and oxygen cycles and thereby influences atmospheric composition. The areal extent of the oceans and the abundance and rapid growth of the phytoplankton confer prime status on oceanic production in the present-day oxygen balance. Quantitatively, oceanic production was even more important in the geologic past. Oxygen-producing algae appeared some 3 b.y. ago, and a highly oxygenic atmosphere probably was attained geologically soon.

Through the interrelated processes of carbon fixation and oxygen production, periods of high oceanic phytoplankton productivity had a somewhat more oxygenic atmosphere, reduced CO_2 pressure, and higher oceanic *p*H. These conditions are recorded by abundant dispersed organic carbon in sediments and petroleum accumulation (both of isotopically light carbon), lime deposition with increasingly heavy carbon isotope ratio, and an abundance of marine filter-feeding, detrital feeding, and carnivorous animal life. Atmospheric conditions were favorable for land animals.

Times of greatly limited phytoplankton production resulted in relative oceanic increase of procaryotes, particularly bacteria, increased CO_2 levels, lowered pH, and oxygen depletion, silica deposition, less organic carbon in sediments, limestones of increasingly light $^{12}C/^{13}C$ ratio and eventual lime solution, and deposition of gypsum, barite, and other sulfates of isotopically light sulfur. The marine food web was drastically tightened, filter feeders and reef-forming invertebrates were largely eliminated, and only the more efficient food gatherers survived. Relative atmospheric depletion of oxygen affected both marine and terrestrial animals, eliminating many, whereas the increased CO_2 pressure stimulated terrestrial plants.

The world-wide contemporaneity of certain geochemical, sedimentological, and marine and terrestrial biologic events is suggested by data obtained in numerous unrelated studies. Changes in phytoplankton quantitative distribution, discussed herein, also coincide with these events and are suggested to have formed the connecting link between them.

CONTENTS

Figure

INTRODUCTION

The time of origin, rates of differentiation and dispersal, and subsequent fluctuations in abundance and diversity of the phytoplankton should be of interest to a wide range of disciplines, for the photosynthetic carbon fixation by these organisms has affected the atmospheric evolution and geochemistry of sediments in addition to supplying the base of the aquatic food pyramid. The photosynthetic protists were ancestral to both the animal and plant kingdoms and through their various life processes repeatedly supplied evolutionary stimuli to coexisting organisms.

Bradley (1963) observed that only as the geologic sciences increased their assimilation of information from the basic sciences of physics, chemistry, and biology would the conceptual aspects of geology and the formulation of general principles surpass the observational aspects. In spite of the remarkable advances made in the study of fossil phytoplankton over the past two or three decades, incorrect statements concerning their nature, fossil record, and evolutionary history abound in the current geologic and biologic literature. Published data concerning both fossil and modern microalgae are dispersed among journals related to a wide variety of disciplines. Even the valuable review articles have generally had a restricted emphasis (such as biochemistry of pigments, quantitative primary production, flagellar structure, or geologic ranges of selected taxa), so the wealth of potentially useful knowledge generally has been ignored for cross discipline application. Palynologists and micropaleontologists commonly deplore the lack of applicable information on modern life histories or ecologic tolerances and emphasize the stratigraphic value of phytoplankton as "index microfossils." Meanwhile, phycologists and protozoologists base phylogenetic classifications and evolutionary theories on morphologic similarities of modern protists and their presumed "primitive" or "advanced" characters, stating that these micro-organisms have persisted without change throughout most of geologic time or that there is unfortunately no fossil record of those lacking a mineralized shell.

The value of cross-discipline application of data from biology to paleontology is well-recognized. Ecologic parameters of modern organisms, determined from distribution and experimentation, have long been used for paleoecological interpretation of depth, temperature, or distance from shore. More recently, water mass distribution, paleocurrents, and paleoclimatic interpretation have been based on such data from marine organisms, and synecological interpretations have been based on the prey-predator, host-parasite, and symbiotic relations of modern counterparts.

Not as widely recognized is the value of paleontologic data to studies of the geochemistry of biogenic carbonate and silicate sediments, the rate of isotopic fractionation of carbon, sulfur, and oxygen, or the qualitative and quantitative evolution of the atmosphere.

Because changes in productivity throughout time seem to correlate with other biological and physical events, these changes are here suggested to be worthy of consideration by other disciplines. Problem solution and the formulation of theories, concepts, and principles, in geology and the related sciences, are dependent upon correct observations. Unhappily, some generalizations concerning fossil micro-organisms were formulated on incomplete or inaccurate primary data yet have come to be regarded as established principles. Examples discussed below in more detail suggest that an expanded interdisciplinary communication shows promise of being equally valuable to the neontologist, the paleontologist, and the geologist.

PHYTOPLANKTON TAXA IN THE GEOLOGIC RECORD

Biologists commonly are unaware of the extent of the geologic record of the protists. Thus Klein and Cronquist (1967, p. 256) suggested that "Lacking a fossil record—a state of affairs which is unlikely to be remedied—we can . . . conclude that phylogenetic speculation on the lower plants is a futile and sterile intellectual game."

The paleontologist also finds difficult the task of keeping abreast of the rapidly expanding subdiscipline of micropaleontology, in particular that part concerned with phytoplankton studies.

Phytoplankton Fossilization

Paleontologists commonly emphasize the importance of "hard parts" in the fossil record. The skeletal materials calcite, calcium phosphate, and silica are considered most durable, chitin and aragonite less so, and "soft parts" preservable only under unusual conditions, perhaps as carbonized impressions. Enclosure in chert, amber, or ice (frozen mammoths) or dehydration (mummification in deserts or caves) are indicated by textbooks as the unusual but nearly exclusive means of preservation of organic remains. For microfossils, resistance to destruction through time is not wholly correlative with mineralization of skeletal parts. Some organic compounds are extremely resistant, and organic-walled bacteria and blue-green microfossils have actually persisted five times as long geologically as any other type of fossil material. To the usual definition of potential fossils as the skeletal carbonates, phosphates, silicates, or chitin must be added these very durable organic structures.

Selective preservation or destruction of the potential fossils may occur in the water column, on the sea floor, or during post-depositional diagenesis, metamorphism, or weathering. Relatively impervious clays that have undergone only slight amounts of heat or pressure contain the greatest variety of unaltered microfossil skeletons, frustules, tests, loricae, cell walls, or resting cysts. Potential fossils have also varied in character and relative abundance through time because of the gradual process of evolution.

Preservation of Carbonates. Either calcite or aragonite may occur in microfossils, although this occurrence is genetically controlled by the organism concerned. Calcite is far more common; it is found in foraminifers and tintinnids, as well as in sponges, some corals, and many other metazoans, including echinoderms, mollusks, arthropods, and so on. In the plant kingdom, calcite occurs in the blue-green and some red algae, coccolithophorids, the probably related discoasters, and charophytes. Aragonite characterizes the foraminiferal superfamily Robertinacea (Loeblich and Tappan, 1964a, p. 387; 1964b, p.C97, C766), green and some red algae, the skeletons of scleractinian corals (Lowenstam, 1954, p. 290), and tunicate spicules (Lowenstam, 1963, p. 150).

The sea contains large amounts of calcium, carbonate, and bicarbonate ions, although calcareous skeletal material has accumulated at a rate of 1 cm/1000 years or more to form great quantities of chalks and limestones. The increase in dissolved CO_2 at depth results in solution of the calcareous shells below the calcium carbonate compensation depth. Depth of this level differs from the Atlantic to the Pacific; it also varied in the recent geologic past in each. The effect of solution is somewhat complicated by the nature of biogenic carbonates. Skeletal material secreted on an organic matrix (as in the coccolithophorids and foraminifers) has greater resistance to solution because of the interspersed organic material. Bacterial destruction of the latter before burial may allow solution or recrystallization of the carbonate compounds. All gradations may occur, from removal by solution or recrystallization to the excellent preservation of detailed structures at the scale of a micron or less (as in fossil coccoliths). Aragonite is more readily dissolved than calcite and may be selectively removed even where the calcite is perfectly preserved. In general, aragonite is best preserved in clays and seldom remains in limes or marls. In no occurrence known to us has an originally aragonitic microfossil been converted to calcite, although this is a common misconception. Fossil tunicate spicules, robertinid or ceratobuliminid foraminifers, or small scleractinian corals may occur as aragonite in sands and clays or their indurated counterparts but are destroyed by solution in highly calcareous deposits rather than being converted to calcite. The dissolved carbonate may then be redeposited as cementing material in the limestone.

Preservation of Silica. Because of the stability of quartz and its resistance to solution by normal seawater, silica is commonly regarded as the most durable skeletal material. Crystalline quartz is highly resistant to destruction; hence an agglutinated foraminiferal test has an excellent chance of preservation. Biogenically deposited silica is in the much less stable hydrated opaline form, however.

The seas are undersaturated in dissolved silica, but fresh waters may contain somewhat more. As organisms can biologically concentrate relatively scarce substances from seawater, silica-secreting animals and plants nevertheless are widely distributed. They illustrate the importance of selective preservation, for very delicate or weakly silicified diatoms may be abundant in the surface plankton and absent from the bottom sediments. Elsewhere, dispersed silica in the sediment and amorphous cherts may represent the remains of the former siliceous skeletons whose source material is no longer recognizable. All amorphous silica (opaline) in modern sediments is said to be of biogenic origin (Lisitsyn, 1967b) and largely due to the skeletal accumulation of diatoms, silicoflagellates, radiolarians, and sponge spicules.

The limiting factor in preservation of siliceous skeletal material seems to be the hydrogen ion concentration (Kolbe, 1957), for the opaline silica readily dissolves under the relatively alkaline conditions (increased *p*H) that accompany a decrease in dissolved CO_2. Diatomites or radiolarian oozes accumulate only if the dissolved CO_2 level is high and the *p*H depressed; the opposite conditions favor deposition and preservation of carbonates. If deposition is rapid or detrital material quantitatively more important, both calcareous and siliceous skeletal elements may be preserved; commonly one or the other is selectively dissolved.

Preservation of Organic Compounds. Acids used in laboratory palynological preparation dissolve the "resistant" silicates, carbonates, and pyrite, one or another of which is commonly regarded as required for preservation, yet leave intact various organic materials. Many plant taxa have cell walls of cellulose, an excellent strengthening material for the living organism but hydrolyzed by bacterial action and therefore rarely preserved in the fossil record. Certain portions of the higher plants (for example, plant cuticle, pollen, spores) are constructed of more resistant organic compounds that persist long after the cellulose has been destroyed. Similarly, many lower plants such as some green algae, charophytes, brown algae, and armored dinoflagellates have easily destroyed cellulosic walls, but some produce more resistant organic cysts (as in the green Prasinophyceae or the dinoflagellates), fungi may have hyphae or spores with chitinoid walls, and blue-greens and some bacteria may have gelatinous sheaths. Certain hydrocarbons have been preserved since the Precambrian (Eglington and others, 1964; Cloud and others, 1965; Meinschein and others, 1964; Barghoorn and others, 1965; Meinschein, 1965; Oró and others, 1965). The hydrocarbons phytane and pristane, which may be produced by alteration of chlorophyll, were identified by geochemical analysis of rocks as old as 3 b.y. (Oró and Nooner, 1967), as were various amino acids (Schopf and others, 1968). Although acid-resistant, organic-walled microfossils can be obtained from rocks of all ages, at least within the last 2 b.y., the exact composition of the resistant material is not known. Possible chemical alteration may be involved in fossilization or during the sample processing, and the small size range prevents obtaining sufficient quantities for exact analysis.

An elemental analysis of the Acritarcha, made by Timofeev (1956), showed 71.88 percent carbon, 7.84 percent hydrogen, 2.16 percent nitrogen, and 18.12 percent oxygen and sulfur. He stated that the similarity of these "spores" from the *Obolus* beds (Tremadocian, Lower Ordovician) from near Leningrad to fossil sporonins indicated a plant origin. No more exact identification can be made of this material, nor has the exact composition yet been determined for modern organic-walled cysts, such as of the dinoflagellates.

Not only do such organic compounds withstand conditions that may selectively dissolve the originally accompanying calcite or silica, but their resistance to acids allows their recovery from a wide variety of limestones, shales, or sandstones, even when these are highly indurated. Organic compounds are susceptible to oxidation, however, and their preservation is limited by the amount of heat or pressure to which the enclosing rocks were subjected. Jekhowsky (1958, p. 1406) stated that organic microfossils may withstand rock temperatures up to about 300° C and remain recognizable.

Slightly metamorphosed sediments that have been subjected to greater temperatures may contain relatively well preserved megafossils, but palynological sample preparation produces only a black cloud of finely comminuted carbon. This could be regarded as a microscopic level of coal formation in which the original structure is completely obliterated. Because of their small size, impressions and molds of the smaller microfossils are seldom obtainable, yet the microfossil must be well-preserved structurally if it is to be recognizable.

Geologic Occurrence of Phytoplankton Taxa

The geologic distribution of the microplankton groups has been summarized in various publications (for example, Deflandre, 1936; Downie, 1967; Loeblich III and others, 1968; Sarjeant, 1967; Tappan, 1968b; Wilson, 1956). The compositional variation in phytoplankton has been suggested above but also shows an evolutionary trend through geologic time (Fig. 1).

Dominance of Organic-Walled Phytoplankton. From the earliest record in the Precambrian through the Paleozoic, the phytoplankton assemblage apparently was exclusively organic-walled, although benthic representatives of various algal phyla were lime-secreting.

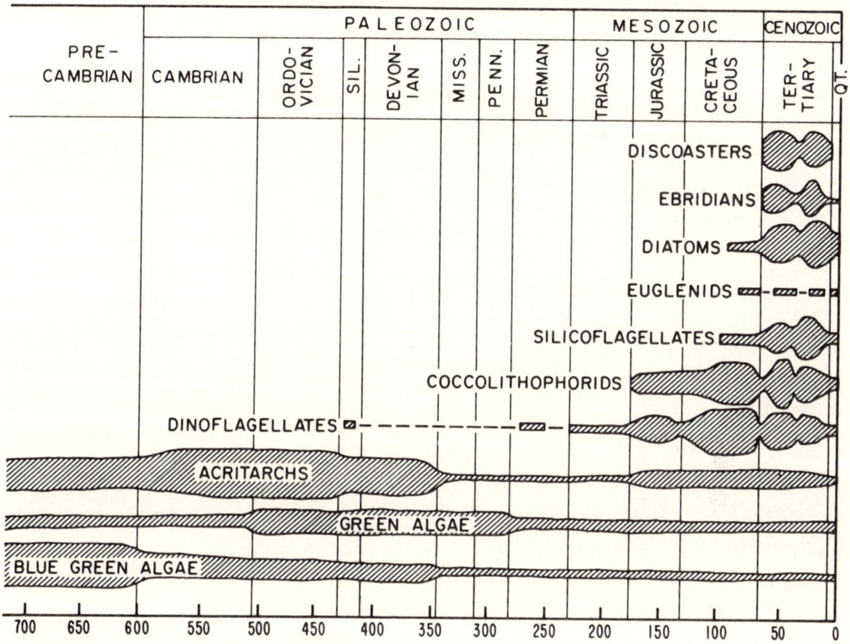

Figure 1. Geologic distribution of important fossil groups of phytoplankton, showing variation in relative and absolute abundance.

The oldest known and simplest modern organisms are the lower protists, or procaryotes, which have no membrane-bound nuclei or other organelles. These include bacteria (Schizophyta) and blue-green algae (Cyanophyta), both known as fossils from the Precambrian to the Holocene, their oldest occurrence reported to date being that in the 3-b.y.-old Fig Tree Series of South Africa (Barghoorn and Schopf, 1966; Schopf and Barghoorn, 1967). Iron-depositing, sulfur-depositing, and even pathogenic bacteria and agglutinating, lime-depositing, and lime-dissolving blue-green algae have been reported widely throughout the geologic column.

Among the most primitive living eucaryotes (organisms whose cells contain distinct organelles such as nuclei, chloroplasts, mitochondria, and so on) are the green algae or Chlorophyta. Modern representatives are highly varied in form and widely distributed. The benthic Dasycladales and Siphonales have a more extensive geologic record in the Phanerozoic than do the planktonic representatives of the green algae. The latter occur sparsely in the Precambrian (Barghoorn and Schopf, 1965); some Precambrian and younger fossil taxa that had formerly been referred to the Acritarcha or other groups are now recognized as Chlorophyta (*Botryococcus, Tasmanites, Leiosphaeridia*). Locally they may be so abundant as to form boghead coal or Tasmanite (white coal).

Prominent in the late Precambrian and lower Paleozoic is the large phytoplankton group of uncertain affinity, known as Acritarcha. Most and perhaps all Acritarcha represent cysts. Some have an attached operculum (Loeblich and Tappan, 1969) or the opening (pylome) left after this was shed (Deunff, 1961; Eisenack, 1958, 1959; Rudavskaya, 1965). Others excysted by a median splitting (Staplin and others, 1965; Loeblich and Tappan, 1969). Acritarcha are typically Paleozoic in distribution, particularly lower Paleozoic. A few stragglers persisted into the Mesozoic and Cenozoic, but their modern affinities are still uncertain. Morphologically similar cysts occur in various algal phyla, particularly the Chlorophyta, Chrysophyta, Xanthophyta, and Pyrrhophyta (dinoflagellates). The nature of the resistant organic compounds in modern cysts (such as those of dinoflagellates) is as yet unknown. Many post-Paleozoic Chrysophyta have siliceous cysts, but these are not universal in modern species. One chrysophyte group, the coccolithophorids, includes representatives with a calcite-covered, nonmotile stage or cyst.

Another group of organic-walled microplankton, the dinoflagellates, appeared in the Paleozoic but became more important in the Mesozoic and Cenozoic. Throughout their range (Silurian to Holocene), dinoflagellates are known from their organic cysts. Some modern dinoflagellates are armored with a cellulosic theca, but naked dinoflagellates may also produce cysts, and both may have an equally good fossil record. Some aberrant living dinoflagellates are attached or parasitic in habit or are endosymbionts, but none of these are known unquestionably as fossils.

Among modern unicellular algae that lack mineralized skeletal material are the Euglenophyta. Particularly characteristic of fresh water, they have a sparse fossil record from the Eocene to present (for example, *Phacus*). So-called flagellates in Cretaceous cherts (for example *Scuticabolus*) are acritarchs (cysts).

Very early in their history, the algae are suggested to have given rise to the colorless osmotrophic fungi, both fungal mycelia and spores having been reported from the late Precambrian to Holocene.

Dominance of Calcareous Phytoplankton. The Mesozoic is characterized by an abundance of the lime-secreting protists. Although benthic calcareous red and green algae and charophytes had been present in the Paleozoic, only the doubtfully assigned mid-Paleozoic calcispheres might possibly represent planktonic calcareous protists. Calcareous zooplankton (Tintinnida) is known from the Ordovician, Silurian, and Devonian but was abundant only in the Late Jurassic to Early Cretaceous, then rare in later periods.

Characteristic calcareous nannoplankton (coccolithophorids and nannoconids) first appeared in the Jurassic (perhaps Triassic), rapidly becoming so numerous as to produce coccolith chalks in the later Mesozoic. By mid-Cretaceous, planktonic foraminifers were also abundant. This plankton assemblage was noticeably depleted at the end of the Mesozoic, but calcareous nannoplankton again diversified and became quantitatively important in the Tertiary. The coccolithophorids then were accompanied by the now extinct discoasters, a probably related group. Coccolithophorids are largely represented by the isolated calcareous platelets that continually are being formed and shed by the cell, but some entire coccospheres occur. A few of the latter seemingly represent a cyst stage. Rare calcareous dinoflagellate cysts occur in the Jurassic, Eocene, Miocene, and Holocene, representing nine species and as many genera. By Miocene time the calcareous phytoplankton groups again began to diminish in importance, although they are still pre-eminent in tropical regions.

Increased Importance of Siliceous Phytoplankton. Siliceous phytoplankton appeared in the late Mesozoic. Like the calcareous phytoplankton, this had been long preceded by siliceous zooplankton (Radiolaria). Marine siliceous phytoplankton includes the diatoms (Bacillariophyta) and some Chrysophyta (silicoflagellates) and Pyrrhophyta (*Actiniscus* and the ebridians). Many living freshwater Chrysophyta have flasklike siliceous cysts. Such cysts occur in freshwater diatomites of the Tertiary. Similar cysts in marine rocks, associated with silicoflagellates and diatoms, were described as a separate family, the Archaeomonadaceae, as they were not known to occur in living marine species. Recently, however, Kozlova and Mukhina (1967) reported chrysomonad cysts in the plankton of the Pacific. In the southern temperate zone, between 56° and 42° S., surface waters contained weakly silicified chrysomonad cysts in numbers of up to 2000 cells per cubic meter. These were obtained on filters and doubtless had been overlooked previously because of their very small size. The weak silicification also allows more rapid dissolution of the cyst so that they are not abundant in bottom sediments. However, between 6000 and 9000 cells per gram occurred in iceberg sediments on the Antarctic Shelf. Other algal taxa, at least of generic level and above, occur in both marine and fresh water; hence the continued separation of the Archaeomonadaceae solely on an environmental basis does not appear to be justified.

Silicoflagellates, centric diatoms, and chrysomonad cysts all occur from mid-Cretaceous on (reports of Jurassic diatoms are questioned) but are far overshadowed in numbers by the calcareous plankton. In the Tertiary, diatoms increased greatly in importance, the Pennales appearing in the Paleocene. The marine plankton is still dominated by the more ancient centric diatoms, whereas the pennate forms have become prominent in the benthos. Diatoms also invaded fresh water by the Eocene, and perhaps as early as Paleocene. Here all benthic diatoms and most planktonic ones are pennates, and only a minor number of freshwater planktonic diatoms are centrics (Proshkina-Lavrenko, 1960).

Silicoflagellates occur rarely in the Cretaceous but are more frequent in the Eocene and Miocene marine diatomites. The nonpigmented ebridians first appeared in the Paleocene, probably derived from the dinoflagellates, although previously classed with the Chrysophyta. They diversified rapidly in the Tertiary to include about 15 genera in the Miocene, then declined as rapidly to only 3 in the present seas (Loeblich III and others, 1968).

EVOLUTION IN PROTISTS

Evolution at the Intracellular Level

Although small in size, protists are not merely simple, primitive organisms. As noted by Hutner (1964), even a bacterium is sufficiently complex that studying its various components could use the entire world supply of biologists and chemists, and the provision of cells for analysis would require the world supply of industrial microbiologists, chemical engineers, and the like. Other protistans are equally or more complex biochemically and far more complex physiologically and morphologically. Because of the biochemical classification based largely on pigments, the great diversity in the plant protists is recognized by botanists in assigning them more than one-half of all plant divisions (phyla) (Papenfuss,

1955). Protozoans are also varied and complex, but the morphologic emphasis in zoology still places all unicellular organisms in a single phylum. That the protists include only 5 percent of the known animal species and 10 percent of the plants is doubtless in part because of their small size and relative neglect during the period when descriptive taxonomy was emphasized. Nevertheless, they early demonstrated both biochemical and morphological diversification, most of the major algal phyla having a very long geologic history.

Usually only the skeleton, test, or cyst of the protist is preserved, whereas modern taxa are based on cytology, biochemistry, and physiology. Hence, evolutionary changes in the protists are commonly overlooked by both neontologist and paleontologist. Corliss (1962, p. 38) stated that ". . . the possibility that thousands of present-day protozoan species are virtually unchanged descendants from perhaps a Cambrian or even pre-Cambrian ancestry rightly dismays the student of evolution and natural selection."

Protistans have been able to remain small and unicellular over a 3-billion-year existence, not because they remained unchanged in a stable environment but because their morphological differentiation at the organelle level (nuclei, plastids, flagella, skeletal composition, and morphology) rather than as tissues or organs and physiological diversification at the biochemical level (pigments, storage products, secretions) allowed exploitation of every available niche (Table 1). Their rapid rate of reproduction, and the variety of its method, allowed recovery from catastrophic depletions of populations and assemblages. The "stability" is also in part "taxonomically imposed." Protistans must persist without the formation of specialized tissues, because they are so limited by definition! Various branches have become compartmentalized multicellular organisms but have then been referred to metazoans or metaphytes. One could as logically state that vertebrates have been virtually unchanged from their origin because they have retained an internal segmented backbone throughout their 500-million-year history.

The Proto-Organism. The earliest life probably arose by chance from an accumulation of abiogenically produced organic molecules. Since the discovery that such molecules could be produced in the laboratory, by the action of various energy sources on reduced gaseous mixtures, the possible biochemical pathways and their timing have been discussed in detail. The water of the oceans and the various components (H_2O, CO_2, H_2, CO, N_2 and so on) of the early atmosphere accumulated by the degassing of the primitive earth (Rubey, 1951, 1955; Abelson, 1966). The organic compounds that were produced from this mixture required a limited temperature range for their preservation, such as would allow the presence of liquid water on the earth's surface. According to estimates by Donn and others (1965), this required temperature was probably attained about 4 b.y. ago. The oldest dated fossil remains yet known are about 1 billion years younger than this, however. The slow accumulation of abiogenically formed molecules, as a sort of warm organic soup, is sometimes regarded as having involved a billion years. Needham (1959) suggested that such a theory neglects the time scale, confusing evolutionary and metabolic events. The rate of molecular recombination is rapid, for an excited atom unloads its energy in 10^{-9} seconds. In fact, "such rapid and worldwide activity would leave few possible metabolic pathways untested for 'viability,' and indeed few not fully exploited from a very early stage. Most would be tested with a frequency which could only be expressed in astronomical figures" (Needham, 1959, p. 195). As the raw materials of the organic world (C, H, O, N, Na, K, and Ca) were available panglobally, the development of living systems, even with the necessary assimilation of less common materials (S, P, Fe, I, and so on) would have been on a historical rather than geological time scale. The earliest proto-organisms could utilize other abiogenically preformed organic molecules for growth, but these would become exhausted within the limited area available to the individual more rapidly than they were replaced, and a more reliable energy source would soon have become necessary. As solar energy was always available, a primitive form of photosynthesis probably was not long delayed (Needham, 1959).

The greater amount of ultraviolet light reaching the earth's surface before the development of an oxygenic atmosphere and protective ozone layer is sometimes suggested as requiring the development of life in protected deeper water. Even a few meters depth would effectively reduce the amount of penetrating ultraviolet, as would atmospheric water vapor. Sagan (1961) estimated that the quantity of ultraviolet reaching the earth's surface would have required a replication period of 10^6 seconds (about 12 days) to avoid an excessive mutation rate. But bacteria and many protists reproduce much more rapidly than this. As noted by Mayr (1963), only a small number of animal species, including metazoans, have a life span of more than two years, and in most it is considerably shorter. Hence, it is probable that the amount of ultraviolet present even in surface waters would not have been excessive for procaryotic organisms or even the eucaryotic protists, although it may have speeded various reactions.

Among the important early chemical reactions were those that resulted in the formation of certain porphyrins, such as the hemes used in cellular respiration, and the very similar photosynthetic chlorophylls. These pigments may have given protection against excessive radiation, as does a sunbather's tan. Interestingly, purple bacterial photosynthetic pigments absorb both the ultraviolet and infrared wavelengths (compare Fig. 2). Possibly these early autotrophs utilized the ultraviolet energy in synthesis; only the

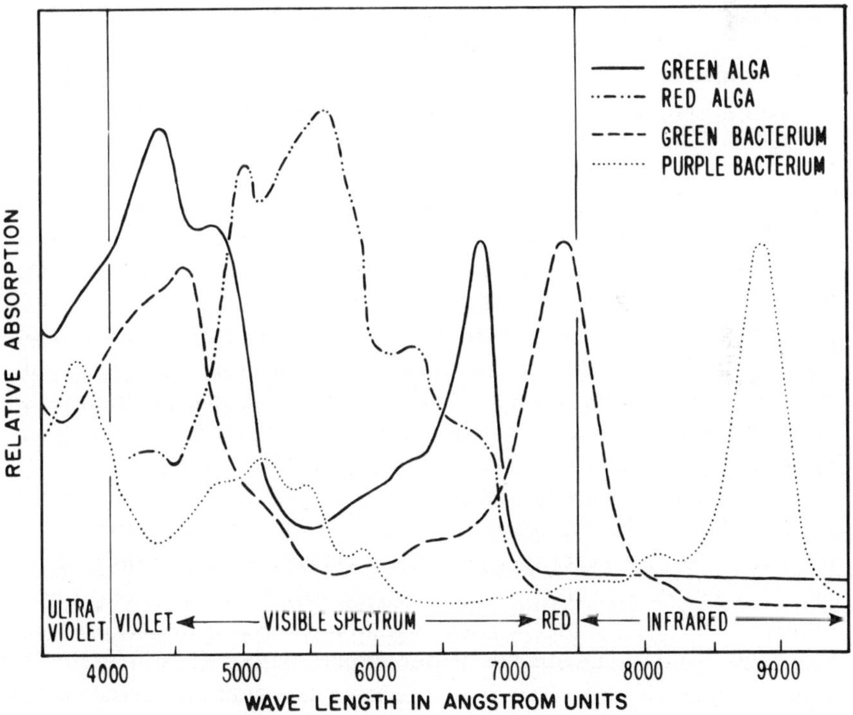

Figure 2. Light absorption by four protists, showing complementarity in visible and infrared regions of spectrum and ultraviolet and infrared absorption by bacteria. Use of longer wavelengths by the latter is correlative to their deeper water, anaerobic habitat. (Based on Stanier and others, 1963, and Haxo, 1960.)

later evolution, involving a slower reproductive rate and different pigments, allowed use of visible light and energy of lower quantal value (Needham, 1959, p. 205).

The primitive organisms, some 4 b.y. ago, probably resembled a modern anaerobic bacterium, obtaining energy by fermentation but gradually developing the complex enzyme sequence found in all organisms. Photosynthetic bacteria, also anaerobic, probably evolved soon afterwards. Those like the present facultatively aerobic bacteria, and the aerobic nitrifying, sulfur, and iron bacteria, probably appeared later, after the blue-green algae had evolved.

With the development of living organisms, a balanced carbon cycle also evolved. Carbon dioxide was converted to organic matter by the autotrophic organisms, and the organic compounds degraded by heterotrophs to return the CO_2 to the atmosphere. The organic materials had to circulate at a rate at least near that of the longest biological generation times, conforming to a biological time scale. With a unidirectional energy source, any system will quickly reach a stable end condition if it does not participate in a cyclic process. Energy acquired from the sun tends to dissipate and allow the changed material to return to its original state. Because this rate of dissipation varies, some organisms can retain energy for considerable periods, overnight, over winter, or considerably longer (for example, as seeds). The biological time scale still remains intermediate between the chemical and geological time scale. Before the rise of autotrophs, growth and reproduction would have been limited to the rate of abiogenic production of organic molecules, as suggested above. Similarly, if all early organisms had been autotrophic, the system would have attained a steady state geared to the slow rate of spontaneous dissipation of energy, resulting in a much less dynamic system than at present. Autotrophic organisms thus depend upon heterotrophs for completion of the carbon cycle, as biological materials do not otherwise rapidly degrade. The early carbon cycle, restricted to quickly growing protistan producers and consumers, probably had a rapid turnover. Rapid cycling traps more solar energy per unit of time, even with a constant total biomass. At present about 0.12 percent of solar energy reaching the earth's surface is utilized by organisms (Needham, 1959, p. 202), but the present biosphere is renewed by photosynthesis in less than 50 years. Dauvillier (1965) estimated that organic matter would have been synthesized by the primitive organisms to the present level of the biosphere in as little as 1500 years, a period of time that is geologically instantaneous. Any out-of-phase part of the cycle could be disastrous to both autotrophs and heterotrophs, hence the importance of self-regulation of population size.

Throughout time some unbalanced activity of living systems has resulted in "stockpiling," as in formation of coal, peat, oil, guano, chalk, chert, and the like.

The Oxygen Producers. An eventual mutation, more than 3 billion years ago, converted the simpler photosynthetic bacterial pigments to true chlorophyll, and the blue-green algae evolved. The bacterial pigments, bacteriochlorophyll and the chlorobium chlorophylls, are less efficient. In reducing carbon dioxide to an organic compound, they utilize other hydrogen donors than water and hence oxygen is not a byproduct of their carbon fixation. The slight change to the chlorophyll pigments of the blue-green algae and all higher plants allowed water dissociation in photosynthesis and production of free oxygen. The change to an oxygenic atmosphere then paved the way for the expansion of aerobic organisms.

Evolution was not only biochemical at this stage. All organisms other than the bacteria and blue-green algae possess membrane-bound organelles (for example, nucleus) within the cell and are termed eucaryotes. As the organelles are similar in all eucaryotic animals and plants, these are believed to have had a common ancestry. Possibly local surface irregularities of the cell membrane were at first advantageous in increasing the surface area for better absorption. Eventually, small portions were pinched off, to remain as membrane-bound organelles within (Robertson, 1962). Seemingly, the process was more complex, as not all organelles have identical membranes. It is nevertheless interesting to note that the bacterial cell membrane contains a variety of enzymes, amino acids, amino sugars, lipids, and proteins such as occur in separate parts of the eucaryote cell. The advantages of intracellular differentiation of organelles were somewhat offset by a loss of metabolic independence. Eucaryotes do not fix atmospheric nitrogen, and many autotrophic protists also require preformed organic compounds (for example, vitamin B_{12}) that are formed by the procaryotes. Supply of these nutrients by diffusion to the cell surface from the surrounding medium would be slow, and the diffusion rate would thus limit metabolic rates (Carlson, 1962). Diffusion could be increased by stirring the medium or by cell movement, both means being possible after the development of flagella. As Carlson (1962, p. 145) stated, "No matter how inefficient the motility mechanism is, it is always possible by moving to achieve a net gain in the collection of nutrient." The similar ultrastructure of the flagella and cilia of all eucaryotic organisms suggest that the survival value of this organelle was discovered prior to the general differentiation of the eucaryotes.

Green algae are generally regarded as the most primitive living eucaryotes. The ancestral eucaryote may have resembled a flagellate green algal monad. However, as the distinctive pigments, flagella, and various cell organelles are not preserved in the fossil specimens, identification of the particular phylum concerned must be based on morphological characters alone. Whether the early acritarchs represent brown or green algal lineages is uncertain, although the two undoubtedly diverged quite early, to give rise on the one hand to the yellow and brown algal phyla, fungi, and the animal kingdom, and on the other hand to the green algae, euglenids, charophytes, and higher green plants. Red algae may have diverged separately from the blue-greens.

Tilden (1933, p. 59, 60) suggested that the primary factor in algal differentiation was a change in the quality of solar energy reaching the organism or a gradual increase in the quantity of light. The different pigments then developed in response to the changing radiant energy. They may merely have allowed wider choice of depth, however. Blue-green and red algae have red and violet phycobilin pigments, characteristic of deeper living species today; the yellow pigments of brown algae allow adaption to stronger light, and the green algae use the strongest bright sunlight.

Because their pigments are responsive to different light wavelengths (Fig. 3), some algae use the deeper penetrating rays and others are depth-limited, resulting in an aquatic depth stratification. The chlorophyll *c* and accessory pigments of brown algae, diatoms, and dinoflagellates are efficient in deeper water, and these organisms are common some meters below the surface. Only the blue-green and green algae have moved into the terrestrial environment, probably because their pigments are best suited to the stronger light levels.

Phytoplankton range from complete autotrophs (certain bacteria and blue-greens) that need only water, carbon dioxide, and atmospheric nitrogen to others that are saprophytic, ingesting dissolved organic material, or heterotrophic, utilizing particulate

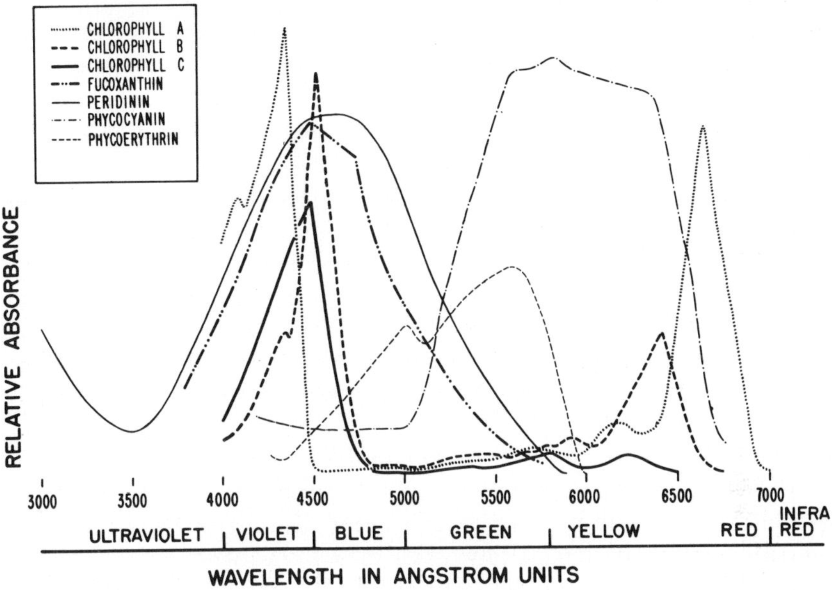

Figure 3. Absorption spectra of phytoplankton pigments. Note that chlorophyll *a* and *b* utilize similar light wavelengths in the red and violet ranges, whereas the phycobilin pigments of blue-green and red algae (phycocyanin and phycoerythrin) utilize light of the green and yellow regions. The xanthophylls and carotenes of the brown algae, chrysophytes, and dinoflagellates make greater use of the blue range. Greater photosynthetic efficiency is obtained by a combination of pigments than by a single pigment. The longer wavelengths of light penetrate more deeply in water, whereas ultraviolet and shorter wavelengths of visible light are rapidly filtered out. (Data compiled *from* Bogorad, 1962, Haxo and Fork, 1959, Healey and others, 1967, Loeblich III and Smith, 1968.)

food. Some may use one or another nutritional mode alternately, photosynthesizing when light is available and consuming dissolved or particulate food when light is insufficient, as in deeper waters. Both the facultative heterotrophic nutrition and the encysted resting stage of some phytoplankton aid their survival when oceanic conditions result in their sinking below the photic zone. The organisms remain viable until turbulence or upwelling currents return them to the lighted levels.

The many reproductive modifications in Protozoa and algae range from asexual fission of haploid organisms, schizogony, or spore formation to sexual conjugation or gamete formation, with the production of resting zygotes. Different methods may alternate regularly, or one or another may be emphasized in response to particular environmental conditions.

Over a sufficient period of time, most groups show undirected testing of all evolutionary possibilities. Most algal phyla independently developed free-swimming flagellate monads, coccoid cells, or palmelloid colonies and filaments and elaborate foliose or siphonous thalli while maintaining their distinctive biochemical and micromorphologic features. The Chlorophyta include all such variations. The Chrysophyta, Euglenophyta, and Pyrrhophyta contain more monad variety but are rarely multicellular. Only the thallus type is known in the red and brown algae, although Fott (1965, p. 125) suggested that the brown algae may have had evolutionary lines similar to those of the green algae; the postulated brown monad and coccoid types became extinct, leaving only the heterotrichous living ones. At this complex level of organization the brown algae vary widely in morphology and life cycles. No siphonate Chrysophyta are known, but the primitive ancestral yellow-brown monad probably gave rise to Xanthophyta, modern Chrysophyta, Bacillariophyta, Phaeophyta, and fungal Phycomycetes, colorless protozoan flagellates, rhizopods, and other animals (Bourrelly, 1962).

RATES OF EVOLUTION

Marine versus Terrestrial Evolution. Evolution is commonly stated to have been more rapid on land than in the seas and more rapid in vertebrates than in invertebrates (Mayr, 1942, p. 218; Simpson, 1949; Rensch, 1960). Evolution has also been extremely rapid at times in the marine realm, however, and major structural diversification leading to the origin of higher taxa has been greater there than in the terrestrial environment.

Microfossils that are morphologically similar to modern bacteria have been described from the 3.1 b.y.-old Fig Tree Series near Barberton, South Africa. But the 1500 or more living species of bacteria are classified and differentiated on the basis of their very diverse physiological characters rather than their morphology!

Archaeosphaeroides barbertonensis Schopf & Barghoorn (1967), from the Fig Tree, resembles the modern blue-green Chorococcales and suggests the presence of photosynthesizing oxygen producers early in the Precambrian. Probable eucaryotic organisms (higher protists), perhaps referable to green algae or fungi, occur in the Gunflint Formation about 2 b.y. old. The Bitter Springs Formation of Australia (1 b.y. old) contains definite green and blue-green algae and fungi, and the Belt Series of Montana contains blue-green algae, fungi, and acritarchs. The supposed foraminifers (Pflug, 1965) of the Belt Series are probably blue-green algae.

Algal diversification followed the appearance of the earliest eucaryotic cell. One-half the phyla of plant protists were well-represented before the beginning of the Paleozoic: the blue-green, green, red, and possibly the brown algae, the fungi, and the Acritarcha. Their diversification in pigments, plastids, nuclei, flagellar form and position, and other features occurred within the marine environment, as did the major later diversification of other groups of algae. All terrestrial vascular plants evolved from the green algae, which they resemble in possessing the pigments chlorophyll *a* and *b*, starch food storage, and cellulose cell walls.

The present ecologic pioneers (bacteria, blue-green, green, and golden-brown algae, and fungi) were undoubtedly the first colonizers of the freshwater and terrestrial environment, early in their history, and have continued to invade first the newly available areas throughout subsequent periods. Representatives of each presently inhabit fresh water and soil, and although these environments are least likely to produce a fossil record, such organisms probably occupied oceans, seas, fresh water, and soils in the Precambrian. The lithology of the Precambrian rocks suggests that mature soils were present then as now.

If the land-plant diversification has been largely limited since the Precambrian to structural modifications of the green algae, have land animals shown a different pattern? Seemingly they have not. The first metazoan probably appeared near the end of the Precambrian, about 600 m.y. ago. The important development of metazoan structure allowed increased size and an exploitation of new evolutionary pathways. The rapidity of this marine diversification is indicated by the presence of representatives of all those phyla and most of the invertebrate classes capable of fossilization, by the mid-Ordovician, only some 120 m.y. after the appearance of the oldest known metazoan. Every animal phylum arose in the marine environment while dependent entirely upon the plant protists for food, energy, and even the oxygen for respiration. The terrestial environment offered no such stimulus for major diversification, although later it was invaded repeatedly by various animal phyla, as it had been by their plant ancestors.

Was diversification below the phylum level more rapid on land? The oldest known vertebrates are of late Cambrian age (reported by Cygan and Koucky, 1963, p. 33, from the Gros Ventre Formation of Wyoming), and only by the Jurassic, some 350 m.y. later, had the last of the vertebrate classes (Aves) appeared. Vertebrate evolution at the class level thus was slower than that of the invertebrates, in contrast to the usually accepted statements!

Diversification of kingdoms and phyla thus took place only in the marine environment and at the class level was more rapid in marine invertebrates than in the vertebrates. The oft-cited more rapid evolution of higher terrestrial organisms is also questioned at the generic or family level. The recognized stratigraphic value of the protistan fossils would seem to be a clear indication of their rapid evolution. Yet Bonner (1962, p. 25), stated that "the living Foraminifera and Radiolaria of today are essentially the same as those of the early seas. These are not progressive groups in an evolutionary sense, but what might be called successful dead ends; dead ends, because they have not changed appreciably, but successful, because they continue to exist throughout the ages."

Foraminifera have been recorded from the Cambrian to the present. Through 1962 some 1200 foraminiferal genera were recognized (Loeblich and Tappan, 1964a, 1964b) and nearly 300 have since been described. This indicates an average evolutionary rate of one new genus every 400,000 to 500,000 years, a somewhat surprising rate for a group that has "not changed appreciably." Furthermore, speciation in the Foraminifera occurred at the average rate of one new species every 20,000 years.

Rensch (1960) indicated that, in general, species last an average of 100,000 to a few million years, genera have a duration about 10 times as great, families about 1.5 times that of genera, and orders 2 times that of families. The Foraminifera include 97 families for the approximately 1500 genera and probably 30,000 species (Loeblich and Tappan, 1964a, 1964b, and unpublished data). The 7 families of planktonic Foraminifera include 56 genera, with an average duration of only 27 million years; the longest duration of any planktonic foraminiferal genus was about 66 million years. The average duration time for all foraminiferal families is approxmiately 150 million years, only 2 of the 97 having representatives from Cambrian to Holocene. These two "long-ranging" families contain a total of 51 genera, only 1 of which itself had such an extensive range. The long-ranging *Bathysiphon* is represented only by a tubular agglutinated test preserving very few features for differentiation. Supposedly long-ranging species or genera are invariably those with fewest diagnostic test characters. Thus a small tubular agglutinated foraminiferal test is referred to the genus *Bathysiphon* or a flasklike one to *Saccammina,* but in view of the cytoplasmic, ontogenetic, and physiologic diversity in living foraminifers, there seems little doubt that the Cambrian and Holocene species of *Bathysiphon* were also quite distinct genetically. A morphologically similar tubular or flasklike organic-walled or agglutinated test or lorica also is found in the ciliate tintinnids, in a variety of algal phyla (here of smaller size), or even among metazoans. As additional criteria for taxonomic differentiation have been recognized (for example, wall mineralogy and microstructure), many of the once "long-ranging" taxa proved to include a variety of easily distinguished, unrelated forms. The reportedly persistent genera and species cited by Rensch (1960) and others actually represent misidentified specimens of a variety of distinct genera, families, and even superfamilies.

Because of the long acceptance of extremely wide generic limits, the Radiolaria have been regarded as a relatively slowly evolving group. Foreman (1963, p. 267) noted that "faulty interpretations of poorly preserved specimens" resulted in assignment of early and middle Paleozoic radiolarians to living genera, whereas well-preserved specimens showed such distinct internal structure as to require their generic separation. Pending generic and familial revisions, discussions of their evolutionary rates would be meaningless. On the other hand, species distinctions are probably more valid. Data from Chediya (1964) indicate a total of 3582 described species of polycystine Radiolaria (Spumellaria and Nasellaria) since the Ordovician, increasing from 87 in the Ordovician and Silurian combined to 888 at present, or an average of one new species every 140,000 years.

The rapid evolution of protozoans is not limited to rhizopods. The ciliate protozoans (Tintinnida) evolved at an over-all rate of one genus every 4.8 m.y., although this was in short periods of expansion, a dozen of these appearing in a period of 20 m.y. at a rate of about 1 in 1.7 m.y.

Phytoplankton Evolutionary Rates. Approximately one-half of the algal phyla had appeared before the end of the Precambrian.

Animal diversification was rapid at the phylum and class levels in the earliest Paleozoic. The last 600 m.y. have seen an exploitation of the various morphological and ecological possibilities of each of the major animal taxa, as well as phytoplankton diversification, evolution, extinctions, and repeated exploitation of varied environments.

Modern and fossil phytoplankton are as yet less well-known than the benthic red, green, and brown algae or the animal protists. However, the rapidly accumulating data indicates that the silicoflagellates evolved at a rate of one genus every 6.6 m.y., the ebridians produced one genus every 3.5 m.y., discoasters one every 5 m.y., and coccolithophorids about one genus every 1 m.y. Based on present data, fossil dinoflagellate cyst genera evolved at a rate of one every 1.4 m.y. This undoubtedly is a low estimate of the genera, as systematic work on this group has scarcely begun. If approximately 300 modern genera based on the vegetative cell are included with the cyst genera, their post-Paleozoic rate of evolution would be about one genus for each 460,000 years. The Paleozoic records of dinoflagellates are not considered here, both because of the questionable nature of some reported dinoflagellates and because of the possibility that many Acritarcha may be dinoflagellates. Although these phytoplankton genera might be regarded as too closely split, speciation rates show a similar distribution. Small (1950) indicated that the 3526 living Pennate diatoms evolved from 3 species in the Paleocene. One new species was produced on an average of every 20,000 years. The evolutionary process was neither a replacement of one species by another nor a transition of one species into another. Instead, the number of species increased in a regular geometric progression. In succeeding periods, new species arose, some persisted from the previous floras, and others disappeared.

When a limited volume of medium is inoculated with a few cells of an algal culture, the population increase follows a regular pattern. A lag phase, during which cells do not reproduce but may increase in size, is followed by a period of rapid cell multiplication, the exponential growth phase. During this phase the number of cells increases in a regular geometric progression. This exponential increase is followed by a phase of declining growth, one where cell number remains stationary, and finally a death phase, after some limiting concentration is reached (Fig. 4). Zeuner (1946) suggested that evolution of all organisms follows a similar pattern.

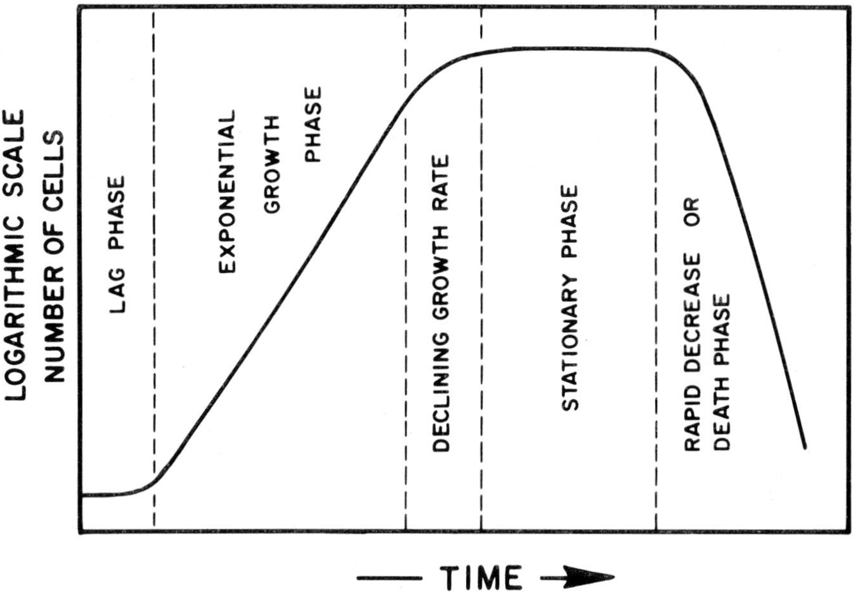

Figure 4. Growth phases of algal protists in limited-volume culture, from inoculation to final death of cells.

A lag phase of relatively insignificant diversification is followed successively by an explosive phase of logarithmic increase (in which speciation exceeds extinction), a climax or stationary phase, and a phase of decline (where extinctions surpass innovations). The explosive phase was regarded as lasting about 50 m.y., regardless of the kind of organism.

Whether or not this concept can be applied to all organisms, it does seem applicable to phytoplankton and may be related to their characteristic asexual reproduction. A slight genetic change in an individual of a species by clonal reproduction could result in populations of two distinct species. Each of these could similarly produce two species. A geometric increase in species could result from a moderate mutation rate and reproduction by binary fission, providing the environment were favorable. Together with the newly evolving species, a fairly constant proportion of old species continues into the subsequent period. Asexual reproduction thus seems conducive to evolution by sudden steps. Small (1950) suggested that the number of newly evolving diatom species was controlled by the number of prior species available as a source rather than by population size or number of generations. In the diatoms, the older and more primitive Centrales passed the lag and explosive phases by the Tertiary and as a group were approaching the stationary phase, although various genera were in different phases. In contrast, the Pennales are now in their explosive phase (Fig. 5).

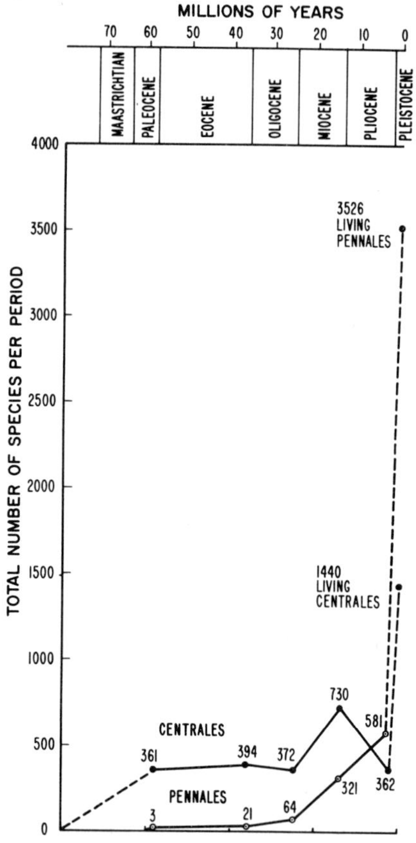

Figure 5. Fluctuating number of species of diatoms throughout the Tertiary (Paleocene to Pliocene). Points marked at end of period for which species are totaled. (Based on data *from* Small, 1950.)

Data compiled by Sarjeant (1967) indicate the average duration of a dinoflagellate cyst genus to be about 42.4 m.y. and that of a cyst family to be about 100 m.y. As was true of the diatoms, the rate of diversification was not steady but showed a lag phase until early Jurassic. The period of taxonomic exponental increase lasted about 24 m.y. (Toarcian through Oxfordian), then was followed by a Late Jurassic decline of 15 m.y. through the Lower Cretaceous Berriasian stage. A renewed exponential diversification began in earliest Cretaceous (comparable to inoculation of new medium or renewal of nutrients in an aging culture), and continued for about 25 m.y. Most of the 28 m.y. of the Late Cretaceous was represented by a declining rate of increase followed by a general sharp reduction at the close of the Cretaceous (Maastrichtian) and in the early Paleocene (Danian). A new expansion occurred in later Paleocene time, exponential increase lasting through early Eocene but succeeded by gradually declining growth to the present (Fig. 6).

Although some 1000 modern dinoflagellate species are known, the hazards of fossilization, collecting, and taxonomic bias suggest

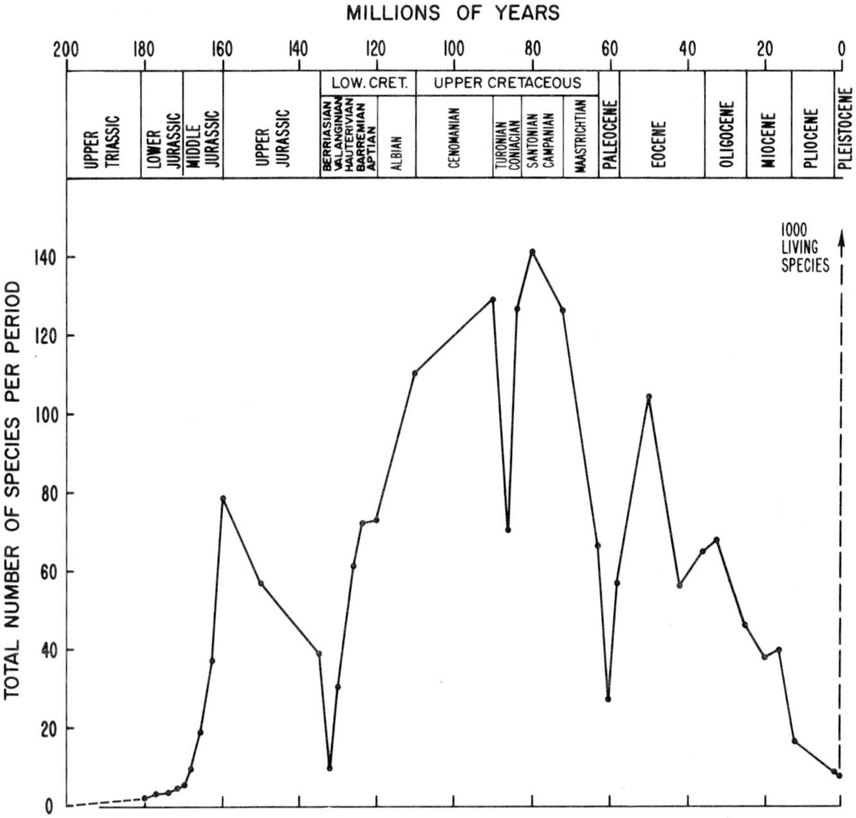

Figure 6. Fluctuating number of species of dinoflagellates throughout the Mesozoic and Cenozoic. Points marked at end of period for which species were totaled (based on data *from* Sarjeant, 1967). Note exponential growth in Middle Jurassic, Late Jurassic decline, exponential increase in Cretaceous, sharp drop at end of Cretaceous, renewed increase in Paleocene and early Eocene, with steady decline since middle Eocene time.

that a relatively small percentage of these might potentially appear in the fossil record. For example, preservation of parasitic dinoflagellates and symbiotic zooxanthellae would be most unlikely, and typically oceanic dinoflagellates do not produce resting stages or cysts. Of the marine species, only the littoral, cyst-producing species could potentially be preserved. Nonmarine dinoflagellates are as yet relatively unknown as fossils, although some 23 genera occur in the modern freshwater environment. It thus seems likely that a rather small percentage represents potential fossils. Yet even though concentrated studies of fossil dinoflagellates have been under way scarcely two decades, Lower and Upper Cretaceous and Eocene microfloras each already include well over 10 percent of the total number of living dinoflagellate species, suggesting that they were in fact more diversified then than now.

As coccolithophorids currently are being studied by many workers, data rapidly accumulate. The general pattern to date

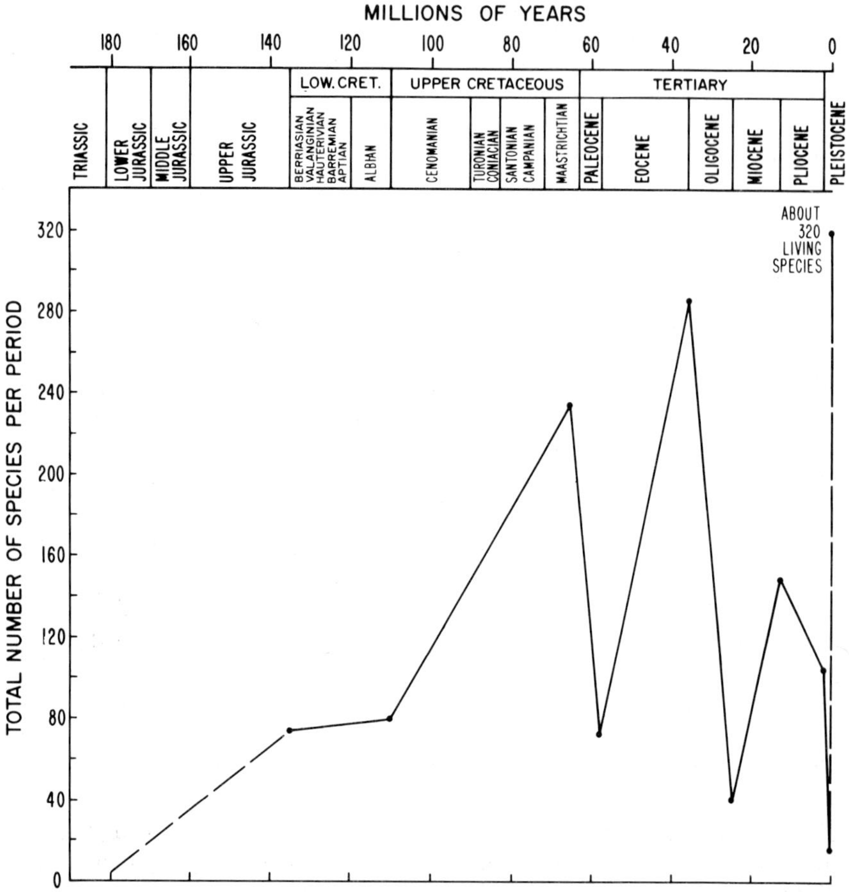

Figure 7. Fluctuating number of species of calcareous nannoplankton (coccolithophorids, etc.) throughout time. Note exponential increase in Jurassic and Cretaceous, sharp decline at end of Cretaceous, rapid increase in Eocene, and post-Eocene decline to the present. The decreased number of Oligocene species and temporary Miocene increase may be real or may result from disagreement as to the stratigraphic limits of the Oligocene. Very high number of species in Cretaceous and Eocene, in comparison to number now living, shows marked decline of this group in later Cenozoic. (Based on Loeblich and Tappan, 1966, and unpublished data.)

suggests a lag phase in the Jurassic and Early Cretaceous (about 75 species each) and an exponential increase (to about 235 species) in the Late Cretaceous, followed by numerous extinctions and a rapid decline. Diversification began again in the Paleocene, but attained only Jurassic levels; then Eocene exponential increase reached a new maximum (about 285 species). The apparently sharp decline in the Oligocene may be in part due to its shorter duration or to boundary uncertainty. Nevertheless, Miocene diversity (about 150 species) remained far below that of the Eocene, and the decline continued through Pliocene and Pleistocene time (Fig. 7). Described Holocene taxa are more numerous, but biological oceanographic studies preceded the study of fossil calcareous nannoplankton. The nearly equal numbers of known Eocene species and living ones again suggests the same relative phytoplankton decline since the Eocene, as in the dinoflagellates.

Ebridians arose in the Paleocene, probably from the dinoflagellates, and rapidly attained their explosive phase. Their numbers are smaller, only 25 species in the Paleocene, and the stationary phase followed in the Eocene. An apparent Oligocene decline, like that of the coccolithophorids, may be a result of stratigraphic bias but was followed by exponential increase in the Miocene to the maximum attained by this group—some 51 species. A precipitous decline followed, to the present four living species (Fig. 8).

Silicoflagellates are known from the mid-Cretaceous on. They increased through the Late Cretaceous to 20 species, then declined for a short time. By the end of the Paleocene 19 were present and 37 in the Eocene. The usual Oligocene decline (to 15 species) was followed by an exponential increase in the Miocene, to a total of 105 species and varieties—production of a new taxon every 100,000 years. A sharp decline followed, to some 5 fossil species in the Pliocene and Pleistocene and the approximately 58 species and varieties reported as living (Fig. 9).

The as yet incomplete data concerning fossil taxa nevertheless indicate distinctly that dinoflagellates, coccolithophorids, centric diatoms, silicoflagellates, and ebridians have already passed an evolutionary peak and that all had a late Cenozoic decline. Data are not yet available to determine whether this decline is still in progress, or whether a post-Pleistocene increase has begun. Of the major phytoplankton groups, however, only the pennate diatoms seem currently to be in an explosive evolutionary phase.

Evolution within the various algal divisions has thus shown inherent patterns that seem unrelated to competition or a one-

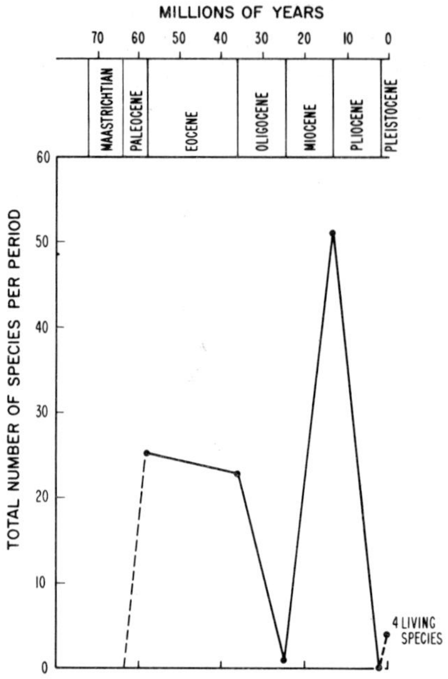

Figure 8. Fluctuating number of species of ebridians throughout the Cenozoic. Note rapid increase in Paleocene, stationary phase in Eocene, decline in Oligocene, exponential rise to climax in Miocene and post-Miocene "death-phase," with only four rare living species. (Based on data *from* Loeblich III and others, 1968.)

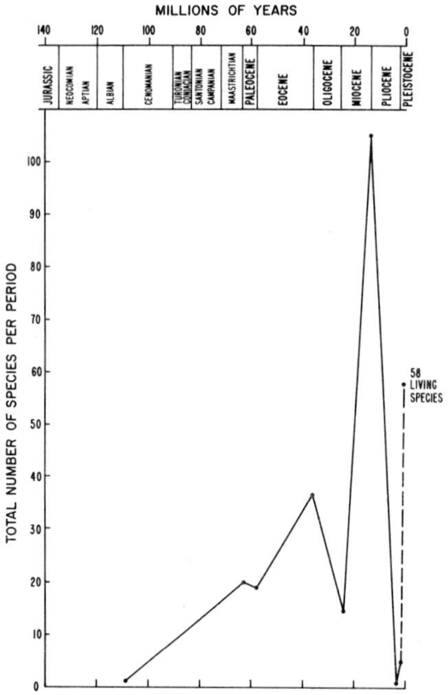

Figure 9. Fluctuating number of species of silicoflagellates throughout time. Note exponential increase from mid-Cretaceous to Miocene, possible decline in Oligocene, and prominent post-Miocene decline, with described living species about half the number of Miocene fossils. (Based on data *from* Loeblich III and others, 1968.)

for-one replacement. The separate groups repeatedly have shown periods of exponential increase, followed by a steady period of varying duration and a precipitous decline that must have been related to external factors (Fig. 10).

Causes of Exponential Evolution. In a newly inoculated laboratory culture, the lag phase before cell division begins is due to the adjustment of the organism to the new medium and its prereproductive growth. The absence of crowding and the availability of light, nutrients, and other requirements soon result in the rapid exponential growth. The spring bloom of phytoplankton in temperate regions resembles the pattern of growth in a limited-volume culture (Fogg, 1965, p. 59). Similarly, the rapid spread of new

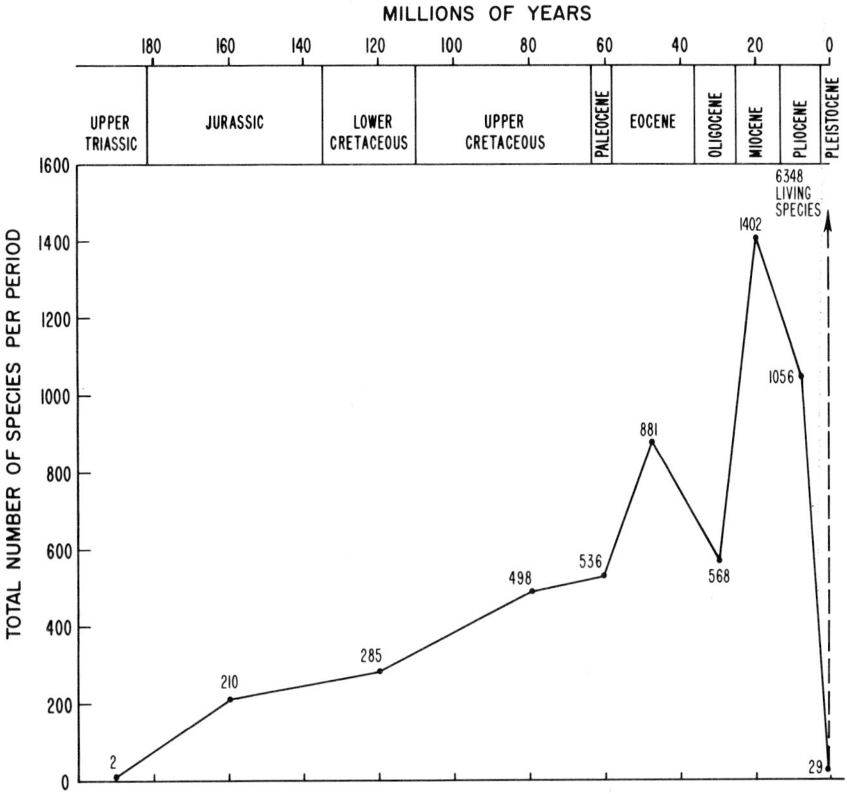

Figure 10. Total described fossil phytoplankton species in the Mesozoic and Cenozoic, by periods. As organic, siliceous, and calcareous groups are combined, fluctuations are smoothed somewhat. Totaling by periods also masks the late Maastrichtian decline of most groups. The low numbers in the Triassic and the post-Miocene decline of all groups of phytoplankton remain quite apparent. (Based on data compiled *from* Small, 1950, Sarjeant, 1967, Loeblich III and others, 1968, Loeblich and Tappan, 1966, and unpublished.)

phytoplankton taxa in the geologic past can be interpreted from such culture data. Exponential increase follows periods of marked reduction of microfloras (absence of crowding) and has accompanied encroaching epicontinental seas (renewed availability of light and nutrients). Similarly, explosive evolution has occurred with the appearance of any important innovation, whether this be environmental, morphological, or physiological. Exponential increase of taxa results from an exploitation of all potentialities of the innovation or environment.

Evolutionary radiation in the Precambrian seems to have been an exploration of the biochemical, physiological, and environmental capabilities of the unicellular organism, whether phototrophic, heterotrophic, or osmotrophic, at first at the level of the procaryote and later at that of the eucaryotic alga and of the protozoan. Exploitation of the various possibilities open to the plant or animal protist has continued in later periods.

The early Paleozoic was marked by the exponential exploitation of another innovation, multicellularity, resulting in the Cambrian diversification of phyla. Later evolution has not included an equally important innovation and hence seems to have slowed. Instead, the Phanerozoic diversification has been marked by the utilization of smaller morphologic or ecologic innovations.

Origin of Protozoa and Metazoa

An oxygenic atmosphere was postulated to have developed slowly during the Precambrian. The slowly evolving and gradually spreading algal flora was considered to be relatively ineffective, as compared to higher plants, and hence the oxygenic time table of Berkner and Marshall (1965) indicated the attainment of only 1 percent of present atmospheric oxygen levels by the beginning of the Paleozoic and 10 percent at the end of the Silurian.

The lengthy Precambrian record of photosynthetic algae and the extent of their diversification and geographic distribution are remarkable in view of the scarcity of suitable rocks and paucity of detailed studies. In addition to the fossil remains, Precambrian biological activity is indicated by preserved hydrocarbons, sedimentary iron ores, and algal stromatolites. The present importance of phytoplankton productivity in the oxygen cycle, the fact that it provided the only source of food for the expanding animal kingdom at the beginning of the Paleozoic or earlier, and the characteristic exponential growth all indicate the need for revision backward of this oxygenic time table.

Exponential Increase and the Precambrian Atmosphere. The exponential increase of phytoplankton in culture is eventually limited by autoinhibiting secretions or quantity of available space, light, or required nutrients (nitrogen, silica, vitamin B_{12}, and so on). Conditions in the present seas differ somewhat from those of unialgal laboratory cultures, but similar growth phases result in exponential growth and periodic phytoplankton blooms and times of decline or stationary growth. Cell doubling time, which results in exponential growth, is constant for each species, independent of the radiant energy available or the temperature (providing the latter is above 10° C). The cells adjust their composition over wide ranges to maintain a steady division rate, independent of other factors, although changes must not be so sudden as to prevent such adjustment (Antia and others, 1963, p. 178).

Reproductive rates are rapid in phytoplankton. In continuous light, Fogg (1965) reported that doubling time for the green alga *Chlorella pyrenoidosa* ranges from 2.6 hours at 39° C to 60.2 hours at 10° C; the xanthophycean *Monodus subterraneus* doubles in 24.3 hours at 25° C or 37.8 hours at 20° C; the coccolithophorid *Cricosphaera carterae* in 8.8 hours at 18° C and the blue-greens *Anabaena cylindrica* 10.6 hours at 25° C, and *Anacystis nidulans* in 2.0 hours at 41° C. In general, growth is most rapid in smaller species and increases with increased temperature to a maximum for each species. The optimum temperature for modern species is commonly between 20° and 25° C, but is as high as 40° C as in certain thermophilic greens (*Chlorella*) and blue-greens (*Anacystis*). If the protists of the Precambrian doubled only once every 24 hours and no cell loss occurred by grazing, the divisions from a single original cell would result in 1 billion cells in a month. Continued exponential growth could allow in 6 months as many individuals as there are atoms on earth (Strughold and Ritter, 1962).

Early oxygen producers of the Precambrian were blue-green algae. Like some bacteria, they may be nearly "perfect organisms," obtaining nitrogen and CO_2 from the atmosphere and requiring little else but water. The present blue-greens are a major group of nitrogen-fixing organisms (Burris, 1966), both in the marine environment (*Nostoc, Calothrix*) and in fresh water (*Anabaena, Chlorogloea*). As few nutrients would be in limited supply, growth would be limited only by available space within the photic zone. At present, grazing zooplankton forms an effective limitation. Harvey (1945) observed that an original population of 100 cells per liter, after six cycles of division, would total 6400 cells per liter. If 1 of every 10 cells were to be eaten between successive divisions, some 413 cells would be thus consumed, but the final population would be only 3400 cells. Prior to such limitation by grazing zooplankton, which did not evolve until oxygen levels were already considerable, net production could have been much greater.

The early carbon cycle consisted of the blue-green and bacterial producers and bacterial reducers, later being modified to include eucaryotic algal autotrophs and facultative heterotrophs and fungal reducers, and eventally protozoan heterotrophs. Incidentally, bacterial activity is even more rapid than that of the phytoplankton, some doubling in number every 15 minutes.

Present phytoplankton productivity is discussed in a later section. It now totals at least 50 percent, and probably 75 percent or more, of the total productivity of the earth's surface. At current rates, all atmospheric oxygen is biologically recycled every 2000 years and the CO_2 of oceans and atmosphere recycled every 350 years (Berkner and Marshall, 1964, p. 112). Much of this now is done by phytoplankton; during the Precambrian and early Paleozoic all photosynthetic oxygen production was due to the algae. How effective would this be in modifying the Precambrian atmosphere?

An amount of 2.3 kg of *Chlorella pyrenoidosa* will produce oxygen at a rate of 37 gm/hr. Oxygen production is also more rapid under somewhat higher temperatures. Starting with 1 kg of a protococcacean alga such as *Chlorella,* exponential growth could have produced the present 1.2 x 10^{21} gm of atmospheric oxygen in less than 2 months (61 days) (Strughold and Ritter, 1962, p. 277).

Exponential growth would have been limited then as now by space, light, and available CO_2. During the early Precambrian algal evolution, higher levels of solar radiation (especially ultraviolet) reaching the earth's surface, greater natural radioactivity levels, extreme tidal action (perhaps related to the formation or capture of the moon), and more active volcanism would have depleted the microfloras locally from time to time. On the other hand, rapid increase and more rapid carbon cycling in the Precambrian seem indicated by the self-sufficient nature and characteristic exponential growth of the algal protists, absence of inhibiting secretions of eucaryotes, absence of grazing zooplankton, and possibly higher temperatures and higher CO_2 concentrations. Stronger tidal actions would have aided cell dispersal and nutrient renewal.

Blue-green algae undoubtedly spread into all available areas geologically instantaneously. The resultant early attainment of high oxygen levels would have been moderated by the inorganic oxidation of reduced compounds (NH_3, CH_4, iron compounds, and the like) in the atmosphere and lithosphere, but this also would have been rapid on a geological time scale.

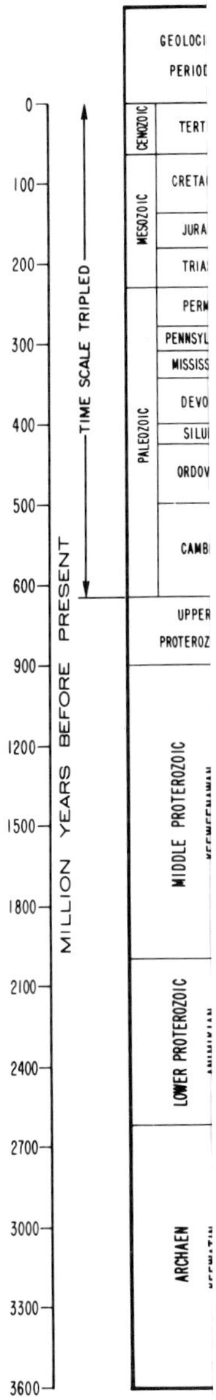

Procaryotic lo
supporting morph
red, acritarch, a
fungal reducers a
tion of multicellu
the advanced rh
temporaneous wit

TAPPAN AND I
Geological Society

An aerobic environment had to be attained before certain other evolutionary advances took place, such as the origin of the protozoans, discussed below. Similarly, both iron and sulfur bacteria use free oxygen in their metabolic processes and are a later development from the anaerobic bacteria, probably arising after the atmosphere already contained considerable oxygen. Both iron and sulfur bacteria were present at least 2 b.y. ago. It seems probable that oxygenic levels near those of the present were also attained by that time. Instead of having only 1 percent of the present amount of oxygen, the Lower Paleozoic atmosphere probably was even more highly oxygenated than ours (Pl. 1).

Origin of Colorless Protists. Modern protozoans are undoubtedly polyphyletic, for they include colorless, heterotrophic or osmotrophic derivations from the green algae, euglenids, dinoflagellates, diatoms, and chrysomonads. The nonphotosynthetic species are morphologically similar to their algal near-relatives but must depend solely upon animal-like nutrition in place of photosynthesis. Many pigmented eucaryotic species can facultatively ingest dissolved or even particulate food, and some can be "cured" of their plastids in the laboratory by irradiation or antibiotic treatment and continue to thrive as protozoans. Interestingly, no blue-greens can grow in the dark, all being strict autotrophs (Hutner and Provasoli, 1964).

In addition to their photosynthetic pigments, members of the plant kingdom are characterized by the presence of cell walls, although thickness or rigidity of the wall is not important to the definition. A cell wall characteristically occurs in bacteria, blue-green algae, and all eucaryotic plants but may be lacking in a few species of various phyla. The sole difference between certain pairs of green algal genera is the presence or absence of such a wall. Some dinoflagellates lack a wall, whereas others have a heavy cellulosic one. The palmelloid Chrysophyta and Xanthophyta in general have naked cells, with only a pellicle instead of a cell wall.

These tendencies of members of a half-dozen living algal phyla to lose both plastids and cell wall indicate a probable source of the naked flagellate protozoans. A single nonphotosynthetic free-swimming chrysomonad-like organism thus is believed to be the ancestral protozoan; the animal-like absence of a cell wall or other durable structure probably would prevent its preservation. Of the known flagellate protists, only the phytoflagellates with cell walls or resistant-walled cysts have a fossil record. Similarly, among protozoans, the more specialized shell-bearing ones are best known geologically, particularly the Rhizopoda (Foraminifera appear in the lower Cambrian and Radiolaria in the Ordovician) and Ciliata (tintinnids occur sporadically from Ordovician on). No evidence as yet suggests that the highly evolved amoeboid or ciliate protozoa were present before the Cambrian; their appearance was perhaps nearly simultaneous with that of the metazoans. Even if nonshelled protozoans were present in the Precambrian, their small size prevents their recognition as carbonized impressions. No undoubted fossil protozoans have yet been obtained from Precambrian rocks.

The algal phyla had a long Precambrian history, and, as today many of them show tendencies toward becoming flagellate consumers rather than producers, one can reasonably assume that they also did so in the Precambrian. It is probable that protozoans did not evolve until atmospheric oxygen levels had become relatively high and ultraviolet radiation was shielded by the ozone layer. Sensitivity of cells to radiation varies, according to their structure and environment. Those most sensitive include embryonic cells, oocytes, lymphocytes, tumor cells, sperm, and so forth, that is, those cells that have an active nucleus and a large ratio of nuclear to cytoplasmic volume and lack an external cell wall. In water, irradiation produces H_2O_2 and the free radicals HO_2 and OH, which react with any available substance. Antioxidants such as catalase may occur in the cytoplasm but are not present in the nucleus of many cells. Catalase also occurs only "in low concentrations in tumor and embryonic cells (large nuclei) as well as in lower organisms that lack cytochrome systems" (Schjeide and others, 1956). Thus any colorless flagellate or fungus that might have developed in the early Precambrian, before the attainment of relatively high oxygenic levels, could not have survived the irradiation, irrespective of its necessity for oxygen. A fossil record is lacking for the shell-less protozoans, but the chitinoid fungal cell wall is preservable. The Precambrian record of fossil fungi thus is another indication of probable high oxygenic levels.

Compartmental Differentiation. The absence of Precambrian Metazoa is generally attributed to lack of sufficient oxygen and their rapid development in the early Cambrian related to the increased oxygen, even if only 1 percent of the present.

Fischer (1965) suggested that the early protozoans and metazoans evolved as small soft-bodied organisms in respiratory dependence upon host plants that provided small oxygenic oases and later became independent by acquiring algal symbionts. But morphologic evidence suggests that benthonic Protozoa and Metazoa were preceded by a wide variety of ancestral free-swimming flagellate protozoans. Symbionts were probably not involved in early metazoan evolution, for those known today are in general rather specialized members of the various algal phyla. They are also more commonly found in benthic organisms (corals, foraminifers, mollusks) and appear adapted to a much smaller degree to free-swimming organisms. Those in which the endosymbiont is referable to the blue-green (cyanellae) and green algae (*Zoochlorella* spp.) occur almost solely in freshwater or terrestrial habitats. Only the dinoflagellate *Zooxanthella* spp. are characteristically marine, and dinoflagellates are unknown before the Silurian. Furthermore, Droop (1963) noted that in the algal-animal symbiosis the primary benefit is always to the algal partner. The relationship is nutritional, the respiratory products of the animal being exchanged for the photosynthetic products of the alga. The early development of animal life dependent upon such specialized endosymbionts seems improbable.

Weyl (1968) suggested that abiotically formed organic molecules accumulated at the tropical oceanic thermocline between 50 and 100 m. due to density gradients. Not only did the first organisms arise there, but development of the blue-green algae led to an accumulation of oxygen at this level, supplying an extensive favorable environment for the development of planktonic eucaryotes and even metazoans, perhaps a billion years before the Cambrian. A benthic habitat and skeletal cover could develop after the higher oxygenic levels finally allowed colonization of the continental shelves at the beginning of the Cambrian.

If, as is suggested here and elsewhere (Tappan, 1968b), the atmosphere attained present oxygenic levels within the Precambrian, why was the Precambrian assemblage limited to protists? A twofold explanation of the sudden appearance of Metazoa seems likely. The selection pressures for small size and rapid reproduction decreased and the innovation of multicellularity was followed by the exploration of its possibilities. Most important, natural selection favors small size when rapid exploitation of an area is important. Large size and multicellularity slow the reproductive process by the insertion of many somatic cell generations. Thus, although phytoplankton blooms appear immediately with favorable conditions, a similar exploitation by multicellular organisms would be impossible. The physical environmental conditions indicated earlier as probable in the Precambrian would have produced a selection pressure toward small size and rapid reproduction. The large populations resulting from exponential growth would have extensively tested various genetic recombinations and the prevalence of asexual reproduction would similarly have favored the clonal retention of different mutations. As the selective pressures of the environment decreased, multicellularity was no longer a disadvantage (Table 2).

The reduction of volcanism and radioactivity in the latter 15 percent of the earth's history is indicated by the greater prevalence of sedimentary rocks in the Phanerozoic and the halflife decay of radioactive elements. Under more benevolent conditions, the development of the animal mode of nutrition would favor an increase in size, for a predator needs to be larger than its prey. Perhaps the many multinucleate protistans, including many rhizopods, opalinids, and even siphonous algae, represent abortive attempts to attain larger size and multicellularity, but, as they failed to develop separating membranes, they did not attain compartmentalization.

Infolding of the cell membranes, or some other mechanism, had long before given rise to the various membrane-bound organelles of the eucaryotic protists. A similar compartmentalization of portions of the more complex cells probably led to multicellularity. Sponges represent an evolutionary side branch, with an origin separate from that of the true metazoans. Multicellularity also arose separately in fungi, green, red, and brown algae, charophytes, and higher plants.

Single-celled organisms in general have continuous growth. Cell enlargement ceases only at reproduction (which involves the entire parent cell), and growth is then renewed by the next generation. In contrast, multicellular organisms are characterized by an adult size limit. Growth is no longer continuous but ceases at maturity. Individual death has become a part of the cycle, only the reproductive cells continuing. For species survival, genetic recombination requires that individuals must be replaced, but production of new individuals requires death of earlier ones to prevent slowing or cessation of the cycle (Needham, 1959).

The cellular differentiation, specialization, and diversion of energy to physical processes other than simple growth and reproduction enable the metazoan to modify an inhospitable environment and reduce the necessity for rapid turnover. Attainment of com-

TABLE 2. DOMINANT GRADES OF EVOLUTION

Approximate time and duration (billions of years)	Form	Generation rate	Energy source	Environment	Processes	Morphologic grade
4.3-3.0 (duration 1.5)	Chemical	10^{-9} sec	Solar radiation radioactivity	Shallow marine and nonmarine	Inorganic reactions	Molecules
3-2.1 (duration 0.9)	Biochemical	Few hours to days	Ultraviolet, infra-red, and visible light (photosynthesis) and organic compounds (heterotrophy)	Shallow marine Freshwater? Land?	Molecular differentiation, development of enzymes, reactions, photosynthesis; dominantly asexual reproduction, haploidy, early sexuality	Procaryotes (bacteria, blue-green algae)
2.1-0.6 (duration 1.5)	Intracellular	Hours to weeks	Visible light (photosynthesis) and organic compounds (heterotrophy)	Shallow marine Freshwater? Land?	Development of organelles (flagella, nuclei, plastids), asexual and sexual reproduction equally important	Eucaryotes (fungi, eucellular protists, multicellular algae)
0.6-present (duration 0.6)	Intercellular	Weeks to years	Plant photosynthesis (visible light), animal heterotrophy, and bacterial and fungal organic reduction	Marine Freshwater Land	Tissue differentiation, delayed growth, specialized reproductive cells and organs lengthened, diploid stage, sexual reproduction dominant	Metazoans and higher plants

plexity to protect the reproductive cells is at the expense of the reproductive rate. Such specialization may be of survival value for the species but not for the individual, for as the somatic cells become interdependent they also accumulate the adverse changes that result in aging, with eventual senescence and death of the individual (Strehler, 1959, p. 126).

Multicellular organisms have different abilities and respond differently to similar pressures than do the protists; hence the establishment of multicellularity would result in rapid exploitation of the new possibilities. As the advantages of multicellularity surpassed those of the protistan grade of organization, the exclusively algal-fungal-protozoan Precambrian biosphere was overshadowed by extensive metazoan diversification in the Cambrian.

Evolutionary Mechanisms

Evolution requires a suitable environment at the proper time and an organism genetically capable of adapting to it. The evolutionary process and the rate of change are influenced by the rate of mutation and effect of a single mutation (factors that supply the raw material for recombination and selection) and the genetic recombinations (by sexual reproduction) and selective processes. Selection may involve isolation of part of the population and may reflect either the physical environment or biological interactions. The relative importance of these factors not only varies from one set of circumstances to another but also differs somewhat between plants and animals in general and between protists and multicellular organisms.

Mutation and Recombination. Mutations occur regularly in all organisms, at a rate per gene of about once in every 10^{-5} to 10^{-7} cells. Some may result from copying error, but the rate is increased by irradiation, high temperatures, or chemical mutagens. The constant mutation rate due to the earth's natural background radiation has allowed most large multicellular organisms to accumulate a great genetic variability. For these, recombination is believed to be more effective in their continued evolution than the incidence of new mutations. However, Haldane (1949) noted that observed mutation rates could alone account for evolutionary changes and should not be neglected merely because they are slow by laboratory standards.

Important differences in the organisms themselves result in the reliance on recombination by metazoan animals, whereas mutations may be of greater importance in plants and in protists in general. Because of their very small size, asexual reproduction, which does not require two parents, is more certain for protistans. Mutational rate is correlative to generations; hence it is more effective in those organisms with short generation time and large, rapidly produced populations. Haploidy also emphasizes the mutational effect, for all genetic characteristics are expressed by the haploid organism or generation. Unfavorable mutations are immediately eliminated, not retained as recessives. Viable mutations may be clonally reproduced for many generations without recombination, and the danger of incompatible mutations in sexual partners is thereby avoided. Leedale (1967, p. 8) noted that because of the complete absence of sexual reproduction in the Euglenophyta, asexual clonal reproduction requires definition of euglenoid "species" on morphology alone, "without recourse to interbreeding capabilities as a criterion."

Although mutation may occur in any cell, only those in the germ cell will affect the next generation. The human adult has about 10^{12} cells, which requires a minimum of 40 cell generations (doubling the number with each division). Thus the chance of an important mutation occurring in the one necessary metazoan cell is relatively slight. In contrast, every mutational change will be in the protistan "germ cell." The effect of the mutation is always expressed for unicellular organisms, and cannot be expressed only in postreproductive senescence; typically the entire protistan organism is involved in reproduction, after which it ceases to exist.

Mutations are commonly stated to be predominantly unfavorable, as in most higher organisms viable possibilities have already been tested. Nevertheless, mutation is the basic source of genetic variability, upon which recombination depends. East (1936, p. 155) suggested that the destructive mutations are more readily observed, whereas numerous constructive mutations may remain unnoticed yet be of considerable importance to the organism.

When organisms are well-adapted to a stable environment, selection favors the reduction of change. Mutations may supply sufficient genetic variabilty in the rapidly reproducing protists and hence asexual reproduction is favored. Sexual reproduction may be infrequently interspersed between many asexual generations or even suppressed entirely.

The foraminifer *Rosalina* reproduces only asexually in the cooler temperatures of the higher latitudes of its range or in the laboratory at temperatures up to 16° to 17° C. This foraminifer seemingly is adapted well to its environment and variation is thereby restricted. In warmer waters, which are generally associated with more changeable conditions, the sexual stage is also present. This can be triggered in the laboratory cultures by raising the temperature to 18° C (Sliter, 1965).

Many plant protists also vary their reproductive process with environmental change. The dinoflagellate *Amphidinium carterae* was maintained in cultures by Vien (1967) in an alternating light-dark cycle at a temperature of 17° C. Sexual reproduction produced immobile zygotes. Only when the temperature was raised to 20° C did the zygote germinate to produce the four new individuals.

Thus the long generation time, numerous cell divisions before maturity, differentiation of somatic and reproductive cells, diploidy with dominant and recessive genes, and smaller populations of large multicellular organisms all tend to place greater evolutionary emphasis on sexual reproduction and recombination. In contrast, the short generation time, rapid turnover of genetic material, haploidy, small size, and large populations result in a greater reliance by the protists on asexual reproduction. The variation induced by mutation then is relatively more effective in determining evolutionary change.

Size in the Protists. Protists usually are regarded as having evolved little throughout geologic time, because of their small size and unicellular state. Large size is advantageous to a multicellular organism; hence many lineages show progressive size increase. Bonner (1965) suggested that small and large organisms have little conflict. Because the ecologic niches for small organisms were filled first and became available only through changed conditions or catastrophes, development of new forms favored an increased size. Large size also requires more food and leads to a decreased average number of individuals. Both variation and evolutionary rate are affected by population size. Large size in animals is correlative also with greater cell differentiation; hence selection for increased size is also for increased complexity.

The small size of protists is not absolute, for the size range is large (Table 3). Adult Foraminifera may range between 20 μ and 110 mm in maximum test size (the continually changing pseudopodial network is here disregarded, although this may extend five or more times the diameter of the shell). Such a 5500-fold size

TABLE 3. SIZE RANGE OF PROTISTS*

		volume (cubic microns)	
	Taxon	Normal range	Extreme limits
Eucaryotes	Protozoa	10,000-50,000	20-150,000,000
	Yeasts	20-50	20-50
	Unicellular algae	5-15,000	5-100,000
Procaryotes	Blue-green algae	5-50	0.1-5,000
	Bacteria	1-5	0.1-5,000
Viruses	Pox virus	0.01	0.00001-0.01
	Polio virus	0.00001	

*Data *from* Stanier and others, 1963.

range is greater than that from the 1-cm goby fish to the 34-m sulfur bottom whale (a range of 3400 times). Other protists have similar broad size ranges. Coccolithophorids range from 5 to 100 μ, diatoms from 2 to 1000 μ, dinoflagellates from 10 to 2000 μ. Although this size range is great, the protists seldom surpass 1 or 2 mm; hence selection must have demonstrated an advantage in small size.

The larger a sphere, the greater is the volume/surface ratio. Food assimilation and gas exchange of a cell are related to the surface area, but metabolic functions are related to volume. As a consequence, diffusion is effective in organisms of 1-mm radius, but greater volume requires either a convoluted surface or a transport system (Bonner, 1962).

The rounded or ovate coccoid or flagellate cells are commonly of lesser size than the rhizopods (for example, Foraminifera and Radiolaria), which attain a greater surface-to-volume ratio by means of their network of pseudopodia. Algal modifications for increased surface include cell elongation, invagination of the surface, and the development of various ridges or other projections.

Sizc minimum is equally limited in organisms. The lower size range of bacteria is controlled by molecular limitations, for in spite of their lack of cytological complexity, a bacterial cell is biochemically complex. Minimum size of higher or eucaryotic protists is related to the minimum size and number of the necessary organelles (for example, nucleus, plastids). Multicellular organisms are in turn limited by the necessary variety of tissues and organs and their minimum number of cells.

Population Size. Unicellularity and hence limited size are of selective value in an environment requiring rapid turnover. Seasonal variations in light, temperature, and perhaps necessary nutrients require similar seasonal population fluctuations of the phytoplankton. If the species were to remain at its lower population size, it would soon disappear, by loss to predators or failure to compensate for those sinking below the lighted zone. Because of the requirement of numerous cell generations during growth, the seasonal population growth of metazoan zooplankton commonly is somewhat behind that of the phytoplankton, thus allowing prior development of its food supply.

Although most habitats have been completely occupied in time, the resident populations fluctuated widely in number in response to various physical and biological factors. A terrestrial desert can accommodate a much smaller biomass than can a tropical rain forest. Similar variations in population size occur in the marine environment, where temperature, nutrients, light, salinity, and biologic interactions replace moisture as the limiting factor. Reduction in number may decrease the variability and result in isolation, tending to accelerate the evolutionary process.

Population size may be enforced upon the organism by the environment, or may result from an internal regulatory mechanism, such as the territoriality shown by higher animals. Crowding may be prevented in some phytoplankton species by their inhibitory excretions. These may repel only other species or may be self-inhibiting. The green alga *Chlorella* excretes the organic substance chlorellin, which is stimulatory to its growth when population levels are low but becomes an inhibitor in high concentration. It also may be an antibacterial agent. Similar excretions (nitrogenous compounds or vitamin B complex) may selectively inhibit or stimulate other species. Some blue-greens (*Anabaena, Microcystis*) excrete substances that are lethal to other algae; and the toxins of some blue-greens and the dinoflagellates *Gonyaulax* and *Gymnodinium* may accumulate to lethal levels in grazing or filter-feeding animals during the periodic blooms, or "red tides" (Saunders, 1957).

Bonner (1965, p. 197) stated that in a variable environment the population size of a small species will increase or drop in response to the slightest external change. Changes in temperature or salinity, regions of upwelling, or the spring influx of nutrients due to increased runoff may result in phytoplankton blooms. Successive changes in these factors result in a seasonal succession of species. Because large multicellular organisms can internally adapt to their environment and somewhat modify its effects, they fluctuate less rapidly in numbers. Smaller organisms are most subject to catastrophic population decreases; hence the selection pressure for higher reproductive rates necessary for repopulation may also result in large standing populations.

Mechanisms of Isolation. When an opportunity for expansion is available, a population with large genetic variability has a better chance of possessing the necessary requirements. Too great a varibility would be detrimental, as even favorable characters could not become established. Although chance selection in the sexual reproductive process results in less variability in a population than the number of theoretically possible combinations, rapid evolution can occur only if the variability is further limited by isolating mechanisms.

Both speciation and the origin of higher taxonomic categories depend upon nearly complete isolation of a portion of the population. Selection then may act upon the limited genetic variability of this particular group of individuals. Isolation of terrestrial organisms is commonly regarded as geographically initiated. Similar ecologic barriers are recognizable for the shallow-water marine benthos but are less evident for pelagic organisms; hence pelagic species have been regarded as geographically wide-ranging and slow to evolve. The exceptional value of pelagic organisms for stratigraphic delineation (graptolites, ammonites, conodonts, planktonic foraminifers, coccolithophorids, dinoflagellates, and other microplankton) indicates that on the contrary they evolved rapidly.

Mechanisms of isolation dominating planktonic metazoan speciation (temperature or depth barriers, small populations, and so on) are not necessarily applicable to protists. The small individual size, large populations, and cosmopolitan distribution of much of the protist plankton, at least within certain limits, suggests that asexual clonal reproduction may be of greater importance in isolation, as indicated earlier. Selective pressure on different clones allows the favored genetic complement to quickly replace that less adapted. Sexual reproduction and cyst formation in neritic species may accompany the onset of unfavorable conditions. Such a benthic resting stage is of little value for the oceanic plankton, where conditions are more stable, and would in most instances be useless because of the greater depths involved and difficulty of returning to the photic zone after excystment.

Phytoplankton may also be affected by geographic isolation resulting from a particular salinity or temperature tolerance, depth stratification, or rate of sinking. Sinking may be retarded by lighter cytoplasmic inclusions or cell sap, varied skeletal morphology, or by turbulence, upwelling, and water temperature stratification. Spatial isolation of individuals may result from nutrient limitation or the effect of inhibiting or toxic excreted metabolites.

INTERACTIONS OF PHYTOPLANKTON AND ENVIRONMENT

Phytoplankton has both evolved and varied quantitatively through time. The record of past microfloras has proven to be far more extensive than was previously suspected, but the stratigrapher must also consider whether the presence or absence of given taxa in particular strata is due to their nonexistence at that geologic moment or to paleoecologic restriction. As recently as a dozen years ago, marine phytoplankton ecology was regarded as a neglected field urgently needing study (Braarud, 1954, p. 178). Since that time, many experimental data have been obtained from laboratory cultures that can be applied not only to paleoecologic interpretation but also to a wide variety of biological, geochemical, and sedimentological problems.

Phytoplankton Autecology

Although exponential growth in cultures may cease because of exhaustion of a nutrient (nitrate, iron, vitamin B_{12}, and so on) or the supply of CO_2 or O_2, reduction of light by self-shading, alteration of *p*H through use of certain constituents of the medium, or the accumulated inhibitory excretions, these data may be difficult to apply in oceanic conditions.

Modern phytoplankton assemblages, like those of all organisms, are affected by physical factors of the environment. The environmental conditions and the particular reactions of the different species are so diverse that each living species has a distinct distribution. The importance of such general physical features as temperature, oxygenation, radiation, salinity, *p*H, lateral and vertical water movements, and abundance and chemical form of nutrients is well recognized; hence they are discussed in detail. Some factors are of particular importance to the phytoplankton however, or are suggested to have geobiologic applications.

Light. All actively photosynthetic organisms are restricted to the photic zone. The various pigments of different algae utilize light of various wavelengths, the filtering effect of seawater resulting in a depth stratification.

Chlorophyll *a* is present in all algal groups. In addition, chlorophyll *b* occurs in the Chlorophyta (green algae), euglenids, charophytes, and higher green plants. Light utilized by these pigments is rapidly filtered out by the upper water layers: hence they are less effective in deeper water. Green plants are therefore more characteristic of fresh water and the terrestrial environment and less important in the sea. The Siphonales, Siphonocladales, and Dasycladales are exclusively marine Chlorophyta, although other marine species occur.

Algal phyla characterized by abundant accessory carotenes and xanthophylls have chlorophyll *c* in place of *b*. The different absorption bands of this pigment allow use of the more deeply penetrating light rays. Thus the brown algae (Phaeophyta), golden-brown algae (Chrysophyta), including chrysomonads, silicoflagellates, and coccolithophorids, the diatoms (Bacillariophyta), and dinoflagellates (Pyrrhophyta) are most characteristic of the oceanic habitat. Diatoms, dinoflagellates, and coccolithophorids also occur in fresh water but are generally subordinate in this environment to the blue-green and green algae.

Depth stratification in the sea results not only from the optimum photosynthetic capabilities of the different pigments but also from photo-oxidation by excessive amounts of light. As the photosynthetic organism is susceptible to destruction by intense light, maximum phytoplankton density commonly lies some meters below the surface. Some species show a vertical diurnal migration in response to light intensity. In tropical areas light penetration is greater, and phytoplankton are abundant to about 150 m in summer, as compared to 30 to 50 m in temperate regions. Light is much lessened or even lacking during part of the year in high latitudes; hence polar species must persist for various periods with almost no photosynthesis. This ability is perhaps as important a factor as temperature in their distribution. Some species survive the winter period in a resting stage, but other species heterotrophically utilize dissolved or particulate organic matter in the sea during periods when light is insufficient.

Culture studies have demonstrated the wide adaptability of many phytoplankton species, whose successive generations are increasingly tolerant of changed temperature or light or normally inhibiting secretions. Adaptation may be rapid. Growth commonly is related to light intensity, yet if other factors remain constant the algae adapt to different light intensities. Fogg (1965, p. 23) noted that during exponential growth *Chlorella* becomes adapted to low or high light in about one division cycle (17 hours).

Phytoplankton that have a partially heterotrophic nutrition may also live in deeper levels of the ocean. Some have even adapted to complete darkness and are soil dwellers, occur in contaminated waters, or live in interstitial waters of the sand in a littoral marine habitat. The chrysophyte *Ochromonas* is capable of heterotrophic, chemotrophic, or autotrophic nutrition, growing most rapidly when utilizing organic substrates in addition to photosynthesis and more slowly when utilizing light and CO_2 alone. This chrysophyte resembles a primitive animal that has retained the ability to photosynthesize if meals are too widely spaced (Fogg, 1965).

Nutrients. Plants manufacture their food by means of photosynthesis. Like any manufacturing process, this requires a constant supply of raw materials. Unlike conditions in a controlled laboratory culture, nutrient depletion is difficult to measure in the sea. Low quantities may indicate a balance between consumption and supply rather than a limited supply.

The most important elements, needed in largest amount by all organisms, are hydrogen, oxygen, carbon, and nitrogen, the latter a constituent of protein and nucleic acids. Sulfur and phosphorus are needed in somewhat lesser quantities for proteins and nucleic acids. Very minor amounts of potassium, magnesium (in chlorophyll), manganese, calcium, iron, cobalt, copper, zinc, and molybdenum are used in certain enzymatic reactions by most organisms. Some organisms have special requirements, as that of diatoms for silicon and of vertebrates for iodine. Although the same chemical elements are involved, various organisms have become adapted to the use of particular compounds, which they obtain in different ways. Water is the most necessary compound, but some bacteria and blue-greens can use molecular hydrogen, and bacteria may even use hydrogen sulfide as a hydrogen donor in photosynthetic carbon reduction. Molecular oxygen is used by all aerobes but cannot be used by the anaerobic bacteria. Photsynthesizers use CO_2 to build a variety of organic molecules, but some also require various preformed organic compounds. Some bacteria use acetic acid as their only carbon source; other plants require certain vitamins and sugar complexes or may even need a wide range of preformed organic molecules. Nitrogen is available to some only in the form of nitrates;

by others ammonia is used, and others assimilate molecular nitrogen from the atmosphere. The differing requirements and tolerances commonly result in a seasonal succession of species. Neritic species also have more extensive requirements than do oceanic ones; the latter have developed methods of trapping minute quantities and would find toxic the increased amounts near shore.

Most required elements are present in sufficient quantities in the seas, although in varying concentrations. The amount of fixed nitrogen and phosphorus, for example, is about 100 times that annually used by phytoplankton. According to Provasoli (1958), assay experiments indicate that many flagellates require three molecules of vitamin B_{12} per cubic micron of volume. The amount may be greater at warmer temperatures by several hundred times. Nevertheless, comparison of the biomass to the lowest measured B_{12} quantities in seawater suggests that the excretion of this vitamin by various bacteria and blue-green algae provides enough in the sea at all times.

In general, the loss of nutrients to the sediments and air is balanced by the influx from land, with the oceanic cycle of use and regeneration resulting in a steady state. Nutrient regeneration in the depths occurs through the heterotrophic metabolism of various forms of animal life, as well as some facultative heterotrophic flagellates (coccolithophorids and dinoflagellates are reported to occur to depths of 4000 m), blue-green algae, diatoms, bacteria, yeasts, and fungi (Wood, 1963). Extensive grazing by zooplankton and other animal life, together with bacterial action, recycles much of this material while still within the photic zone.

Excreted compounds were suggested above as a factor in isolation of phytoplankton individuals. These external metabolites may be of considerable quantitative importance in supplying dissolved organic material to the sea (Fogg, 1966). Some chlamydomonads excrete more than 25 percent of their photosynthetic products under high light intensity (Coleman, 1961). These substances may attract other organisms, to result in a form of mutualism or near-symbiotic relationship, as in certain diatoms and bacteria, or tintinnids and diatoms. Lucas (1955, p. 143) noted that external metabolites of some dinoflagellates stimulate the feeding movements of barnacles and the pumping rates of oysters. Salmon and other fish appear to recognize streams by the external metabolites in their waters.

The excretions may also repel other organisms, being inhibiting (antibacterial secretions of some blue-greens) or highly toxic, as are those of various blue-greens and dinoflagellates (Nielsen, 1955; Redfield and others, 1963). Bacteria and algae also can adapt to accumulations of various antimetabolites and antibiotics. Successive subcultures of the blue-green *Anacystis* subjected to gradually increased amounts of streptomycin produce strains that are resistant to 50,000 times the quantity formerly tolerated (Fogg, 1965).

$CaCO_3$ and Silica. The concentration of $CaCO_3$ and silicic acid may limit skeleton-secreting phytoplankton. Diatom growth in cultures ceases when the silica supply is exhausted (Busby and Lewin, 1967). The quantities in the ocean are not as low, however, and organisms can greatly concentrate small amounts. Dissolved CO_2 and *p*H determine the preservation or regeneration by solution of this skeletal material if the supply is limited.

Protistan skeletal deposition is largely an internal process. No organism can form crystals of shape foreign to inorganic crystallography unless the mineralization occurs within an organic membrane, where the concentration of the mineral is increased above that of the seawater. This organic component or covering similarly is a protection from skeletal solution during life of the organism.

The ornate skeletons of opaline silica characteristic of the diatoms, silicoflagellates, ebridians, and the dinoflagellate *Actiniscus,* as well as those of radiolarians or thecamoebians, are internally formed.

The diatom silica shell is deposited within a membrane-bounded vesicle and is entirely enclosed by organic material (Reimann and others, 1966). The degree of silicification in different species is genetically controlled, as the same plankton sample may contain both heavy-shelled species (for example, *Coscinodiscus*) and weakly silicified ones (for example, *Chaetoceros*).

Different foraminiferal taxa in an identical environment will selectively form tests of calcite or aragonite deposited on an organic membrane. The mineralogical form is controlled by internal cell chemistry rather than by the external environment. The crystal form of coccoliths is also organically controlled. The ornate heterococcoliths have a characteristic crystal structure, in which crystals have continued to grow laterally, overlap, and become highly modified. In some, the basic crystal form remains identifiable; in others, such as *Emiliania huxleyi,* the coccolith is constructed of numerous radially arranged units in which the crystal form is unrecognizable. These units are shown by X-ray diffraction analysis to represent separate single crystals of calcite, in the form of a hollow crystal, with C axes in the plane of the coccolith (Watabe, 1967). The coccoliths were formed by deposition of calcite on an internal organic "mold" and later pushed to the cell exterior (Isenberg and Lavine, 1964). Holococcoliths may occur in alternate generations of some species. They consist of unmodified hexagonal or rhombohedral crystals of calcite, deposited externally on the surface of organic scales that were previously internally formed. Possibly because of the small size of the crystals and their more open arrangement, or the absence of organic components, these holococcoliths rapidly disintegrate and few occur in bottom sediments (Parke and Adams, 1960).

Lime is more nearly saturated in the ocean than in fresh water; hence calcite-depositing phytoplankton (coccolithophorids) and protozoans (foraminifers) are most characteristic of the marine environment. Coccolithophorids also occur in fresh water but are less abundant than the silica-secreting Chrysophyta (chrysomonads) and Bacillariophyta (diatoms). The freshwater skeleton-secreting Protozoa similarly are siliceous (thecamoebians and heliozoans) rather than calcareous.

Plankton samples taken in the present oceans may contain both siliceous (diatoms, silicoflagellates, radiolarians) and calcareous components (coccolithophorids, foraminifers), but the recent bottom sediments and the sedimentary rocks exposed on the continents commonly are dominated by one or the other to a greater extent than would be expected on the basis of present distributions. This reflects selective preservation rather than original availability of these compounds.

Temperature. High-latitude species are tolerant of low temperatures and high, constant salinity (over 31# 0/00 salinity with a1 mean of about 33.88 0/00 salinity). The Antarctic contains quite distinctive species, but the Gulf Stream at present modifies the North Atlantic and Arctic regions. Diatoms are most characteristic of the cold-water assemblages, as only a few dinoflagellates and probably no coccolithophorids meet the tolerance requirements. The latter may occur seasonally in higher latitudes, due either to annual influx or to reseeded populations.

Tropical species must withstand high temperatures, and many can tolerate variable salinity. Tropical assemblages are dominated at present by coccolithophorids and dinoflagellates. Only a few pelagic diatoms are truly tropical, although temperate species may range into the tropics. In the Pacific Ocean, diatom distribution appears to be more closely related to nutrient supply or *p*H than to temperature alone, for the gradual decrease in diatom populations from the poles toward the lower latitudes changes at the equator. From about 5° S. to 6°N., diatoms again occur in large numbers. Both large, heavily silicified forms and some lightly silicified species (*Nitzschia* spp., *Rhizosolenia*) occur in the plankton, but the latter dissolve readily and are not represented in the sediments. Kozlova and Mukhina (1967) regard the increased phosphate content of the waters as responsible for the diatom abundance locally in low latitudes.

Most cosmopolitan species, which may be circumglobal, are neritic and have a wide temperature tolerance. Although zooplankton can live at deeper levels in low latitudes and near the surface in high latitudes to maintain similar environmental temperatures, phytoplankton must remain in the upper lighted zone. Important cosmopolitan species today include many diatoms (for example, *Chaetoceros compressus* and *Skeletonema costatum*), some dinoflagellates (for example, *Exuviaella baltica* and *Prorocentrum micans*), and the coccolithophorid *Emiliania huxleyi* (Samayda, 1958).

Water Depth and Distance from Shore. Phytoplankton species may be confined to coastal and inshore regions (neritic) or be present in the open ocean, where coastal influence is minimal. These habitats are not simple, and the distribution may in fact result from a combination of many different factors (Smayda, 1958). Coastal species are strongly influenced by their proximity to shore as this is reflected by depth, supply of nutrients, and growth factors. Other species live in the open sea and are restricted to areas free of coastal influence. Some are indifferent to coastal influence and have fewer requirements and greater tolerance so that they may exist in either environment.

Environment and life cycle may also be interrelated. Some coastal planktonic species have a sedentary stage, form resting spores, or have a dormant pelagic cyst during the life cycle. Such stages are related to temperature or salinity control and commonly coincide with oceanographic factors that cause accelerated sinking rates. The benthic or encysted stages must be followed by a return to the photic zone, as a result of changes in water density or turbulence, in order to allow reseeding of the population (Braarud, 1962); hence such species are in general restricted to relatively shallow depths. Not all coastal forms are yet known to possess benthic stages or cysts, but increasing numbers of these now are recognized. In contrast, the continuously pelagic species in the open-sea environment are independent of the bottom, although a few may have a planktonic resting stage.

Nearly all major phytoplankton groups include cyst-forming coastal species, and their recognition would have an important paleoecologic application. Typically, oceanic diatoms and coccolithophorids are well-represented as fossils in deeper water deposits, such as the coccolith-nannoconid-tintinnid limestones of the Alpine Jurassic-Cretaceous, or Tertiary oceanic cores. In contrast, all dinoflagellates in the fossil record were of more neritic habit, for their only preservable structures are the resting stages. Ancient water depths or original distance from shore may thus be postulated from the enclosed phytoplankton assemblage. Similarly, Holocene open-ocean deposits contain no organic microplankton remains, although both calcareous and siliceous skeletons of other taxa may be abundant. We examined for dinoflagellates numerous samples from the JOIDES (Joint Oceanographic Institutions' Deep Earth Sampling) ocean coring program off the eastern coast of Florida. Dinoflagellates proved to be abundant in material from those holes on the continental shelf and slope nearest to the present coast, where cyst-producing species could exist. None was found in material from the deeper offshore locations on the Blake Plateau. Coccolithophorids and planktonic Foraminifera occurred in all locations. Many oceanic cores from other areas lack dinoflagellate remains, yet obviously their absence from the deep-water sediments does not indicate conditions unfavorable to microplankton in general nor local extinctions of dinoflagellate taxa—only the absence of species that produce preservable organic cysts.

Flotation and Sinking. Braarud (1962) suggested that means of vertical distribution differed for motile and passive phytoplankton. Although motile species might have a morphologic adaptation for flotation in addition to a means of locomotion, the morphologic or physiologic adaptation is of greater relative importance for the nonmotile species.

Many phytoplankton species (diatoms, some coccolithophorids, and the like) are nonmotile and tend to sink in the water. Spines and modifications of cell shape have been suggested as possible adaptations to aid flotation. However, the smallest phytoplankton cells may be retained in the photic zone by water turbulence as well as by flotational mechanisms or reduced specific gravity

(Fogg, 1965). The supposed adaptations for flotation may be of nutritional value, however, in increasing the cell surface-to-mass ratio. This facilitates absorption of necessary nutrients. If turbulence is the controlling physical factor, size and surface area are of prime importance. Yentsch (1963) observed that sinking is best retarded if the ratio of surface area to volume is large, and hence isolated small cells have greater flotation capability than aggregates of cells. The colonial or palmella stage of a coccolithophorid has a lower surface-to-volume ratio than do the isolated cells and has an increased sinking rate. Bernard (1963) noted that *Cyclococcolithus* sank at a rate ranging from 4 to 260 m/hour, according to the size of the cell cluster. The sinking may be of some nutritional value, as access to additional nutrients may be obtained as the cell moves vertically through the water. In other phytoplankton, sinking rate is related to the general physiological condition of the cell, for in the laboratory it is greatest in old cultures, in nutrient-rich media, and in darkness. Gas vacuoles may be formed within some cells, particularly in the blue-green species. Others accumulate fats and oils in the cell. The green alga *Botryococcus* has a reduced specific gravity because of the large fat percentage of the cell, but in most other phytoplankton such high fat and oil production is related to abnormal cellular metabolism due to nutrient depletion and is not an adaptation for flotation. The chlorophyte affinities of *Botryococcus,* demonstrated by Blackburn (1936, p. 851-853), are generally accepted (Papenfuss, 1955; Silva, 1962; Fogg, 1965), although elsewhere it has been included with the Xanthophyta.

Phytoplankton Productivity

Present Productivity. The oceans now cover 70 percent of the earth's surface, of which 2 percent is occupied by the littoral flora. As phytoplankton occupies well over two-thirds of the global area, it is responsible for the major part of the organic production in the sea (Braarud, 1954, p. 178). Oceanic production was estimated by Riley (1944) as fixing between 44 and 208 billion tons of carbon annually (2 to 10 times the estimated production of all land areas). Nielsen and Jensen (1959, p. 89) regarded Riley's estimate as high, and, basing productivity estimates on a world-encircling sampling cruise, suggested that annual net oceanic production was about 12 to 15 billion tons of carbon, approximately equal to that of the land. Ryther (1959) noted that the single observations for each station made by Nielsen and Jensen rarely if ever included seasonal maxima. Detailed measurements suggested that average annual production in the relatively barren Sargasso Sea was about six times as great as Nielsen and Jensen had stated, and even twice as high as their estimate for the oceans as a whole. Ryther concluded (1959, p. 608) that the seas were at least more than twice as productive as the land. Yentsch (1963, p. 172) quoted recent estimates of oceanic production as 2 to 3 times that of the land. Strickland (1960, p. 95) summarized problems of obtaining accurate productivity estimates but concluded that annual oceanic production was at least 20 billion tons of carbon, and perhaps several times greater. Phytoplankton production thus is between 50 and 90 percent of the earth's total. Both complex food webs and biologic recycling of atmospheric oxygen is largely dependent upon phytoplankton photosynthesis.

The importance of the phytoplankton in global productivity, in spite of the tiny cell size, results from the great abundance. Smayda (1966) recorded diatom numbers in the Gulf of Panama of 4.6 million cells per liter of seawater. Lackey (1964) reported 1.8 million cells per liter of the dinoflagellate *Gymnodinium flavum* during a bloom in Mission Bay, San Diego, with a total protist cell count per liter of 3.5 million. In 1947, *G. breve* occurred in Florida in numbers up to 60 million per liter. Coccolithophorid records of 115 million cells per liter report a resultant "white water" similar to the "red water" or "red tides" of the dinoflagellates and blue-green algae. Bloom visibility depends upon both the cell count and their relative size (biomass), about 6000 cells/ml, of *Peridinium triquetrum*, but only 20 cells of *Noctiluca*, being necessary for visible discoloration of the sea.

In general, the smaller species (nannoplankters) with greater proportional surface area assimilate more carbon per unit of biomass than do the larger diatoms; hence qualitative counts are equally as important as quantitative ones. Fogg (1965) stated that more than one-half of the photosynthesis is by algae small enough to pass through a net of 35 μ mesh.

Exact production figures cannot be obtained by presently known methods. Although the standing phytoplankton crop can be measured locally, the smaller and more delicate, yet most productive, species are commonly overlooked; measurement of chlorophyll *a* content may be complicated by the presence of detrital chlorophyll and disregards the important photosynthetically active accessory pigments of marine phytoplankton, especially chlorophyll *c* and the carotenoids (for example, peridinin and fucoxanthin) of the more abundant marine brown and yellow groups, coccolithophorids, diatoms, and dinoflagellates. Rate of turnover is more important than biomass at a given time, but attempts to measure oxygen production, the uptake of nitrates and phosphates, or use of ^{14}C-labeled compounds all suffer from technical difficulties related to bacterial increase, disruption of the normal cycle by enclosure in bottles, organic recycling during the measurements, and isotope discrimination by the organisms for organic carbon or calcium carbonate (Strickland, 1960; Raymont, 1966). Distribution in nature is also extremely spotty, plankton samples a few meters apart routinely differing by 20 to 50 percent (Verduin, 1964). Seasonal succession is an additional factor, temperate areas commonly having spring and fall blooms and lower summer and winter productivity, whereas polar areas may have a single peak of production. Completely different assemblages may be involved through the year. Diurnal variations and depth stratification add further complications.

Diversity and abundance are not in general correlative. Hulburt (1963) observed that when growth conditions were poor, diversity was greatest. The few specimens represented a considerable number of species. As conditions became favorable to growth, one or two species rapidly gained dominance to the exclusion of others. Some maxima include a greater variety. Hart (1966) recorded a bloom of 2,315,000 cells per liter, in which ten species were present in more than 5000 cells each; of these four had more than 150,000 cells each.

Areas of richest oceanic productivity were identified by Hart (1966) as temperate coastal waters during spring (and sometimes fall) maxima, subpolar and polar waters that showed exaggerated seasonal effects, and warmer seas where turbulence maintains nutrient levels in the euphotic zone, as in upwelling areas of the western coasts of continents, the coasts of monsoon areas, current boundaries, or the wake streams of islands.

High latitudes are commonly characterized by large standing crops, but, because of the short duration of the growing period, the total annual productivity is not as high as in tropical regions where light is present over longer periods and penetrates to greater depths. The smaller nannoplankton of the lower latitudes represents a less impressive standing crop, but its more rapid growth rate results in a more rapid cycling. A close balance of production and consumption by grazing zooplankton also might result in a small standing crop even with rapid cycling and high productivity (Raymont, 1966). The true relative productivity may be indicated by the increase in bacterial count from the polar regions to the equator. Heterotrophic bacteria are indicators of organic matter, their abundance in warmer water being dependent upon the greater productivity. The bacterial maximum abundance lies about 10 to 40 m vertically below the phytoplankton maximum. Sinking and dead cells are utilized by bacterial growth and contribute to the rapid nutrient cycling (Wood, 1963).

Productivity Changes in the Past. Just as an accurate quantitative indication of present productivity is difficult if not impossible to obtain, so also is exact quantitative production of past geologic time difficult to determine. Present abundance and rate of increase of modern microfloras can merely indicate probabilities in past cycling, with the added difficulty that relative abundance and the over-all composition of microfloral assemblages have also changed through time.

Nevertheless, as indicated earlier, quantitative changes in dominant assemblages are recognized. The organic-walled Paleozoic phytoplankton was eventually replaced by the Mesozoic calcareous nannoplankton, and this in turn by siliceous Cenozoic microfloras. The changes were relatively abrupt; in each case the formerly abundant dominant component was nearly completely eliminated far in advance of the rise to prominence of the new group.

Acritarchs were extremely abundant for about 150 m.y., in the Ordovician, Silurian, and Devonian, but dropped sharply in both numbers and variety toward the end of the Devonian. Possible fluctuations within the lower Paleozoic need further documentation, but the acritarchs are generally scarce after the Devonian. Locally the green algal *Tasmanites* and related forms (*see* Wall, 1962, and Parke, 1966, for discussion of the relationships of *Tasmanites* and the modern *Pachysphaera*) were abundant in late Paleozoic shallow waters, but most phytoplankton were scarce throughout the late Paleozoic and Triassic.

The appearance in the Jurassic of calcareous nannoplankton with the accompanying newly important organic-walled dinoflagellates thus was not a replacement of one competing group by another but a new radiation into a long-vacated niche. This microflora was dominant for some 120 m.y. through the Cretaceous and then it, too, had a sharp decline. Microfloras were extremely sparse in the earliest Paleocene. The siliceous plankton (particularly the diatoms) rose to importance in the late Paleocene and Eocene, accompanied by a renewal of dinoflagellates and calcareous microplankton, including the new discoasters.

Geographically, stratigraphically, and lithologically varied material suggests that the general distribution of the major phytoplankton groups now has been accurately delineated, although many additional assemblages have yet to be described, countless taxa require detailed examination and characterization, and quantitative studies have yet to be initiated.

We indicated above that most of the major groups of plant protists appear now to be in a phase of decline, their present variety and abundance being much below their respective levels in the Late Cretaceous and early and middle Cenozoic. As phytoplankton productivity is of primary importance in the global economy today, it must have been relatively far more important at earlier times of more extensive seas and maximum abundance. Such marked changes in microfloras and thus in total productivity in the geologic past must have drastically affected the carbon and oxygen cycles and should therefore be recognizable by changes in associated biota and contemporary sediments.

Geochemical and Sedimentological Aspects of Productivity

Rock-forming properties of animals and plants are too well-known to need extensive elaboration, although the importance of various protists and their progressive changes in skeletal composition has not always been recognized. In the Paleozoic, phytoplankton fossil material occurs almost exclusively as organic carbon. Where the floras were abundant, sediments contain much dispersed organic material; locally tasmanite, boghead coal, and petroleum accumulated. The calcium carbonate deposition of the Precambrian and Paleozoic was dominantly due to benthonic algae (such as blue-greens and dasyclads) and invertebrates. A change from this shallow-water lime deposition to that of deep water in the Mesozoic accompanied the expansion of the lime-secreting planktonic protists, particularly coccolithophorids, but also including tintinnids and planktonic foraminifers. The resultant oceanic lime deposits were less likely to be exposed, eroded, and recycled than were the shallower water skeletal accumulations of benthic organisms. The rise of siliceous phytoplankton in the Late Cretaceous and its elevation to major importance in the Tertiary resulted in and from other extensive changes.

Geochemical aspects of phytoplankton production are not as widely recognized. Photosynthesis is the motivating force that turns the carbon and oxygen cycles, both of primary importance to all life on earth. With this trapping of solar energy for reduction of atmospheric CO_2 to organic compounds has also come the present oxygen-rich atmosphere.

A correlation was suggested (Tappan, 1968a, 1968b, and herein) between phytoplankton productivity and changes or fluctuations in atmospheric composition, oceanic *p*H, the deposition or solution of lime and silica, occurrence of sedimentary iron ores, bedded chert and gypsum, black shales, carbon and sulfur isotope variation in limestones and sulfates, and the accumulation of petroleum. Reduced phytoplankton populations on a world-wide basis are correlative with periods of maximum turnover of the biosphere, with widespread extinctions of some taxa and radiation of others.

pH and CO_2. Rubey (1951) stated that seawater varies little in its dissolved constituents, except for salinity and the amounts of bicarbonate and carbonate ions. These latter are largely controlled by hydrogen ion concentration (*p*H) and the atmospheric CO_2.

The present oceans and atmosphere together contain about 1.47 $\times 10^{20}$ grams of CO_2, more than 98 percent of this being dissolved in the ocean. Attainment of equilibrium with the atmospheric CO_2 is dependent upon the rate of turnover of the deeper waters, estimated to require between 2000 and 5000 years. These same interrelated factors, *p*H, $CO_3^=$, and HCO_3^-, are also among the most critical for a wide variety of organisms, and all are related to the photosynthetic process by its effect on the carbon dioxide-oxygen ratios.

The ocean normally is slightly alkaline, with a *p*H between 7.5 and 8.4, commonly between 8.1 and 8.3 near the surface and to about 100 m depth. Where dissolved CO_2 is decreased in quantity, as by a high rate of photosynthesis, the *p*H is elevated to about 8.4 and the number of carbonate and bicarbonate ions in solution is increased. The amount of dissolved CO_2 is higher below the photic zone, due to respiration and oxidation of organic matter. The oxygen is decreased and the *p*H reaches a minimum of about 7.5 at a depth ranging between 200 and 1200 m. Below this oxygen minimum zone, perhaps as a function of hydrostatic pressure

(Park, 1966), oxygen and *p*H gradually increase to a deep-sea maximum between 3500 and 4000 m. Similarly, if atmospheric CO_2 pressure is increased, dissolved CO_2 increases and *p*H drops.

Carbonate and Sulfate Deposition and Isotope Fractionation. If the rate of photosynthesis were to be considerably higher than at present, as was suggested for the lower Paleozoic or the Cretaceous, the amount of CO_2 in the oceanic reservoir would decrease, the *p*H would be elevated, and more extensive deposition of calcium carbonate would probably occur. On a smaller scale this does occur today, photosynthesis by algal symbionts stimulating the calcification by reef corals (Goreau, 1961, p. 270). Algal photosynthesis also causes external deposition of lime (for example, charophyte or siphonalean thalli), even where this is not internally secreted by the species. A drop in photosynthetic production (for example, late Paleozoic, end of Cretaceous) would have the opposite effect, as CO_2 would accumulate, the higher CO_2 pressure of the atmosphere and ocean resulting in slightly lowered *p*H and carbonate solution. Kennett (1966) noted that in the Ross Sea, Antarctica, foraminiferal distribution is controlled by a shallow calcium carbonate solution boundary at a depth of about 500 m. He suggested a relation to low temperature and high salinity and possibly to the higher concentrations of CO_2 resulting from restricted photosynthetic activity of phytoplankton because of aphotic conditions and an ice cover much of the year. A similar relation of CO_2 pressure and carbonate solution occurs on some carbonate reefs where high rates of respiration at night result in partial solution of lime deposited during the daylight.

Pytkowicz (1967) calculated the relative roles of carbonates and silicates in the CO_2 concentration in the oceans. Extensive solution of calcium carbonate at depth results in CO_2 regeneration, although the rate is too slow in comparison to that of the present overturn of the deeper waters to allow saturation to be attained. However, carbonates were demonstrated to be far more effective than silicates in the buffer mechanism of the oceans over a measurable time.

Tappan (1968b) suggested that heavy phytoplankton production in the late Cambrian, Ordovician, and Silurian probably resulted in a higher oxygen level in the atmosphere than at present. This seems to be supported by the great quantities of carbonates then deposited (for example, Arbuckle Group, Trenton Group, Niagara Group). The pre-Devonian low oxygen pressure postulated by Berkner and Marshall (1965) is controverted by these rocks, for the higher CO_2 and lowered *p*H that would have accompanied the low oxygenic levels would have prevented their accumulation.

Because of the linkage of photosynthetic carbon fixation, CO_2 pressure, oxygen production, *p*H, and dissolved $CO_3^=$ and HCO_3^-, times and areas of heavy phytoplankton production in the past generally have been associated with limestone production, petroleum accumulation, and oil shales (Ordovician, Silurian, Early Devonian, Cretaceous, Eocene, Miocene); those of low production had relatively little carbonate deposition but had increased bacterial activity, sulfate production, and lime solution (*see* Fig. 11).

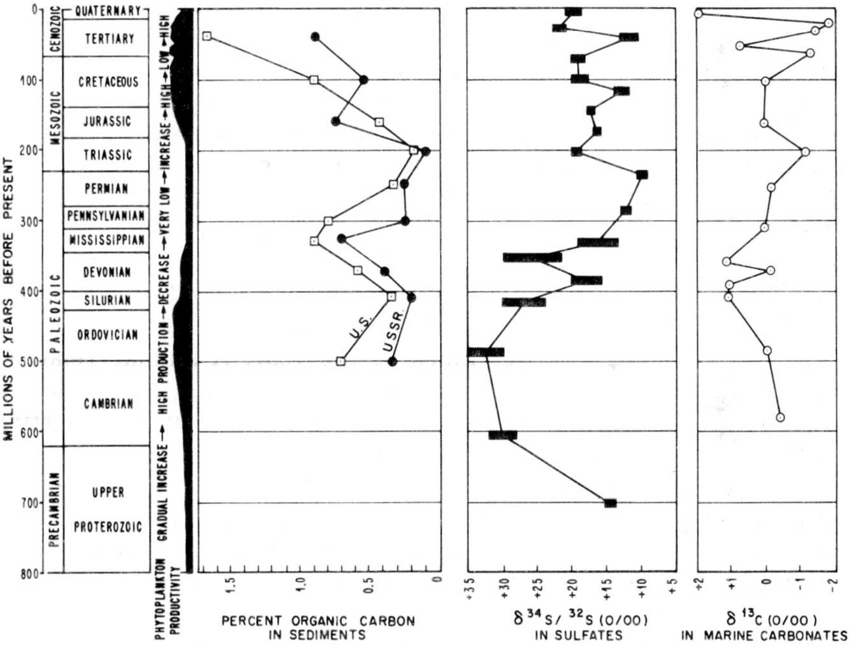

Figure 11. Correlation of phytoplankton productivity, percentage of organic carbon in sediments, and isotope ratios of sulfur in sulfates and carbon in carbonates. Note similarity of downward fluctuations of curves in Devonian, Permo-Triassic, Cretaceous, and mid-Tertiary, except where detail is obscured by averaging of entire period for organic matter in sediments. From top of figure, total phytoplankton distribution indicated as blackened area, based on Figures 5 to 10 and unpublished data. Percentage of organic carbon in sediments from Russian Platform and United States averaged by periods (data *from* Ronov, 1958). Sulfur isotope ratios in sulfates (data *from* Holser and Kaplan, 1966). Carbon isotope ratios for marine carbonates (data *from* Keith and Weber, 1964).

Photosynthetic carbon fixation is accompanied by isotope fractionation, with organic carbon enriched in ^{12}C as compared to the carbon of the oceanic reservoir. Phytoplankton-produced organic carbon is more enriched in ^{12}C than that of benthic algae but less than that of terrestrial plants (Sackett, 1964, p. 173). The organic carbon in shales and petroleum is thus ^{12}C-enriched, whereas marine limestones that are in equilibrium with the isotope ratio of the ocean reservoir are enriched in ^{13}C. Extensive changes in total productivity would modify the rate of this fractionation and hence the carbon isotopic ratio in the oceanic reservoir, even before the rise of carbonate-secreting phytoplankton or of terrestrial plants.

In an examination of isotope ratios of about 100 Australian carbonates and coals, Jeffery and others (1955) reported Cambrian limestones as isotopically light, a single Ordovician sample heavier, Silurian light, and Devonian carbonates heaviest, followed by increasingly lighter ones to the Lower Permian. Cretaceous limes were again heavy, and Tertiary ones (mostly Miocene) were lighter. Periodically increased volcanism and rock weathering related to orogenesis were suggested to have returned carbon to the system and thereby modified the isotopic ratios of atmospheric and oceanic carbon. This was in turn shown by ratios of succeeding carbonate deposits.

Keith and Weber (1964) analyzed more than 500 additional carbonates, of various ages and continents, and concluded that the range in samples of a given age was large but that no consistent or marked change in isotope composition occurred throughout time. Later, Weber (1967, p. 2347-2348) graphically represented the fluctuating isotope ratios for marine and freshwater carbonates, noting that peaks occurred around the Silurian, Mississippian, Cretaceous, and Holocene, following, respectively, an early Paleozoic rise, an early Mesozoic rise, and early Cenozoic rise. Data were regarded as insufficient for the Carboniferous. Similar fluctuations were noted for freshwater carbonates, with a late Paleozoic decrease, Permo-Triassic increase, and late Mesozoic-Tertiary decrease in ratios. The freshwater ratios were suggested to correlate with the development and distribution of land plants, whose abundance resulted in relatively light fresh water carbonates.

Unless considerable organic debris was present in the freshwater limestones, the preferential use of the light isotope by plants would seemingly have had an opposite effect, leaving an excess of the heavier isotope in the reservoir and resulting in heavier limestones in fresh water as in the marine environment (Tappan, 1968b). Re-examination of the data of Keith and Weber shows that times of increasing isotopic weight in the marine limestones are correlative with periods of high phytoplankton productivity and decreasing ratios with periods of reduced abundance. Thus Keith and Weber (1964, Fig. 3) show that the average $\delta^{13}C$ in marine limestone gradually increased from the Precambrian through the Cambrian, Ordovician, and Silurian. The trend reversed in the Devonian and Carboniferous, which were progressively lighter, the Permian was again heavier, Triassic lighter, Jurassic and Cretaceous increasingly heavy, and Tertiary lighter. Evaluation of their basic data (p. 1813-1814) for epochs within the Tertiary shows light Paleocene carbonates, heavy Eocene ones, and increasingly light ones in the Miocene and Pliocene. Thus, although no steady change throughout time is apparent and additional isotope data are needed for certain critical periods, isotopic changes closely correlate with phytoplankton productivity, abundant organic fixation of ^{12}C leaving more ^{13}C in the reservoir and resulting in heavy limestones, and reduced productivity allowing the accumulation of more ^{12}C. An extended period of low productivity, as in the later Paleozoic or at the Cretaceous-Tertiary transition, might result in increased CO_2 pressure, lowered *p*H, carbonate solution, and the return of heavier carbon to the reservoir.

Rubey (1955, p. 646) observed that increased CO_2 in the atmosphere and oceans would lead to a decrease in dissolved carbonate and bicarbonate ions and allow the accumulation of calcium ions until gypsum as well as calcite became insoluble. Perhaps the extensive bedded gypsum of the midcontinent Permian and the presence of barite and other sulfates can be genetically correlated to the reduced marine productivity of the late Paleozoic.

Biogenically formed H_2S and free sulfur are enriched in ^{32}S during reduction by the sulfate reducing bacteria; residual dissolved sulfate in seawater is thereby ^{34}S-enriched. Holser and Kaplan (1968) noted that Permian sulfates of western North America and Germany were isotopically light and suggested that the conversion of sulfides to sulfates utilized an amount of oxygen equal to 90 percent that of the present atmosphere, extending the marine oxygen minimum zone, and causing late Paleozoic extinctions.

Numerous samples of all ages and areas, examined by Nielsen and Ricke (1964, p. 589) and Holser and Kaplan (1966, p. 120), indicate that isotope variation of sulfates is far greater with geologic age than between those of different areas or environments. Isotopically light sulfates in the Cambrian were followed by heavier ones in the Ordovician and Silurian, lighter sulfates in the Early Devonian, heavier in the Middle Devonian, then progressively lighter in Late Devonian, Carboniferous, Permian, and Early Triassic. Somewhat heavier Middle Triassic sulfates were followed by lighter ones in the Late Triassic, then gradually heavier ones to the present, except for sharp reductions in mid-Cretaceous and late Oligocene. Other than at the Cretaceous-Tertiary boundary, for which no sulfur data were given, periods of heavy sulfate production coincide with abundance of phytoplankton and light sulfates with reduced productivity (Tappan, 1968b).

Reduced quantities of phytoplankton would also have meant lesser amounts of the antibiotic metabolites secreted by various chrysophytes and dinoflagellates. These substances normally inhibit bacterial growth, but in their absence the teeming bacterial population would have had only limited quantities of organic matter for oxidation. The bacteria would then have turned to oxidation of sulfides and free sulfur. The process of sulfide oxidation that began as an effect of reduced productivity would further add to the oxygen depletion and carbon dioxide accumulation.

Relation of Siliceous Deposition to pH and CO_2 Pressure. Because radiolarian and diatomaceous oozes are found in northern seas where lime is not deposited or has been secondarily removed, they have been regarded as characteristically deep-water residual deposits. Their siliceous skeletons are regarded as "less soluble," accumulating wherever calcite is dissolved and detrital dilution is unimportant. Unfortunately, like many other widely held beliefs, this is not entirely true. Quartz undoubtedly is less soluble than calcite, but organically deposited SiO_2 is always in the opaline form and readily dissolves under alkaline conditions. Neritic diatoms commonly are smaller and less heavily silicified than the holoplanktonic species and thus are less likely to be preserved. Some weakly silicified diatoms such as *Nitzschia* may be abundant in the plankton but never appear in sediments, whereas the coarser, larger, and more heavily silicified species accumulate in greater quantities in the sediment than the population abundance would suggest (Mukhina, 1966).

Calvert (1964, p. 323) stated that stagnant water and laminated diatomaceous sediments are correlative in some regions but that similar deposits formed in deep basins of the Gulf of California that were not stagnant, being in open communication with the Pacific Ocean. Diatomite distribution was correlated to the depth of the oxygen minimum layer in the water column. The oxygen minimum zone is generally attributed to biological control, oxygen being consumed by organisms and decaying detritus. Where the oxygen minimum zone intersected the basin floor, benthic populations were restricted almost entirely to polychaete worms; rarely shrimp also occur.

Although oxygen is doubtless at a minimum, the abundance of silica-depositing organisms and the accumulation of siliceous deposits may depend upon additional factors. Low oxygenation alone should result in highly organic deposits rather than laminated siliceous ones. As oxygen consumption is controlled by biological activity, its product, CO_2, is similarly controlled. The occurrence of calcareous deposits in some regions and of siliceous ones in others seems closely related to the amount of dissolved CO_2 and the resultant *p*H variation. An additional indication of the importance of CO_2 concentration rather than O_2 is given by the widely noted association of diatomites and radiolarian-rich rocks with volcanism. This correlation probably is not due merely to increased supply of silica from volcanic ash, for the organisms deposit silica even with relatively low concentrations. A more important factor undoubtedly was the added quantity of CO_2, which is second only to H_2O as a prominent constituent of volcanic gases. The atmosphere change due to volcanic eruptions could result in sufficiently lowered *p*H to allow extensive diatom or radiolarian accumulations even at considerable distances from the volcanism itself.

Where dissolved CO_2 is relatively high, as at present in polar areas, *p*H is lower, silica is less soluble, and diatomaceous, silicoflagellate, and radiolarian oozes accumulate. Greater amounts of gases may be held in solution in colder water, but this applies to oxygen as well as to the CO_2; hence siliceous accumulations cannot be attributable solely to oxygen deficiencies. Conditions favoring the deposition and preservation of silica have an opposite effect on calcium carbonate.

In warmer water where less CO_2 is held in solution, or where it is depleted by an unusually high rate of photosynthesis, *p*H is elevated and silica becomes more soluble. In lower latitudes the abundance of photosynthesizing coccolithophorids and other rapidly growing nannoplankton coincides with lowered solubility of CO_2.

Accumulation of siliceous remains is prevented, whereas coccolith and foraminiferal oozes do occur. Where the poorer oxidation at depth again reverses the situation neither calcareous nor siliceous skeletal material can accumulate; only red clay remains in the deeper regions of the warmer seas.

Diatoms and other silica-secreting organisms occur widely in the plankton but are generally most abundant in high latitudes, decreasing in numbers toward the tropics. Diatoms are presently abundant in the Antarctic, but scarce both at the surface and in the sediments in the subtropical region of the Pacific (between 20° and 40° of latitude). The moderate increase in abundance in the equatorial zone, 5° S. to 6° N., was stated to result from greater nutrient supply due to upwelling (Kozlova and Mukhina, 1967). But upwelling of the deeper waters also renews the supply of carbon dioxide and lowers the *p*H and, as suggested above, would allow silica to accumulate in the sediments. Figure 12 indicates a closer (inverse) correspondence of diatoms in the sediment to the *p*H than to the temperature, or to the supply of phosphorus or silica. If nutrients or temperature were the dominant factor, Arctic production and accumulation should more closely match that of the Antarctic and the low-latitude increase should be greatest at the equator. Instead, the *p*H increase at about 15° S. is correlative to the lowest diatom numbers, the decrease in *p*H from 0° to 12° N. is accompanied by an increase in diatom accumulation in the sediments, and the lower diatom production and lesser accumulation of siliceous deposits in the Arctic is related to its higher *p*H.

To form their frustules, the diatoms effectively concentrate dilute quantities of dissolved silica from the undersaturated seas. In silica-deficient cultures, diatom activity so diminishes the quantity in solution as to allow silicon in the glass container to be dissolved (Lisitsyn, 1967a). The genus *Nitzschia* reportedly decomposes koalin, using the silica and liberating hydrated alumina (Vinogradov and Boichenko, 1942). Although laboratory experiments have shown diatoms to have an absolute silicon requirement, a complete absence of silica is unlikely to occur in nature. Even where silica is available and deposited by the organism, however, it may not occur in the sediments unless conditions are equally favorable to its preservation.

Kolbe (1955) observed that although foraminifers were abundant, diatoms were scarce in Atlantic sediment cores as com-

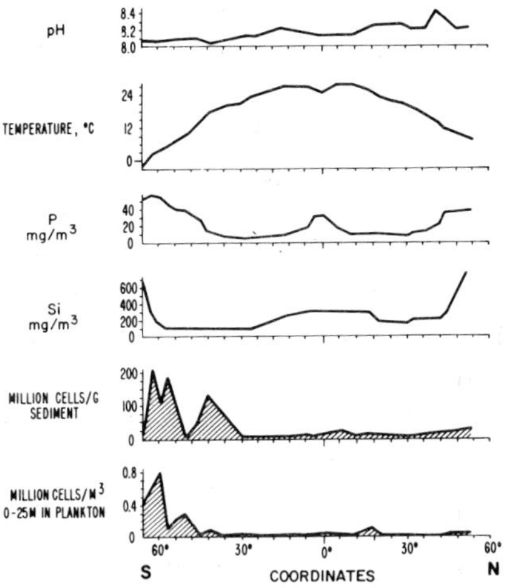

Figure 12. Quantitative latitudinal distribution of diatoms in the plankton and in sediments of the Pacific Ocean, related to variation in dissolved silica (Si), nutrients (P = phosphorus), temperature, and *p*H. Although diatom distribution is commonly related to nutrients, temperature or oxygen, quantitative differences in northern and southern regions correlate better (inversely) to *p*H, being abundant in plankton and sediments with lower *p*H. (Redrawn *from* Kozlova and Mukhina, 1967.)

pared to the Pacific. Specimens present were of coarsely textured species and poorly preserved, due to differential solution. Kolbe stated (p. 163) that low *p*H is detrimental to the preservation of foraminifers and does not affect diatoms, whereas high *p*H values have the opposite effect. He noted that the maximum abundance of diatoms in cores in the Romanche Deep coincided with the maximum occurrence of sediment particles less than 6 μ, but suggested no explanation. Perhaps the fine-grained sediment cover prevented continued solution at the sea floor.

The extensive marls and chalks of the Gulf Coast Cretaceous indicate a decreased CO_2 pressure and higher *p*H, probably related to extensive coccolithophorid and dinoflagellate production. This, in turn, resulted in the almost complete absence of siliceous microfossils. Siliceous remains occur only rarely, and then in silts or clays of lower calcareous content. In the California Cretaceous, siliceous sediments are more common and calcareous ones a rarity, although the presence of rudistids suggests that this is not entirely temperature-related. The California Paleocene and lower Eocene include many coccolith- and planktonic Foraminifera-bearing calcareous sediments, such as the Santa Susana and Lodo Formations, that resemble the Gulf Coast Cretaceous in paleoecologic aspect. Middle and late Eocene radiolarian beds and Miocene diatomites again suggest more extensive oceanic changes than those due to temperature alone. Alternating diatomaceous and foraminiferal beds in the Modelo Formation of Los Angeles demonstrate the general incompatibility of siliceous and calcareous remains, an increase in one being accompanied by a decrease in the other.

Sedimentary Iron Ore. The occurrences of sedimentary iron ore and sulfur deposits are related to bacterial activity. Iron bacteria (*Sphaerotilus* and *Gallionella*) occur in brackish, marine, and hypersaline water as well as in fresh water and may be of greater importance in the marine environment than is generally recognized (Wood, 1963). Precambrian iron ores may have been due to the abundance of iron bacteria coexisting with blue-green algae. After eucaryote algae appeared, their bactericidal secretions were inhibiting. However, bacterial iron deposition again characterized later periods of phytoplankton scarcity (for example, the Permian ferruginous oölitic limestone of the Pechora River basin, which contains fossil bacteria). Kutznetsov and others (1963) stated that marine iron ores are largely limited to pre-Jurassic strata. Commonly associated with coals, they formed along low flatlands that were repeatedly submerged. Such an environment was probably poor in oxygen and seems consistent with the low phytoplankton production of the Carboniferous through the Triassic and the resultant bacterial expansion.

Organic Carbon in Sediments. Large amounts of carbon occur as dispersed organic matter in sedimentary rocks. Ronov (1958) stated that nearly 10,000 times as much organic carbon is present in the disseminated state as occurs in the petroleum reservoirs. Organic carbon determinations were made for some 26,000 samples from the Russian Platform, ranging in age from late Precambrian to Quaternary, and results compared to data from United States oil provinces. Quantitative fluctuations occurred in all rock types but were of greatest magnitude in clay. Peaks of accumulation characterized the Ordovician, Mississippian, and Jurassic-Lower Cretaceous, with secondary peaks in the lower part of the Upper Devonian (Frasnian), in the Upper Permian (Kazanian), and in the Tertiary. Organic carbon was much less abundant in the Cambrian, Silurian, Permian, Triassic, and Quaternary. A positive correlation was found between the maximum disseminated carbon and the important coal and oil deposits. Similarly, the maxima and minima were identical in the U.S.S.R. and the United States, recording periodic increases and retardations in the intensity of biotic activity. Ronov suggested that periods of abundance were relatively calm tectonically, characterized by marine transgressions and a preponderantly humid temperate climate. Minima characterized times of regressions and local aridity.

The periods of abundant dispersed organic matter also coincide with times of high phytoplankton productivity or those in which blue-green algae were abundant. Blue-green stromatolites are characteristic of the Precambrian and early Paleozoic but also are locally common in the Mississippian, Pennsylvanian, and Permian of the mid-continent region, the Triassic, and parts of the Tertiary. Reduction of phytoplankton is commonly followed by increase of blue-green algae, just as occurs in aging mixed cultures.

Maximum productivity is not always correlative with high oxygenation. Under stagnant conditions, many cells may sink below the lighted zone and be unable to return. Complete decay does not always follow, as much plant organic matter is quite stable under anaerobic conditions. Thus it may accumulate as oil shales or other dark organic sediments. The areally extensive Devonian-Mississippian black shales have been suggested to have resulted from decaying benthonic seaweeds, although the latter are no longer identifiable as such. However, as such benthic plants must be within the photic zone, their accumulation would have necessitated very extensive seas of less than 30 m depth. The great quantities of organic matter suggest that decay was incomplete, yet brown algal thalli consist of cellulose and alginic acid and are not particularly resistant.

Abundant phytoplankton, or perhaps blue-green algae with resistant cell walls probably supplied the major organic component of the black shales. Dissolved and dispersed organic matter in the seas may also have coagulated or been assimilated by bacteria and deposited with the remains of the microalgae.

Ronov (1958) suggested that the maximum abundance of dispersed organic matter in sediments indicated a reducing environment and a low Fe_2O_3/FeO ratio. Poorly oxygenated lower levels may result from excessive microplankton abundance, even without oxygen depletion by decay. When productivity reaches "red tide" proportions, respiration of the plant cells at night may so decrease the level of dissolved oxygen as to result in anaerobic conditions at depth, particularly if vertical mixing is limited. Healthy algal cells respire at a rate of 5 to 15 percent that of gross photosynthesis at optimum light intensity. Respiration rates are relatively slow during the lag and early exponential growth phases but increase as the cells age. Nutrient deficiency increases respiration rates until they nearly equal photosynthesis, although it cannot long exceed a 2:1 ratio, as the plankton must survive approximately 24 hours of respiration to about 12 of photosynthesis (Strickland, 1960, p. 69). The respiration rate may also increase by about 10 percent for each 1° C rise in temperature (Harvey, 1955).

Normally the rate of spring oceanic production approximately balances respiration at about 10 m depth. Depending upon water transparency, light penetration, and season, the equilibrium level may range from as high as 1 or 2 m down to 20 or 30 m. Where dense algal production occurs in highly polluted waters, light cannot penetrate the dense growth and the water is anaerobic below the surface (Lackey, 1964). Perhaps the Devonian-Mississippian black shales represent marine conditions similar to those now present in Lake Erie, where dense blue-green algal blooms occur in numbers of 7 million cells per liter; the decaying algae accumulate to a thickness of 3 feet on the beach (Sperry, 1967, p. 353). The coincident accumulation of Devonian-Mississippian cherts and black shales supports the correlation of poor oxygenation, increased CO_2 pressure, and decreased pH in at least part of the water column with lime solution and silica accumulation at the sea floor.

Petroleum. Phytoplankton is probably the major source of the organic compounds in petroleum. Abundant microfossils indicate high phytoplankton productivity at most times of important petroleum accumulation.

Some 95 percent of net oceanic production, in excess of that used in respiration, is consumed by plankton feeders in the upper 200 m of the ocean (Nielsen and Jensen, 1959, p. 111); hence the small fraction buried on the sea floor might seem unlikely to result in important accumulations. Perhaps none is presently being formed in such quantity. During periods of extensive epicontinental seas and lowered continental elevation, temperatures were more equable and the thermocline less pronounced. These factors would allow a more rapid sinking of cells through the water column. The warm climate and greater nutrient influx, related to proximity of the shore in inland seas, would have favored exponential growth of phytoplankton much as occurs in a newly inoculated laboratory culture. Thus greater productivity would be combined with greater settling rates.

Among the less abundant nutrients in the water are the nitrogen compounds that are known to be fixed by both bacteria and blue-greens. Bactericidal compounds secreted by various phytoplankton may have inhibited these nitrogen-fixing groups to result in a relative scarcity of this necessary compound. Culture studies indicate the interesting effects of such nutrient limitation.

During exponential growth in culture, *Monodus subterraneus* has a 70 percent protein cell content, a high amount of chlorophyll and nucleic acids, and a low quantity of carbohydrates and fat (Fogg, 1965). In old, nitrogen-deficient cultures, the cell protein content is less than 10 percent, chlorophyll and nucleic acid content are similarly low, and the quantity of reserve polysaccharides and fats is high. Fats thus accumulate in large amounts in nitrogen-deficient cultures of green algae, diatoms, and Xanthophyta. Although the algal phyla do differ in their usual storage compounds, the difference due to varied growth conditions for a single species may be greater than between the major taxa grown under similar conditions.

Extensive epicontinental seas would favor phytoplankton production with increased cell fat content, as well as an increased sinking rate. Phytoplankton bactericidal secretions would reduce bacterial count and activity, and organic matter could accumulate at the sea floor without destruction by oxidation. Coincidence of extensive seas, heavy phytoplankton growth, and petroleum accumulation characterizes the lower Paleozoic, Cretaceous, and lower Tertiary.

Possibly bacteria were important producers of organic matter during phytoplankton minima, although they would not contribute to oxygen production. Heterotrophic bacteria at present produce oceanic organic matter in an amount equal to 0.5 to 1 percent that formed per cubic meter by photosynthesis. As their available depth is up to 50 times that of the euphotic zone, bacterial production may even equal that of phytoplankton (Strickland, 1963, p. 183). Although much of this production is at the expense of dissolved organic matter, the latter is estimated to comprise 40 percent of the total dispersed organic matter (Saunders, 1957). Parsons and Strickland (1962, p. 211) state that the dissolved organic matter in the sea is 10 times the amount of particulate matter and 50 times that of living cells. Phytoplankton cells may excrete between 35 and 40 percent of their organic matter during growth (Antia and others, 1963; Fogg, 1966), and as the oceans contain some 1.5 kg of dissolved organic matter per square meter of surface, nearly 100 times the total annual production of the phytoplankton (Wood, 1963), the present amount probably represents a long-term accumulation. Bacterial organic production, perhaps using some land-derived detritus, may have been of primary importance in the late Paleozoic seas, as phytoplankton was everywhere scarce, whereas land-derived spores and other detrital organic matter were abundant.

Synecology

Paleoecologic interpretations have commonly been dominated by consideration of physical factors of the environment, which change relatively slowly. Advancing or retreating seas may proceed with geologic rapidity yet allow accommodation by even the attached and sessile invertebrates. Organisms migrate on a biological time scale instead of a geologic one. Less easy to determine for a fossil assemblage are the various biological interactions that may, in fact, have played a far more important role. Loss of a key species in a highly complicated ecologic food web may cause its complete breakup, but even an important synecologic event may be difficult to recognize because of the incomplete nature of the fossil record. Elsewhere the suggestion was made (Tappan, 1968a, 1968b) that changing phytoplankton abundance throughout time had a marked quantitative effect on the atmospheric and oceanic CO_2 and O_2 and on the *p*H of the seas, in addition to a more direct effect on the food web. This postulated megafluctuation in productivity levels, through its effects on the composition of atmosphere, oceans, and sediments, would have had widely ramifying effects on the contemporary biosphere. An abundance and variety of grazing and filter-feeding animals would accompany high productivity, whereas low productivity would result in rapid elimination of less efficient food gatherers, as well as those susceptible to changed *p*H or variation in dissolved carbonate and bicarbonate ions. Reduced productivity and its effect on CO_2-O_2 levels would similarly have selectively affected terrestrial organisms. The record of terrestrial biologic events might serve to check the assumed relations.

Environmental Aspects of Evolutionary Progress

Evolution has various aspects. One of these is increased specialization, shown on a large scale by the change from the procaryote with relatively homogeneous structure to the eucaryote with special organelles enclosed in the cell and thence to the multicellular organism whose various cells have particular characters and functions. This aspect of evolution of the protists has been discussed above. The other and quite distinct aspect of evolutionary progress is said to be that in which organisms have successfully adapted toward greater control of the environment or toward independence of the environment (Huxley, 1942, p. 562). Evolutionary progress thus is regarded as a replacement of one species or group by another that is better at the same game. Huxley commented that eurypterids were better swimmers than trilobites and fish better than either. Early reptiles were clumsy and heavy, but later ones adapted for running. Lower marine organisms have blood or body fluids similar to seawater, whereas the higher fish keep a chemically constant internal environment, and birds and mammals control the internal temperature as well.

Such a replacement of less well adapted species by better adapted ones is suggested by the usual phylogenetic tree but is not upheld as a cause-effect relationship by the paleontologic record. Relative swimming ability probably had little importance in the decline of the trilobites or eurypterids. Either was superior in this respect to the barnacles or to the horseshoe crab, yet these latter arthropod groups long have persisted. Similarly, the internal chemistry of a fish may be biologically controlled, but protistan species living in exactly the same seawater concentrations "control" their intracellular environment so as to cause the deposition of opaline silica and of calcium carbonate as aragonite or as hexagonal or rhombohedral calcite; others may deposit iron carbonate, iron sulfide, or molecular sulfur. Using the same raw materials, CO_2 and H_2O, some protists store their excess photosynthetic products as starch, others as fats or oils, by their "control" of their internal environment.

Regardless of the taxonomic level of the organisms present at any stage in geologic history, the organic world repeatedly and rapidly exploited every possibility open to it, being limited only by its capability for structural adaptation to the available environments.

The statement is commonly made (for example, Huxley, 1942, p. 562) that "There is no certain case on record of a line showing a high degree of specialization giving rise to a new type. All new types which themselves are capable of adaptive radiation seem to have been produced by relatively unspecialized ancestral lines." This seems to be somewhat circular reasoning, in that "specialization" is generally defined by the end members that did not persist. As a matter of fact, consideration of the entire organic world suggests that all adaptive radiation has followed from the chance occurrence of an exceptionally high degree of "specialization" that afforded new ways of solving the universal problems of growth and reproduction. The first blue-green alga was certainly a highly specialized bacterium, with a distinctive pigment complex that allowed use of a different energy source. The first eucaryote was only a highly specialized blue-green whose various cell components had become localized in separate organelles. The first flagellate certainly attained a high degree of specialization in the complexity of the elongate fibrils, grouped to form a flagellum that could propel its few cubic microns a distance of a few millimeters. Yet by modifications of this specialized organelle, the descendants of the primitive flagellate developed a wide variety of ciliary structures. These occur in almost all higher organisms (for example, the ciliated epithelia of the vertebrate respiratory tract). Or, to consider a vertebrate example *Archaeopteryx* might be regarded as only a highly specialized end member of a reptilian lineage, if modern birds were unknown. In fact, each new type of specialization arose as an atypical development whose possibilities were then explored prior to further differentiation in other directions.

The paleontologic record has shown the successive rise and diversification of various groups and the gradual or precipitous decline of some with no immediate replacement. Mammals did not replace reptiles at the end of the Mesozoic merely because progressive changes that gave them dominance resulted in extinctions of one and radiation of the other. Extinctions or marked reductions of major groups have more commonly occurred prior to the radiation of their successors. The vacant niche provided an opportunity for expansion, rather than arising due to the expansion pressure. "Survival of the fittest" is better likened to a long-distance steeplechase than to a prize fight; depending upon the nature of the hurdles, either the tortoise or the hare may prove the winner.

Huxley (1942, p. 564) defined evolutionary progress as "a raising of the upper level of biological efficiency, this being defined as increased control over and independence of the environment." If adaptability to or independence of the environment were the important aspect, evolutionary progress would seemingly have retrogressed, for bacteria are the most widely adapted and adaptable organisms. Each succeeding evolutionary development in one direction has led away from adaptation in another. An increase in size required compartmentalization and transport systems but was gained at the expense of cellular independence and reproductive rate. Specialization for mobility and the heterotrophic use of preformed organic material was at the expense of economy in energy expenditure and autotrophic independence. Temperature regulation was at the expense of temperature indifference. Conquest of the terrestrial environment was at the expense of areal geographical distribution, for land areas are more restricted than the oceans, and their environmental conditions are so varied that much is habitable only by a few particularly adapted species.

What motivating force resulted in the repeated evolutionary bursts and subsequent extinctions that have characterized geologic history? None of the hypotheses previously proposed in explanation of extinctions during or closing the Devonian, Permian, Triassic, Cretaceous, and Pleistocene, or of the radiations in the Cambrian and those following the times of extinctions, explained the coincidence of marine and terrestrial animal extinctions with the radiations of terrestrial plants nor the delayed expansion of the animal successors. Possibly the phytoplankton fluctuations supply such a connecting link.

Phytoplankton Periodicity as an Evolutionary Stimulus

Aquatic Realm. The fluctuations in abundance of phytoplankton throughout time and the effect of changing productivity on CO_2 and O_2 levels and *p*H have been discussed. How did these changes, that are indicated by the phytoplankton fossil record, affect other organisms?

Margalef (1967) described a modern phytoplankton succession. The first stage, characteristic of enriched upwelled water, contains species with high surface-to-volume ratio, such as small flagellates, coccolithophores, and small dinoflagellates. Cell division is rapid, cell density between 100 and 1000 cells/ml, and total productivity high. Food chains are short and inefficient, and filter feeders are abundant. During the succession, cell size shows a progressive increase; production/biomass and rate of cell division decrease. Under poor nutritional conditions, large dinoflagellates become dominant; some are heterotrophic and many toxic. Coccolithophorids are larger and more elaborate (for example, *Rhabdosphaera* and *Syracosphaera*). Cell division is less frequent; cell density is less than 10/ml. Food chains are long, complex, and more efficient, to utilize the reduced productivity to the fullest. Filter feeders are scarce, macrophagous predators common.

Although seasonal successions are not recognizable in the fossil phytoplankton record, the characteristic features of high and low productivity can be observed. In the Jurassic and Early Cretaceous, small coccolithophorids, nannoconids, and dinoflagellates suggest the assemblage of the early stage of succession characteristic of high productivity. The only fossil abundance of ciliates known was at this time, the calpionellids (tintinnid ciliates) being characteristic of the coccolith-nannoconid limestones of the Tethyan region. Tintinnids became scarce by mid-Cretaceous, then fossils are unknown until the Eocene. Late Cretaceous reduced productivity is suggested by the larger and more complex coccolithophorids, large dinoflagellates, and complex food chains dominated by predators. As noted by Margalef (1967, p. 557), complex food chains are based on stability, which in the sea is not compatible with high productivity. Loss of any component of the chain would have disruptive effects on the remainder. Decreased quantities would also result in decreased populations farther along the chain and eventually mean extinction.

Because of food selectivity by some organisms, the qualitative nature of the food supply is also important. An abundant total microflora is not equally available to all grazing species. Protozoa and small zoöplankton, or larval metazoans, can only consume the smaller prey, and even adult organisms cannot utilize all available material. *Mytilus californianus* feeds on very small suspended organic particles, including dinoflagellates. However, the mussel lacks the enzyme cellulase and cannot digest the cellulosic cell wall of the armored dinoflagellate (Moore, 1958, p. 107). This inability to digest cellulose may result in starvation of the grazing predator in the midst of large populations of some green algae or armored dinoflagellates. Many zooplankters, larvae and adults, preferentially utilize as food coccolithophorids, silicoflagellates, or naked dinoflagellates in marine waters, whereas their freshwater counterparts utilize diatoms or chrysomonads, even when other species are numerically more important. This may be because of their lack of cellulose.

Food supplies affect differently the benthonic and pelagic organisms. Benthonic organisms respire at a slower rate and grow more slowly than the planktonic ones and hence require less food (Harvey, 1955). The rapid evolution of planktonic organisms and their similarly rapid extinctions may be in part related to their greater rate of growth, higher respiratory rate, and hence greater food requirements than those of the benthonic organisms.

The relative rate of sinking of organic material and its consumption in the upper levels may have a selective effect on the benthonic filter feeders and detritus feeders. The decrease in marine phytoplankton at the end of the Devonian, which continued on a nearly world-wide basis until Jurassic time, was accompanied by the rapid decline and extinction of a variety of organisms, suggesting that survivors were those most efficient in food gathering or that had lessened food requirements. The phylogenetic series of *Myalina* species, described by Newell (1942, p. 47, Fig. 12) shows a steady increase in relative shell height (perpendicular to the hinge axis) through the Mississipian, Pennsylvanian, and Permian. The evolutionary trend shown by this filter-feeding organism may well have been an adaptation for increased gill area, to aid in filtering organic matter from the water, as the available quantities decreased. These pelecypods also were adapted to shallow water and tolerant of brackish and even fresh water. The abundant brachiopods of the early Paleozoic, restricted to the definitely marine waters with a decreased organic content in the later Paleozoic, eventually starved because of their less efficient filtering systems. Most became extinct at the end of the Devonian. The suggestion (Rudwick, 1965, p. H207), that *Prorichthofenia,* and perhaps to a lesser extent other productoids, may have abandoned the suspension feeding system and become deposit feeders or even carnivores, might explain their Devonian development and persistence into the later Paleozoic. The replacement of cystoids by the crinoids, with their more efficient food-gathering methods, coincided with decreased productivity in the mid-Paleozoic, as did the eventual nearly complete replacement of Pelmatozoa by the motile bottom-feeding echinoderms.

With decreased productivity, the increase in dissolved CO_2 resulted in lowered *p*H, the latter being particularly detrimental to reef-forming organisms. At the Permo-Triassic boundary many of the latter disappeared, including the Tabulata, tetracorals, and stony bryozoans (trepostomes and cryptostomes). At the end of the Cretaceous the rudistid pelecypods met a similar fate.

Inhibitory and toxic excretions of some algae (notably certain blue-greens and dinoflagellates) may not only repel but actually kill feeding organisms. The well-known occurrence of catastrophic fish mortality related to dinoflagellate blooms may be due to one or more causes: toxic excretions, the abundant cells clogging the fish gills, or the excess of respiratory activity over photosynthesis, reducing the dissolved oxygen below the critical level.

Oxygen is a requirement of all animal life, but some are better adapted than others to low oxygen levels. ". . . Most marine organisms are physiologically better adapted to wide variations in oxygen tension than many land organisms. . . . In most aquatic organisms, respiratory consumption of oxygen is essentially independent of the oxygen tension down to some lower limit of the latter. Below this limit, the respiratory consumption may decrease with decreasing oxygen tension, until the tension becomes so low that the organism can no longer survive" (Richards, 1957, p. 211-212). Marine worms persist even in the poorly oxygenated depths, and shrimps and other arthropods may live there in part but obtain sufficient oxygen by means of vertical diurnal migrations. Thus oxygen content is not a primary factor in limiting life and growth in the ocean except under completely anaerobic conditions, where bacterial metabolism utilizes combined oxygen (sulfates, nitrates, and so on). Decreased populations with depth are not exactly correlative with oxygen content and are more closely related to temperature, light, and food (Richards, 1957, p. 213).

Oxygen supply is a more important factor for vertebrates however, and notable changes in oxygenation might exert a greater pressure on these than on the invertebrates. The earliest vertebrates yet known are of latest Cambrian or Early Ordovician age, a time of extensive productivity and hence good oxygenation. Fishes had an explosive evolution in the Silurian and Devonian and by the end of the Devonian were present in all freshwater deposits as well as marine ones. Could their extensive migration into fresh waters have been related to decreased marine productivity? Widespread extinctions of fish at the end of the Devonian might suggest this. Romer (1966, p. 317) stated that "In the space of a single period [Devonian], almost every fish type present in the Lower Devonian was wiped out except for a few acanthodians."

The evolution of the amphibians from the crossopterygian fish also may have been a result of the selective pressures of decreased oxygen supply. Because of the lowered solubility of gases in warm water, oxygen in shallow waters may be completely exhausted at night in the present tropics, as in South America or India, due to the heavy demands of respiration by organisms. The increased CO_2 level due to organic decay further stimulates respiration in vertebrates. Under such conditions, fish must come to the surface to breathe air, and otherwise would drown from lack of oxygen. Modern lungfish occur in such swamps, where even the indigenous snails are air-breathing pulmonates. Pearse (1936, p. 29, 49) suggested that lack of oxygen rather than the drying up of swamps is probably more important in restriction of the local biota to air-breathing taxa. Romer (1966, p. 318) noted that in the fresh-water Upper Devonian deposits of Scaumenac Bay, 90 percent or more of the specimens of fossil fish appear to have lungs. Lungfish continued to be abundant throughout the late Paleozoic and Triassic. Amphibia arose in the Devonian and had an explosive evolution in the Mississippian and Pennsylvanian. The coincident expansion of lungfish, extinctions of other early fish, and the migration of the amphibian descendants from the aquatic to terrestrial habitat may all have been related to the reduction in marine phytoplankton. The extensive black shales indicate poor oxidation, and the interbedded cherts suggest elevated CO_2 and lowered *p*H. Although shallow, brackish-water deposits of the Carboniferous locally contain such abundant green algal remains (*Botryococcus*) as to form boghead coal, they suggest a comparison with the oxygen-deficient shallow waters of the South American tropics; the late Paleozoic marine sediments are nearly barren of phytoplankton.

The newly evolved amphibians could obtain oxygen from the atmosphere but were largely restricted to marshes or coastal regions, for their reproduction and early developmental stages were still aquatic.

Terrestrial Realm. Obviously, the food supply of terrestrial organisms is not directly related to phytoplankton production, yet extinctions and evolutionary bursts of land animals have, in general, been remarkably coincident with those of the seas. The changed atmospheric proportions of CO_2 and O_2, which are dependent upon the oceanic reservoir and its productivity, may supply the previously unrecognized connecting link between these organic realms (Tappan, 1968b).

Poor oceanic and shallow-water oxygenation was suggested above as the major environmental pressure related to the origin of the amphibians. As the effects of low marine productivity accumulated, the poorly oxygenated Carboniferous coal swamps became less habitable and the Amphibia declined. Development of a reptilian grade of organization allowed migration of some from these lowlands, but continuation of the oxygen depletion resulted in the extinction of most of the amphibians by the end of the Paleozoic and of many more by the end of the Triassic. All modern Amphibia (Lissamphibia) have been regarded as descendants of the Paleozoic microsaurs. Interestingly, these scaleless amphibians, such as frogs and salamanders, can respire through the skin as well as through the lungs. In fact, a frog can obtain sufficient oxygen through the skin to remain alive after its lungs are removed (Pearse, 1936, p. 90). Such an ability may well have allowed their survival, although the larger and more impressive-appearing scaled and armored relatives succumbed.

Because they were more independent of the aquatic habitat, reptiles became the dominant vertebrates of the Permian. Although some of these disappeared at the end of the Permian and others in the Triassic, the group persisted into the mid-Mesozoic, where favorable conditions then allowed explosive radiation. Phytoplankton abundance in the Jurassic and Cretaceous lowered the CO_2 levels and increased the atmospheric oxygenation, resulting in lowered *p*H and extensive marine carbonate deposition. These more oxygenated waters allowed the reinvasion of an aquatic habitat by various reptilian lines and also saw an explosive evolution of the bony fish (Holostei and Teleosti). But the Late Cretaceous saw another marked decrease in phytoplankton. Coccolithophorids and dinoflagellates were both greatly reduced, many taxa becoming extinct, and the changed atmospheric and oceanic conditions caused extensive carbonate solution. This change seemingly is also recorded in the vertebrate evolution as most of the dominant Mesozoic reptile groups became extinct at the close of the Cretaceous.

Although the fossil evidence suggests that greater fluctuations probably occurred, with even a reduction of productivity to one-half or one-third, the resultant changes in the CO_2 and O_2 balance might have far-reaching effects on the terrestrial biota.

Increased CO_2 elevates the respiration rate of terrestrial vertebrates, although oxygen levels can be reduced about one-third before having a notable effect on breathing. Heavy exertion increases oxygen consumption about tenfold. The present atmosphere contains about 20.9 percent oxygen and 0.03 percent CO_2. A poorly ventilated room may have up to 0.5 percent of CO_2. If CO_2 reaches a level of 3 percent with oxygen reduced to 17 percent, the breathing rate is notably increased, and 6 percent CO_2 with 13 percent oxygen results in panting and exhaustion. The maximum that can be tolerated for a short period is about 10 percent CO_2, regardless of the amount of oxygen, and this results in unconsciousness (Haldane and Priestley, 1935).

Oxygen requirements are greater for the rapidly growing young animals and for large organisms in general. As a result, the more active carnivorous dinosaurs and large bulky sauropods both succumbed. Just as a minor group of amphibians adapted to the rigorous conditions and could survive the Paleozoic crisis, so also a few reptiles were better adapted to survive that of the Mesozoic. These were the crocodiles, snakes, and turtles, whose metabolic rate is periodically much lowered by the habit of hibernation or estivation.

Mammals had appeared earlier, at the end of the Triassic, but all were small and relatively few are known. Although the metabolic rate of warm-blooded animals is higher, other features left them better adapted to survive the Mesozoic. Distribution of blood to outer tissues and peripheries is better, and their smaller blood corpuscles with greater relative surface have an increased oxygen-carrying capacity (Pearse, 1936, p. 109). The warm Mesozoic climate also made few demands on metabolic heat production. Even with these advantages, those that persisted after the Cretaceous were only the small primitive marsupials and placentals. Even most of the multituberculates of the Paleocene were small (Romer, 1966).

The higher productivity of the early Tertiary allowed expansion into the vacant ecologic niches by the survivors of the Mesozoic. Mammalian radiation into all available areas, including the aquatic realm, followed the pattern previously set by the reptiles.

We have suggested that a direct correlation existed between phytoplankton productivity and expansion in the marine realm and among terrestrial animals. An inverse relation is shown by terrestrial plants. Because plants utilize CO_2 in photosynthesis, increased CO_2 levels are stimulatory instead of harmful. Extensive plant diversification has always been related to high CO_2 levels, beginning with the Precambrian origin and diversification of the algal groups, to the spread of land plants in the mid-Paleozoic. Although vascular plants appeared in the Silurian and Early Devonian, the more primitive Psilopsida died out within the Devonian. The larger Lycopsida (scale trees), Sphenopsida (*Calamites*), and Pteropsida (ferns) diversified extensively in the Late Devonian and Carboniferous. The abundance of these plants in the late Paleozoic coal swamps has been generally regarded as an indication of high oxygen levels. The coal floras indicate warm, humid conditions similar to modern tropical rain forests, but under cloudy conditions, such as characterize these swampy regions, photosynthesis is much reduced and oxygen production may be balanced by respiration. These climatic features, together with the accumulation of large quantities of unoxidized organic matter as coal and petroleum, are indicative of higher CO_2 levels and reduced oxygenation in the Permo-Carboniferous. Rapid diversification and increased abundance of the coal-swamp floras coincided with the late Paleozoic low phytoplankton productivity and hence stimulatory high CO_2 levels.

Perhaps the rise of the gymnosperms in the Devonian and their late Paleozoic diversification were related to their production of seeds. The contained food reserves for the early development of the new plant supplied a distinct advantage over the spore formation when photosynthetic rate was slow under the cloudy, warm skies. Angiosperm diversification was slow during the Cretaceous, but the high CO_2 levels closing the Mesozoic again supplied the plant kingdom with an evolutionary stimulus.

EPILOGUE

The remarkable world-wide contemporaneity of a continuous sequence of biological, geochemical, and sedimentological changes, briefly outlined above, seems scarcely coincidental. Our suggestion that these were related through a previously unconsidered link may induce others to present supporting or contradicting data that we failed to consider. Additional distributional data for the fossil microplankton are needed for the Paleozoic (for example, phytoplankton abundance in certain parts of the Silurian and Devonian), although already its general patterns of increase and decrease are known. Similarly, confirmation of the suggested geochemical relations requires more detailed isotope studies of the Cretaceous-Tertiary boundary.

We may have seemed overenthusiastic in relating so many of the Phanerozoic events and changes in the earth's outer crust to the actions and interactions of tiny green bits of protoplasm in the seas on what might be regarded as circumstantial evidence. Correlation of data from modern laboratory cultures and present oceanic distributions to the admittedly incomplete fossil record, in interpretation of conditions in the geologically distant past, has this common failing with all geologic theory, for no eyewitnesses survived.

ACKNOWLEDGMENTS

Plankton studies by Helen Tappan (Loeblich) were supported in part by National Science Foundation Grant GB-3684 and American Chemical Society Petroleum Research Fund Grant 3657-A2. We are greatly indebted to Dr. W. W. Hay, University of Illinois, Dr. Ralph Lewin and Alfred Loeblich III, Scripps Institution of Oceanography, Drs. E. C. Olson and J. William Schopf, University of California at Los Angeles, Dr. Robert M. Kosanke, U.S. Geological Survey, Denver, and Drs. S. R. Silverman and Dietrich H. Welte, Chevron Oil Field Research Company, for critical reading of the manuscript and for many helpful discussions. Figures were drafted by J. C. Keeser and T. G. Naves, Chevron Oil Field Research Company.

REFERENCES CITED

Abelson, P. H., 1966, Chemical events on the primitive earth: Natl. Acad. Sci. Proc., v. 55, p. 1365-1372.

Antia, N. J., McAllister, C. D., Parsons, T. R., Stephens, K., and Strickland, J. D. H., 1963, Further measurements of primary production using a large-volume plastic sphere: Limnology and Oceanography, v. 8, p. 166-183.

Barghoorn, E. C., Meinschein, W. G., and Schopf, J. W., 1965, Paleobiology of a Precambrian shale: Science, v. 148, p. 461-472, 9 figs.

Barghoorn, E. S., and Schopf, J. W., 1965, Microorganisms from the Late Precambrian of Central Australia: Science, v. 150, p. 337-339, 8 figs.

——1966, Microorganisms three billion years old from the Precambrian of South Africa: Science, v. 152, p. 758-763, 12 figs.

Berkner, L. V., and Marshall, L. C., 1964, The history of growth of oxygen in the earth's atmosphere, *in* Brancazio, P. J., and Cameron, A. G. W., *Editors,* The origin and evolution of atmospheres and oceans: New York, John Wiley & Sons, p. 102-126.

——1965, History of major atmospheric components: Natl. Acad. Sci. Proc., v. 53, p. 1215-1226, Figs. 1-10.

Bernard, Francis, 1963, Vitesses de chute en mer des amas palmelloides de *Cyclococcolithus.* Ses conséquences pour le cycle vital des mers chaudes: Pelagos, Bull. Inst. Océanog. Alger, v. 1, p. 5-34, 5 figs.

Blackburn, K. B., 1936, A reinvestigation of the alga *Botryococcus Braunii* Kützing: Royal Soc. Edinburgh Trans., v. 58, p. 841-854, 4 figs., 1 pl.

Bogorad, L., 1962, Chlorophylls, *in* Lewin, R. A., *Editor,* Physiology and biochemistry of algae: New York, Academic Press, p. 385-408.

Bonner, J. T., 1962, The ideas of biology: New York, Harper and Brothers, 180 p., 24 figs.

——1965, Size and cycle: An essay on the structure of biology: Princeton, New Jersey, Princeton Univ. Press, 219 p., 24 figs.

Bourrelly, Pierre, 1962, Chrysophycées et phylogénie *in* Beiträge zur Physiologie und Morphologie der Algen: Stuttgart, Gustav Fischer, p. 32-36, 2 figs.

Braarud, Trygve, 1954, Ecology of marine phytoplankton: VIII Congr. Internat. Bot. Rap. Commun., Sect. 17, p. 178-185.

——1962, Species distribution in marine phytoplankton: Oceanog. Soc. Japan Jour., 20th Anniversary Volume, p. 628-649.

Bradley, W. N., 1963, Geologic laws, *in* Albritton, C. C., Jr., *Editor,* The fabric of geology: Geological Society of America 75th Anniversary Volume, p. 12-23.

Burris, R. H., 1966, Biological nitrogen fixation: Ann. Rev. Plant Physiology, v. 17, p. 155-184.

Busby, W. F., and Lewin, Joyce, 1967, Silicate uptake and silica shell formation by synchronously dividing cells of the diatom *Navicula pelliculosa* (Breb.) Hilse: Jour. Phycol., v. 3, p. 127-131.

Calvert, S. E., 1964, Factors affecting distribution of laminated diatomaceous sediments in Gulf of California, *in* van Andel, Tj. H., and Shor, G. G., Jr., *Editors,* Marine geology of the Gulf of California: Tulsa, American Assoc. Petroleum Geologists, p. 311-330, 14 figs.

Carlson, F. D., 1962, A theory of the survival of motility, *in* Bishop, D. W., *Editor,* Spermatozoan motility: Am. Assoc. Adv. Sci. Symposium 72, p. 137-146.

Chediya, D. M., 1964, Sistematika E. Gekkelya i evolyutsionnoe rasvitie radiolyariy [Systematics of E. Haeckel and evolutionary development of the radiolaria]: Paleontol. Tadzh., Akad. Nauk Tadzhik. SSR, Tadzh. otdel, vses. Paleontol. obshchest., p. 86-135.

Cloud, P. E., Jr., Gruner, J. W., and Hagen, H., 1965, Carbonaceous rocks of the Soudan Iron Formation (early Precambrian): Science, v. 148, p. 1713-1716, 1 fig.

Coleman, A. W., 1961, The role of modern culture methods in the study of algal life cycles: Quart. Rev. Biology, v. 36, p. 247-253.

Corliss, J. O., 1962, Taxonomic procedures in classification of Protozoa, *in* Ainsworth, G. C., and Sneath, P. H. A., *Editors,* Microbial classification: Cambridge, Cambridge Univ. Press, Twelfth Symp. Soc. Gen. Microbiology, Royal Inst. London, April 1962, p. 37-67, 1 fig.

Cygan, N. E., and Koucky, F. L., 1963, The Cambrian and Ordovician rocks of the east flank of the Big Horn Mountains, Wyoming: Wyoming Geol. Assoc. and Billings Geol. Soc., First Joint Field Conf. Northern Powder River Basin, Wyoming and Montana, Guidebook, p. 26-37, 3 figs., 1 pl.

Dauvillier, A., 1965, The photochemical origin of life: New York, Academic Press, 193 p.

Deflandre, Georges, 1936, Les flagellés fossiles. Aperçu biologique et paléontologique. Role géologique: Actual. Sci. Ind. 335, 98 p., 135 figs.

Deunff, Jean, 1961, Un microplancton à hystrichosphères dans le Tremadoc de Sahara: Rev. Micropaléontologie, v. 4, p. 37-52, Pls. 1-3.

Donn, W. L., Donn, B. D., and Valentine, W. G., 1965, On the early history of the earth: Geol. Soc. America Bull., v. 76, p. 287-306, 2 figs.

Downie, C., 1967, The geological history of the microplankton: Rev. Palaeobotany and Palynology, v. 1, p. 269-281, 3 figs.

Droop, M. R., 1963, Algae and invertebrates in symbiosis: Symbiotic Assoc., 13th Symp. Soc. Gen. Microbiology, Royal Inst. London, p. 171-199, 3 figs.

East, E. M., 1936, Genetic aspects of certain problems of evolution: Am. Naturalist, v. 70, p. 143-158.

Eglington, G., Scott, P. M., Belsky, T., Burlingame, A. L., and Calvin, M., 1964, Hydrocarbons of biological origin from a one-billion-year-old sediment: Science, v. 145, p. 263-264.

Eisenack, Alfred, 1958, *Tasmanites* Newton, 1875 und *Leiosphaeridia* n. g. als Gattungen der Hystrichosphaeridea: Palaeontographica Abt. A, v. 110, p. 1-19, Pls. 1-2, Figs. 1-3.

——1959, Neotypen baltischer Silur-Hystrichosphären und neue Arten: Palaeontographica, Abt. A, v. 112, p. 193-211, Pls. 15-17, Figs. 1-11.

Fischer, A. G., 1965, Fossils, early life, and atmospheric history: Natl. Acad. Sci. Proc., v. 53, p. 1205-1215, 2 figs.

Fogg, C. E., 1965, Algal cultures and phytoplankton ecology: Madison, Wisconsin Univ. Press, 126 p., 31 figs.

——1966, The extracellular products of algae: Ann. Rev. Oceanography Marine Biology, v. 4, p. 195-212, 3 figs.

Foreman, H. P., 1963, Upper Devonian radiolaria from the Huron Member of the Ohio Shale: Micropaleontology, v. 9, p. 267-304, Pls. 1-9, 4 figs.

Fott, Bohuslav, 1965, Evolutionary tendencies among algae and their position in the plant kingdom: Preslia, v. 37, p. 117-126.

Goreau, T. F., 1961, On the relation of calcification to primary productivity in reef building organisms, *in* The biology of *Hydra*: Miami Univ. Press, p. 269-285.

Haldane, J. B. S., 1949, Suggestions as to quantitative measurements of evolution: Evolution, v. 3, p. 51-56.

Haldane, J. S., and Priestley, J. G., 1935, Respiration: Oxford, Clarendon Press, 493 p., 124 figs.

Hart, T. J., 1966, Some observations on the relative abundance of marine phytoplankton populations in nature, *in* Barnes, H., *Editor,* Some contemporary studies in marine science: London, George Allen and Unwin, Ltd., p. 375-393.

Harvey, H. W., 1945, Recent advances in the chemistry and biology of sea water: London, Cambridge Univ. Press, 164 p., 29 figs.

——1955, The chemistry and fertility of sea waters: London, Cambridge Univ. Press, 224 p.

Haxo, F. T. 1960, The wavelength dependence of photosynthesis and the role of accessory pigments, *in* Allen, M. E., *Editor,* Comparative biochemistry of photoreactive systems: New York, Academic Press, p. 339-360, figs. 1-16.

Haxo, F. T., and Fork, D. C., 1959, Photosynthetically active accessory pigments of cryptomonads: Nature, v. 184, p. 1061-1062, 3 figs.

Healey, F. P., Coombs, J., and Volcani, B. E., 1967, Changes in pigment content of the diatom *Navicula pelliculosa* (Bréb.) Hilse in silicon-starvation synchrony: Arch. Mikrobiol., v. 59, p. 131-142, 7 figs.

Holser, W. T., and Kaplan, I. R., 1966, Isotope geochemistry of sedimentary sulfates: Chem. Geology, v. 1, p. 93-135.

——1968, Oxygen balance in the Permian atmosphere, *in* Abstracts for 1966: Geol. Soc. America Spec. Paper 101, p. 97.

Hulburt, E. N., 1963, The diversity of phytoplankton populations in oceanic, coastal, and estuarine regions: Jour. Marine Research, v. 21, p. 81-93, 5 figs.

Hutner, S. H., 1964, Introduction, *in* Hutner, S. H., *Editor,* Biochemistry and physiology of Protozoa, v. 3: New York, Academic Press, p. 1-7.

Hutner, S. H., and Provasoli, Luigi, 1964, Nutrition of algae: Ann. Rev. Plant Physiology, v. 15, p. 37-56.

Huxley, Julian, 1942, Evolution: The modern synthesis: New York, Harper and Brothers, 645 p.

Isenberg, H. D., and Lavine, L. S., 1964, Comparative biology of mineral deposition, *in* Adams, J. P., *Editor,* Current practice in orthopaedic surgery, v. 2: St. Louis, C. V. Mosby Co., p. 177-219, Figs. 94-126.

Jeffery, P. M., Compston, W., Greenhalgh, D., and de Laeter, J., 1955, On the carbon-13 abundance of limestones and coals: Geochim. et Cosmochim. Acta, v. 7, p. 255-286, 2 figs.

Jekhowsky, B. de, 1958, Méthodes d'utilisation stratigraphiques des microfossiles organiques dans les problèmes petroliers: Inst. Français Pétrole Rev., v. 13, p. 1391-1415, Pls. 1-3, 21 figs.

Keith, M. L., and Weber, J. N., 1964, Carbon and oxygen isotopic composition of selected limestones and fossils: Geochim. et Cosmochim. Acta, v. 28, p. 1787-1816.

Kennett, J. P., 1966, Foraminiferal evidence of a shallow calcium carbonate solution boundary: Science, v. 153, p. 191-193, 2 figs.

Klein, R. M., and Cronquist, A., 1967, A consideration of the evolutionary and taxonomic significance of some biochemical, micromorphological, and physiological characters in the thallophytes: Quart. Rev. Biology, v. 42, p. 105-296, Figs. 1-26.

Kolbe, R. W., 1955, Diatoms from equatorial Atlantic cores: Repts. Swedish Deep-Sea Expedition, 1947-1948, v. 7, Sed. cores from North Atlantic, no. 3, p. 151-184, Pls. 1, 2.

——1957, Fresh-water diatoms from Atlantic deep-sea sediments: Science, v. 126, p. 1053-1056, 2 figs.

Kozlova, O. G., and Mukhina, V. V., 1967, Diatoms and silicoflagellates in suspension and floor sediments of the Pacific Ocean: Internat. Geology Rev., v. 9, p. 1322-1342, 13 figs.

Kuznetsov, S. I., Ivanov, M. V., and Lyalikova, N. N., 1963, Introduction to geological microbiology, *in* Oppenheimer, C. H., *Editor;* Broneer, P. T., *Translator*: New York, McGraw-Hill Book Co., 252 p.

Lackey, J. B., 1964, The ecology of plankton algae, *in* Jackson, D. F., *Editor,* Algae and man: New York, Plenum Press, p. 205-220.

Leedale, G. F., 1967, Euglenoid flagellates: Englewood Cliffs, New Jersey, Prentice-Hall, Inc., 242 p., 176 figs.

Lisitsyn, A. P., 1967a, Basic relationships in distribution of modern siliceous sediments and their connection with climatic zonation: Internat. Geology Rev., v. 9, p. 631-652, 13 figs.

——1967b, Basic relationships in distribution of modern siliceous sediments and their connection with climatic zonation: Internat. Geology Rev., v. 9, p. 842-865, 7 figs.

Loeblich, A. R. III, and Smith, V. E., 1968, Chloroplast pigments of the marine dinoflagellate *Gyrodinium resplendens*: Lipids, v. 3, p. 5-13, 6 figs.

Loeblich, A. R., Jr., and Tappan, Helen, 1964a, Foraminiferal facts, fallacies and frontiers: Geol. Soc. America Bull., v. 75, p. 367-392, 13 figs.

——1964b, Treatise on invertebrate paleontology, Part C, Protista 2, Sarcodina, chiefly "Thecamoebians" and Foraminiferida: Geol. Soc. America and Kansas Univ. Press, 2 vols., 900 p.

——1966, Annotated index and bibliography of the calcareous nannoplankton: Phycologia, v. 5, p. 81-216.

——1969, Acritarch excystment and surface ultrastructure with descriptions of some Ordovician taxa: Revista Española Micropaleontologia, v. 1, no. 1, p. 45-57, Pls. 1-4, 1 fig.

Loeblich, A. R. III, Loeblich, L. A., Tappan, Helen, and Loeblich, A. R., Jr., 1968, Annotated index of fossil and recent silicoflagellates and ebridians, with descriptions and illustrations of validly proposed taxa: Geol. Soc. America, Mem. 106, 319 p., 21 figs., 53 pls.

Lowenstam, H. A., 1954, Factors affecting the aragonite: calcite ratios in carbonate-secreting marine organisms: Jour. Geology, v. 62, p. 284-322, Figs. 1-15.

——1963, Biologic problems relating to the composition and diagenesis of sediments, *in* Donnelly, T. W., *Editor,* The earth sciences, problems and progress in current research: Chicago, William Marsh Rice Univ. and Chicago Univ. Press, p. 137-195, 14 figs.

Lucas, C. E., 1955, External metabolites in the sea: Deep-Sea Research, v. 3, Suppl. Papers in Marine Biology and Oceanography, p. 139-148.

Margalef, Ramon, 1967, The food web in the pelagic environment: Helgolander wissenschaftliche Meeresuntersuchungen, v. 15, p. 548-559.

Mayr, Ernst, 1942, Systematics and the origin of species from the viewpoint of a zoologist: New York, Columbia Univ. Press, Columbia Biological ser. no. 13, 334 p., 29 figs.

——1963, Animal species and evolution: Cambridge, Massachusetts, Belknap Press of Harvard Univ. Press, 797 p., 19 figs.

Meinschein, W. G., 1965, Soudan Formation: Organic extracts of early Precambrian rocks: Science, v. 150, p. 601-605, 10 figs.

Meinschein, W. G., Barghoorn, E. S., and Schopf, J. W., 1964, Biological remnants in a Precambrian sediment: Science, v. 145, p. 262-263.

Moore, H. B., 1958, Marine ecology: New York, John Wiley & Sons; London, Chapman and Hall, Ltd., 493 p.

Mukhina, V. V., 1966, Siliceous organisms in suspension and in the surface layer of Indian Ocean bottom sediments (from data obtained on the 36th cruise of the Research Vessel Vityaz'): Oceanology, v. 6, p. 660-667, 5 figs.

Needham, A. E., 1959, The origination of life: Quart. Rev. Biology, v. 34, p. 189-209.

Newell, N. D., 1942, Late Paleozoic pelecypods: Mytilacea: Kansas Univ. Pub., Kansas Geol. Survey, v. 10, p. 1-115, 15 pls., 22 figs.

Nielsen, E. S., 1955, The production of antibiotics by plankton algae and its effect on bacterial activities in the sea: Deep-Sea Research, v. 3, Suppl., p. 281-286.

Nielsen, E. S., and Jensen, E. A., 1959, Primary oceanic production, the autotrophic production of organic matter in the oceans: Galathea Rept., v. 1, p. 49-136.

Nielsen, H., and Ricke, W., 1964, Schwefel-Isotopenverhältnisse von Evaporiten aus Deutschland; Ein Beitrag zur Kenntnis von $\delta^{34}S$ im Meerwasser-Sulfat: Geochim. et Cosmochim. Acta, v. 28, p. 577-591.

Oró, J., and Nooner, D. W., 1967, Aliphatic hydrocarbons in Pre-Cambrian rocks: Nature, v. 213, p. 1082-1085, 2 figs.

Oró, J., Nooner, D. W., Zlalkis, A., Wikstrom, S. A., and Barghoorn, E. S., 1965, Hydrocarbons of biological origin in sediments about two billion years old: Science, v., 148, p. 77-79.

Papenfuss, G. F., 1955, Classification of the algae: A century of progress in the natural sciences 1853-1953: San Francisco, California, Acad. Sci., p. 115-224.

Park, Kihlo, 1966, Deep-sea *p*H: Science, v. 154, p. 1540-1542, 1 fig.

Parke, Mary, 1966, The genus *Pachysphaera* (Prasinophyceae), p. 555-563 *in* Barnes, H., *Editor,* Some contemporary studies in marine science: London, George Allen and Unwin, Ltd.

Parke, Mary, and Adams, Irene, 1960, The motile (*Crystallolithus hyalinus* Gaarder & Markali) and non-motile phases in the life history of *Coccolithus pelagicus* (Wallich) Schiller: Jour. Marine Biological Assoc. U. K., v. 39, p. 263-274, Pls. 1-4.

Parsons, T. R., and Strickland, J. D. H., 1962, On the production of particulate organic carbon by heterotrophic processes in seawater: Deep-Sea Research, v. 8, p. 211-222.

Pearse, A. S., 1936, The migrations of animals from sea to land: Durham, North Carolina, Duke Univ. Press, 176 p., 4 figs.

Pflug, H. D., 1965, Foraminiferen und ähnliche Fossilreste aus dem Kambrium und Algonkium: Palaeontographica, Abt. A, v. 125, p. 46-60, Pls. 10-12, Figs. 1-4.

Proshkina-Lavrenko, A. I., 1960, K evolyutsii diatomovykh vodorosley [On the evolution of diatoms]: Byull. Mosk. Obshch. Ispyt. Prir., otd. biol., v. 65, no. 5, p. 52-62.

Provasoli, L., 1958, Nutrition and ecology of Protozoa and algae: Ann. Rev. Microbiology, v. 12, p. 279-308.

Pytkowicz, R. M., 1967, Carbonate cycle and the buffer mechanism of recent oceans: Geochim. et Cosmochim. Acta, v. 31, p. 63-73, 2 figs.

Raymont, J. E. G., 1966, The production of marine plankton: Advances Ecol. Research, v. 3, p. 117-205.

Redfield, A. C., Ketchum, B. H., and Richards, F. A., 1963, The influence of organisms on the composition of sea water, *in* Hill, M. N., *Editor,* The Sea, Pt. 2: New York, John Wiley & Sons, p. 26-77.

Reimann, B. E. F., Lewin, J. C., and Volcani, B. E., 1966, Studies on the biochemistry and fine structure of silica shell formation in diatoms. II. The structure of the cell wall of *Navicula pelliculosa* (Bréb.) Hilse: Jour. Phycology, v. 2, p. 74-84, 21 figs.

Rensch, Bernhard, 1960, Evolution above the species level: New York, Columbia Univ. Press, Columbia Biological Series no. 19, 419 p., 113 figs.

Richards, F. A., 1957, Oxygen in the ocean: Geol. Soc. America Mem. 67, v. 1, p. 185-238, 11 figs.

Riley, G. A., 1944, The carbon metabolism and photosynthetic efficiency of the earth as a whole: Am. Scientist, v. 32, p. 129-134.

Robertson, J. D., 1962, The membrane of the living cell: Sci. American, v. 206, p. 64-72, illus.

Romer, A. S., 1966, Vertebrate paleontology (3d ed.): Chicago, Chicago Univ. Press, 468 p., 443 figs.

Ronov, A. B., 1958, Organic carbon in sedimentary rocks (in relation to the presence of petroleum): Geochemistry (translation of Geokhimiya), p. 510-536, Figs. 1-11.

Rubey, W. W. 1951, Geologic history of sea water. An attempt to state the problem: Geol. Soc. America Bull., v. 62, p. 1111-1147, Figs. 1-4.

——1955, Development of the hydrosphere and atmosphere, with special reference to probable composition of the early atmosphere: Geol. Soc. America Spec. Paper 62, p. 631-650.

Rudavskaya, V. A., 1965, Gistrikhosfery Kembriya yuzhnoy chasti Sibirskoy platformy: Paleofitologicheskiy Sbornik, Vses. Neft. Nauchno-Issled. Geol.-Razved. Inst. (VNIGRI) Trudy, vyp. 239, p. 95-107, Pls. 1-2.

Rudwick, M. J. S., 1965, Ecology and paleoecology, *in* Moore, R. C., *Editor,* Treatise on invertebrate paleontology, Pt. H, Brachiopoda: New York, Geological Society of America, and Lawrence, Kansas Univ. Press, p. H199-H214.

Ryther, J. H., 1959, Potential productivity of the sea: Science, v. 130, p. 602-608.

Sackett, W. M., 1964, The depositional history and isotopic organic carbon composition of marine sediments: Marine Geology, v. 2, p. 173-185, Figs. 1-6.

Sagan, Carl, 1961, On the origin and planetary distribution of life: Radiation Research, v. 15, p. 174-192.

Sarjeant, W. A. S., 1967, Stratigraphical distribution of fossil dinoflagellates: Rev. Palaeobotany Palynology, v. 1, p. 323-343.

Saunders, G. W., 1957, Interrelations of dissolved organic matter and phytoplankton: Bot. Rev., v. 23, p. 389-409.

Schjeide, O. A., Mead, J. F., and Myers, L. S., Jr., 1956, Notions on sensitivity of cells to radiation: Science, v. 123, p. 1020-1022.

Schopf, J. W., and Barghoorn, E. S., 1967, Alga-like fossils from the early Precambrian of South Africa: Science, v. 156, p. 508-512, 8 figs.

Schopf, J. W., Kvenvolden, K. A., anJ Barghoorn, E. S., 1968, Amino acids in Precambrian sediments: an assay: Natl. Acad. Sci. Proc., v. 59, p. 639-646, 4 figs.

Silva, P. C., 1962, Classification of algae, *in* Lewin, R. A., *Editor,* Physiology and biochemistry of algae: New York, Academic Press, p. 827-837.

Simpson, G. G., 1949, The meaning of evolution. A study of the history of life and of its significance for man: New Haven, Yale Univ. Press, 364 p.

Sliter, W. V., 1965, Laboratory experiments on the life cycle and ecologic controls of *Rosalina globularis* d'Orbigny: Jour. Protozoology, v. 12, p. 210-215, 5 figs.

Small, James, 1950, Quantitative evolution XVI. Increase of species-number in diatoms: Ann. Botany, n. ser., v. 14, p. 91-113, 4 figs.

Smayda T. J., 1958, Biogeographical studies of marine phytoplankton: Oikos, v. 9, p. 158-191, 7 figs.

——1966, A quantitative analysis of the phytoplankton of the Gulf of Panama. III. General ecological considerations, and the phytoplankton dynamics at 8° 45′ N, 79° 23′ W from November 1954 to May 1957: Bull. Interamer. Trop. Tuna Comm., v. 11, p. 355-612.

Sperry, Kathleen, 1967, 1. The battle of Lake Erie: Eutrophication and political fragmentation: Science, v. 158, p. 351-355, 2 figs.

Stanier, R. Y., Doudoroff, M., and Adelberg, E. A., 1963, The microbial world (2d ed.): Englewood Cliffs, New Jersey, Prentice-Hall, Inc., 753 p.

Staplin, F. L., Jansonius, Jan, and Pocock, S. A. J., 1965, Evaluation of some acritarchous hystrichosphere genera: Neues Jahrb., Geologie u. Paläontologie Abh., v. 123, p. 167-201, Pls. 18-20, Figs. 1-13.

Strehler, B. L., 1959, Origin and comparison of the effects of time and high-energy radiations on living systems: Quart. Rev. Biology, v. 34, p. 117-142.

Strickland, J. D. H., 1960, Measuring the production of marine phytoplankton: Canada Fisheries Research Board Bull., no. 122, p. 1-172.

——1963, Significance of the values obtained by primary production measurements, *in* Doty, M. S., *Editor,* Conference on primary productivity measurement, marine and freshwater, Hawaii Univ. 1961, Proc., p. 172-183.

Strughold, H., and Ritter, O. L., 1962, Oxygen production during the evolution of the earth's atmosphere: Aerospace Medicine, v. 33, p. 275-278.

Tappan, Helen, 1968a, Three faces of evolution: Biologic, atmospheric and lithologic, p. 218 *in* Abstracts for 1966: Geol. Soc. America Spec. Paper 101.

——1968b, Primary production, isotopes, extinctions and the atmosphere: Palaeogeography, Palaeoclimatology, Palaeoecology, v. 4, p. 187-210, 1 fig.

Tilden, J. E., 1933, A classification of the algae based on evolutionary development, with special reference to pigmentation: Bot. Gaz., v. 95, p. 59-77, 1 pl.

Timofeev, B. V., 1956, Early Paleozoic flora of the Baltic provinces and main stages of its development: XX Congreso Geologico Internacional, Mexico, Resumenes de los Trabajos Presentados, p. 187-188.

Verduin, Jacob, 1964, Principles of primary productivity: Photosynthesis under completely natural conditions, *in* Jackson, D. F., *Editor,* Algae and man: New York, Plenum Press, p. 221-238.

Vien, Cao, 1967, La reproduction sexuée chez un péridinien libre, l'*Amphidinium carteri*: Jour. Protozoology, v. 14, Suppl., p. 45-46.

Vinogradov, A. P., and Boichenko, E. A., 1942, Razlozhenie kaolina diatomovymi vodoroslyami [Decomposition of kaolin by diatoms]: Akad. Nauk SSSR, Doklady, v. 37, p. 135-138.

Wall, D., 1962, Evidence from recent plankton regarding the biological affinities of *Tasmanites* Newton 1875 and *Leiosphaeridia* Eisenack 1958: Geol. Mag., v. 99, p. 353-362, Pl. 17, 2 figs.

Watabe, N., 1967, Crystallographic analysis of the coccolith of *Coccolithus huxleyi*: Calcified Tissue Research, v. 1, p. 114-121, 9 figs.

Weber, J. N., 1967, Possible changes in the isotopic composition of the oceanic and atmospheric carbon reservoir over geologic time: Geochim. et Cosmochim. Acta, v. 31, p. 2343-2351, 2 figs.

Weyl, P. K., 1968, Precambrian marine environment and the development of life: Science, v. 161, p. 158-160.

Wilson, L. R., 1956, Composite micropaleontology and its application to Tertiary and near-Recent stratigraphy: Micropaleontology, v. 2, p. 1-6.

Wood, E. J. F., 1963, Heterotrophic microorganisms in the oceans: Oceanography Marine Biology Ann. Rev., v. 1, p. 197-222.

Yentsch, C. S., 1963, Primary production: Oceanography Marine Biology Ann. Rev., v. 1, p. 157-175.

Zeuner, F. E., 1946, Dating the past, an introduction to geochronology: London, Methuen and Co., Ltd., 444 p., 24 pls., 103 figs.

The Use of Dinoflagellates and Acritarchs for Zonation and Correlation of the Navarro Group (Maestrichtian) of Texas

JAMES B. ZAITZEFF
U.S. Naval Oceanographic Office, Washington, D.C.

AUREAL T. CROSS
Michigan State University, East Lansing, Michigan

ABSTRACT

Eighty-one species of dinoflagellates and acritarchs have been identified from two sections of the Navarro Group in Texas. Forty-one species were distinguished from the Olmos and Escondido Formations from a 1170-foot core section in Frio County. Nearly all of those species and more than 30 additional taxa were identified from surface sections of the Neylandville Marl-Nacatoch Sand, Corsicana Marl, and Kemp Clay composited from several localities near Austin. The ranges of 56 of these entities are plotted to show zonation of the Navarro Group into three parts and tentative correlation of the Austin and Frio County sections. Twenty species found in the Neylandville-Nacatoch rocks near Austin were not found in the Frio County section, indicating that the base of the Navarro in Frio County is younger than the lower part of the Navarro in the Austin area, probably about early middle Corsicanan in age. An unconformity is demonstrated at the top of the Neylandville Marl-Nacatoch Sand in the Austin sections by the termination of range of these 20 species and by the first appearance of 15 additional species in the lowermost Corsicana beds. The ranges of nine forms terminate at approximately the position of the base of the Kemp Clay, to mark the base of the upper of the three acritarch-dinoflagellate zones.

The termination of range, in this area, of *Palaeohystrichophora infusorioides* and the peak development of five other species at successively younger time-correlative levels in both sections furnish a basis for further subdivision of the Navarro and time-line correlations between the two sections about 135 miles apart.

Pollen-spore/microplankton ratios in the Austin section indicate relative water depths or distance from shore, or both. Comparison of palynomorphs from other areas previously studied indicates that the Corsicana Marl and Kemp Clay, and comparable age strata of the Olmos and Escondido Formations, are probably about the same age as the Red Bank Formation of New Jersey and older than the Upper Moreno Formation of California.

CONTENTS

INTRODUCTION

Purpose and Scope

The purpose of this study is to report on the occurrence of dinoflagellates and acritarchs from selected outcrop sections and cores of subsurface rocks of the Navarro Group of east-central and southwestern Texas and to demonstrate that vertical successions of distinctive dinoflagellate and acritarch assemblages can aid in the biostratigraphic zonation and correlation of rocks comprising the Navarro Group in this area.

Two essentially complete sections of the Navarro Group were analyzed. The first, a composite section of Navarro rocks, is in the vicinity of Austin, Texas, and is based on several outcrop localities and supplemented by sections from two cores. The total thickness of this composite section is approximately 550 feet (Fig. 10). The second section analyzed is a subsurface cored sequence consisting of Escondido and Olmos rocks from the northwestern part of Frio County, Texas. The cored section is approximately 1170 feet thick. The Navarro is 1320 feet thick in this well. The Frio County well is about 135 miles southwest of Austin.

Previous Work

There are relatively few published works describing dinoflagellates from Upper Cretaceous and Tertiary rocks from North America. The only description of major Maestrichtian assemblages from North America to date is Drugg's recent report (Drugg, 1967) of the palynomorphs (paleomicroplankton, pollen, and spores) from the Upper Moreno Formation (Maestrichtian-Danian) of California. One must rely primarily on European and Australian works for assemblage comparisons. Tasch and others (1964) described an Albian assemblage from Kansas. Other pertinent North American works that include some descriptions and illustrations of dinoflagellates, acritarchs, or their assemblages are Evitt and others (1967), Leopold and Pakiser (1964), Singh (1964), although the latter two deal with materials from Cretaceous rocks older than Maestrichtian, and Stanley (1965), in which several species are described from the Paleocene of South Dakota, which appear to be conspecific with Navarro forms. European and Australian Mesozoic dinoflagellate studies are more numerous, and these were used extensively here.

SAMPLE LOCALITIES

Sections in the Vicinity of Austin, Texas

Corsicana Marl. Travis County; Jones' Crossing section. 0.25 mile downstream (east) from highway bridge (U.S. 71) over Onion

Creek. Section begins at lowest exposed stratum in creek bed and extends eastward up steep, high bluff. Kemp Clay present at top of section (bright yellow clay). Total thickness 90 feet.

Kemp Clay. Travis County; A. B. Beddow Ranch section. Bluff of Colorado River about 1 mile due north of Garfield, Texas. Base of section is about 50 feet above highest part of the Jones' Crossing section. Total section 40 feet.

Travis County; type locality of the Littig Glauconitic Member, near Littig, Texas. Upper Kemp Clay and Kincaid Formation collected from cut in west side of road leading south from Littig. Cut is 1.85 miles, by road, from railroad crossing in Littig. Upper Kemp Clay also collected in brick clay pit on east side of road leading south from Littig. (Road leading into pit is 1.75 miles from railroad crossing in Littig.)

Neylandville Marl. Travis County; Gilleland Creek section. Taylor Marl-Neylandville Marl. Section measured on southwest bank of Gilleland Creek 3.35 miles south-southwest of Manor on State Highway 973. Section begins in bank of creek 200 feet west of middle bridge (in three-bridge sequence) and passes southwest beside highway along northwest side of road cut.

Frio County Section. Humble Oil and Refining Company, Frank Doering #1 well, located 13.5 miles northwest of Dilley, Texas, in Frio County. Top of the Navarro at 3050 feet. Bottom of the Navarro at 4370 feet. Total cored sequence 3102 to 4272 feet.

GEOLOGY

Nomenclature and Stratigraphic Relations

The Navarro Group, formerly classified as a formation, includes the youngest Cretaceous rocks in the Gulf Coast, the upper part of the Gulfian Series (Figs. 1, 2). The existence of Upper Cretaceous rocks in Texas and to which the name "Navarro beds" was applied was first reported by Shumard (1861, 1863). His type section was located somewhere in Navarro County and may have been a composite of several sections, but he did not specify the location(s) of the type locality, although Chatfield Point and Corsicana are frequently mentioned as fossil localities in his text, and one or both may be portions of the type locality.

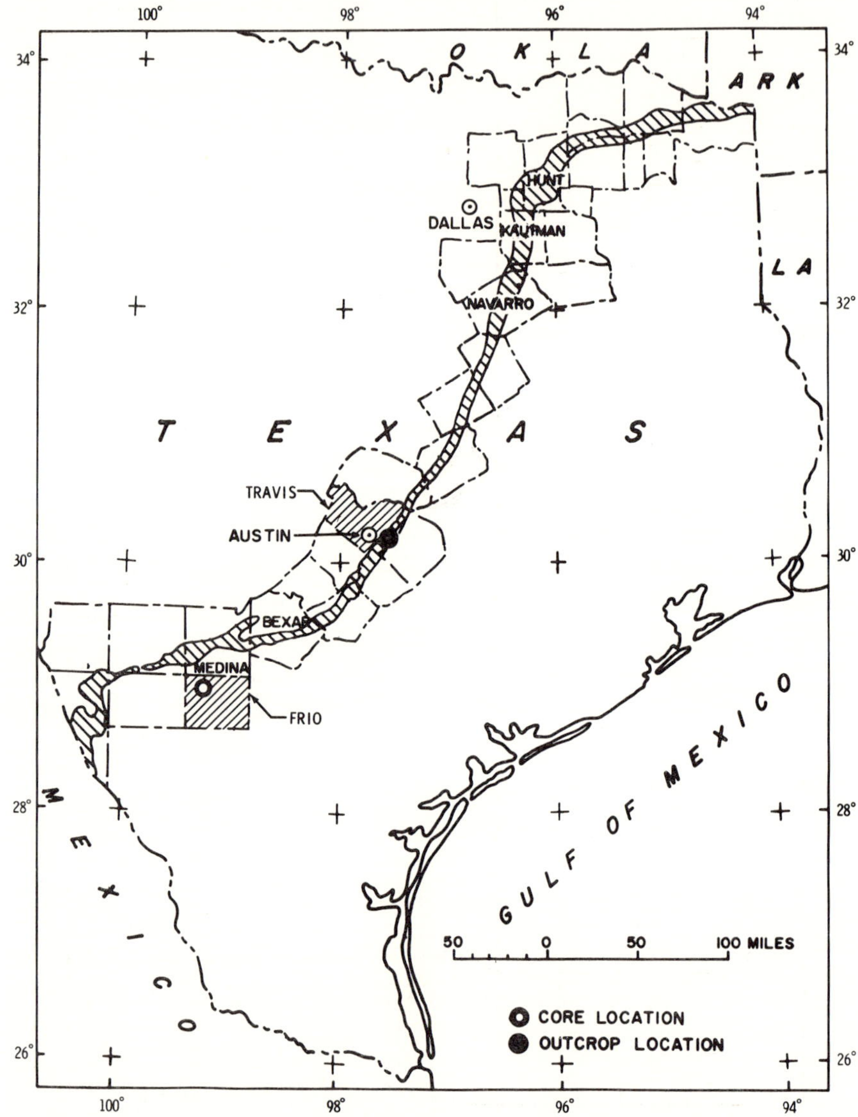

Figure 1. Map showing the belt of outcrop of the Navarro Group in Texas and locations of the Austin, Texas, and Frio County, Texas, sections.

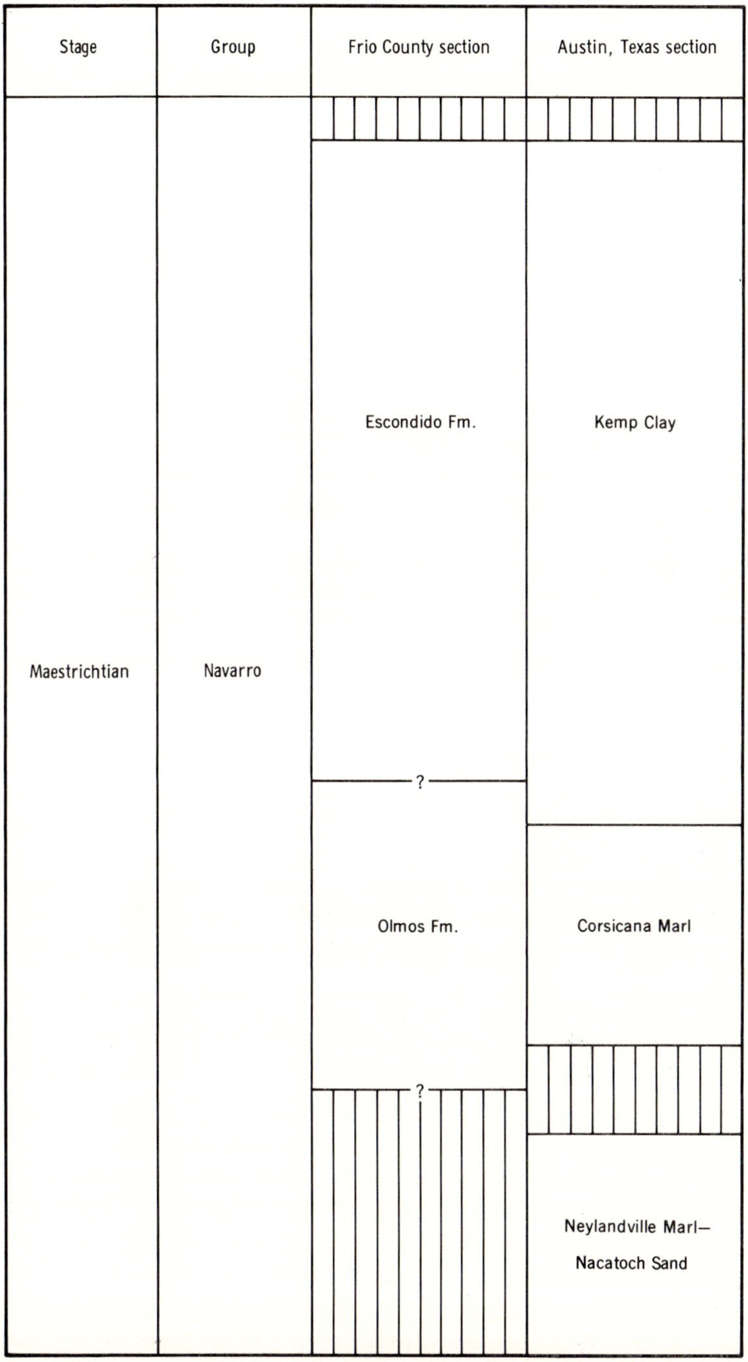

Figure 2. Correlation of the recognized formations or members in the wells or sections studied, Upper Cretaceous Navarro Group, Frio County, Texas, and near Austin, Texas.

The Navarro Group has been subdivided into four formations in Navarro, Kaufman, and Hunt Counties, where it is most completely developed. These formations are, in ascending order: Neylandville Marl, Nacatoch Sand, Corsicana Marl, and Kemp Clay (Adkins, *in* Sellards and others, 1932). This classification was adopted for use by the U.S. Geological Survey and appears on the geological map of Texas issued by the Survey in 1937.

Southwest of Navarro County and westward (Fig. 1), the Kemp Clay changes into rocks of different lithology to which the name "Escondido Formation" is given. The lithologic change is gradational from the Kemp and the separation of these two formations is rather arbitrary.

Invertebrate fossils near the base of the Navarro Group in Bexar and Medina Counties and near the base of the Escondido Formation farther west in Medina County indicate that the age of basal beds there is approximately equivalent to the middle of the Navarro in Navarro, Hunt, and Kaufman Counties in northeastern Texas. This indicates the existence of an unconformity at the base of both the Navarro Group and the Escondido Formation in the Bexar and Medina Counties area. This unconformity between the Navarro and the underlying Taylor Marl continues westward between the Anacacho Limestone and the overlying Escondido Formation, but in Maverick County (Rio Grande Embayment area) the stratigraphic gap represented by the unconformity is, in part, filled in by nonmarine beds, originally known as the "coal series," now defined as the "Olmos Formation." In Maverick County, the Olmos is overlain by the Escondido. The Olmos overlies the San Miguel (Senonian), which may be at least in part equivalent to the Anacacho Limestone farther east.

The Navarro Group of central and northeastern Texas and the Escondido Formation of the southwestern part of the state are unconformably overlain by various overlapping Tertiary formations of Midway or Wilcox age. The amount of time represented by the hiatus between the uppermost Navarro or Escondido and the overlying Tertiary is unknown, but it is probably a long interval.

At different places on the outcrop, Navarro rocks of different ages are directly overlain by overlapping Tertiary strata. Rocks of the upper part of the Escondido Formation are probably younger than the uppermost beds of the Navarro Group in central and northeastern Texas. The differences of age of the youngest deposits of the Navarroan at different localities are due in part to varying amounts of material having been removed by erosion from the top of the Cretaceous at different places, some now exposed in the outcrop, and in part to an earlier retreat of the sea in some areas than in others. The problem as to where the present pre-Tertiary subcrop surface is an erosional surface developed on Navarroan strata and where the Tertiary may conformably overlie depositional surfaces of different ages of Navarroan strata may eventually be resolved by refining such zonation or time-line correlation as is presented here in an introductory way.

General Features of the Navarro Group

The Navarro Group is exposed at the surface in a belt up to 23 miles in width (Fig. 1). The marls, clays, and sands comprising the Navarro Group weather to gray and black soils and subsoils that blanket and partly conceal the formations, or alter their appearance, except where there are fresh exposures. The exposures of greatest thickness occur along banks and bluffs of streams, and occasionally good outcrops are also found in gullies, road cuts, and railroad cuts, along ditches, and near water tanks. Because of the mantle of weathered materials, differentiation and mapping of the various formations and accurate stratigraphic placement of outcrop samples often become difficult and sometimes impossible. The surface distribution of the four formations of the group has been differentiated and mapped, although in a highly generalized fashion.

The sediments composing the marine rocks of the Navarro Group were deposited in water probably not exceeding 100 fathoms (Stephenson and others, 1942) but for the most part deep enough to escape appreciable disturbance by wave action. The rocks consist dominantly of massive bedded marls, chalky marls, clays, and sands with subordinate indurated concretionary masses or layers cemented with calcareous carbonate. Bentonite forms a minor, although well-distributed, part of the group. The aggregate thickness of the Navarro sediments now exposed anywhere in the outcrop is estimated to be 550 to 750 feet, but downdip in the subsurface, the group thickens appreciably to more than 1000 feet.

In Maverick County, the Escondido Formation, equivalent to the Kemp Clay in age, attains thicknesses at the outcrop of 750 feet. The Escondido here consists of dark clays and marls, interbedded with limestones, shales, and sandstones of quite variable thickness and areal distribution. In the area of outcrop of the

Olmos Formation in Maverick County, it ranges in thickness from a few feet to 500 feet. The formation consists largely of nonmarine clays, shales, and sandstones and seams of coal. Along Olmos Creek, in Maverick County, rocks of irregularly stratified sandstones and clays containing ferruginous concretions and silicified wood are exposed. The small beds of sand and clay usually are not constant laterally but are interfingering lenses.

METHODS OF STUDY

Sample Preparation Technique

Preparation of the Navarro shales and calcareous clay shales for microscopic analysis were carried out in a procedure similar to standard palynological preparation techniques for the study of acid-insoluble micro-organisms.

To release the organic entities from the enclosing matrix, 5-gram aliquots of samples were treated successively with HF, HCl, Schulze's solution, and NH_4OH, with intermittent washings, and then dispersed in a dilute solution of Alcojet,[1] which acts to some extent to remove additional humic material and silt-sized particles, followed by a specific gravity separation in zinc chloride solution at 1.90 specific gravity. The residues were then stained with Safranin O. Residues are stored in an aqueous solution of glycerin and phenol.

Sample Analysis

Four slides of each residue were made. One entire slide was examined in detail and types were recorded and established. A total of 5000 dinoflagellates and acritarchs were observed in some instances on the first slide. A second slide was used to insure random sampling, for the determination of relative abundance of the individual species. Relative abundance of species was determined by the recording of not less than 200, where specimens were rare, and a minimum of 500 where specimens were abundant. Also recorded was the relative abundance of pollen and spore groups or genera such as bisaccates, *Proteacidites, Classopollis,* triporates, tricolpates, *Ephedra,* and *Sphagnum.*

---

[1]Alcojet is a patented detergent manufactured by Alconox Inc., New York, New York.

ZONATION AND CORRELATIONS

Dinoflagellate-Acritarch Assemblages Selected

Of the various palynomorphs differentiated in the study of the plant microfossils of the Navarro (Zaitzeff, 1967), paleomicroplankton appeared to be most promising for zoning the marine Navarro rocks and correlating constituent strata. Some other palynomorphs, such as dispersed pollen, spores, and detritus of larger plants also are of value, particularly if their relative frequency in the total microfossil flora is considered or if their natural affinities are known. The present study excludes these terrestrial palynomorphs except for several groups or particular kinds of pollen and spores used to aid in analyzing environmental conditions of sedimentation by pollen-spore/microplankton ratios as shown in Figure 11.

Table 1 is a list of 56 species of acritarchs and dinoflagellates deemed useful for zonation and correlation out of the total microplankton flora. Each of these entities is illustrated on Plates 1 to 6, and the stratigraphic range of each is plotted for both the Austin and Frio sections in Figure 3.

Zonation of the Austin Section

Eighty-one species of dinoflagellates and acritarchs have been recognized in rocks of the Navarro Group in the Austin section in this study. The vertical distributions of 56 selected species are recorded on the range chart (Fig. 3). The Neylandville and Corsicana Marls contain a number of species that, based on data presented here, appear to have restricted ranges, whereas the Kemp Clay contains many species that are common to the Corsicana Marl and Neylandville Marl as well. The Kemp has few if any restricted forms. Several of the 14 species that continue from the Neylandville Marl through the Corsicana Marl and Kemp Clay have known occurrences in the Tertiary rocks from other areas. The dinoflagellate-acritarch zones are based on the qualitative nature of the assemblages. The characteristics of the zones are discussed below.

Zone A. Zone A includes the Neylandville Marl in the sampled area. There are 40 species recognized in the Neylandville Marl, 18 of which are also common to the Corsicana Marl and Kemp Clay. The top of Zone A is marked by the uppermost (youngest)

TABLE 1. LIST OF ACRITARCHS AND DINOFLAGELLATES*

1. *Baltisphaeridium* sp. 2
2. *Cymatiosphaera radiata* O. Wetzel 1933
3. *Palambages* Forma A
4. *Palambages deflandrei* Gorka 1963
5. *Hystrichosphaeridium* sp. 1
6. *Hystrichosphaeridium* sp. 2
7. *Hystrichosphaeridium patulum* Davey & Williams 1966
8. *Palaeostomocystis* sp. 1
9. *Polysphaeridium* sp. 4
10. *Polysphaeridium* sp. 2
11. *Tanyosphaeridium* sp. 1
12. *Tanyosphaeridium* sp. 2
13. *Cleistosphaeridium* sp. 1
14. *Polysphaeridium* sp. 1
15. Forma F sp. 1

16, 17. Forma A sp. 1

18. Forma A sp. 3
19. Forma A sp. 2
20. *Callaiosphaeridium* cf. *asymmetricum* (Deflandre & Courteville) Davey & Williams 1966
21. *Systematophora* sp. 1
22. *Cordosphaeridium fibrospinosum* Davey & Williams 1966
23. *Cordosphaeridium ? inodes* Klumpp 1953
24. *Achomosphaera ramulifera* (Deflandre) Evitt 1963
25. *Hystrichosphaera ramosa* var. 1
26. *Hystrichosphaera ramosa* (Ehrenberg) Davey & Williams 1966
27. 27a. *Cyclonephelium* sp. 1
28. *Cannosphaeropsis fenestrata* Deflandre & Cookson 1955
29. *Cyclonephelium divaricatum* Williams & Downie 1966

30-30b. Forma B sp. 1

31. *Tenua* sp. 1
32. *Tenua* sp. 2
33. *Areoligera senonensis* Lejeune-Carpentier 1938
34. *Areoligera* sp. 2
35. *Dinogymnium nelsonense* (Cookson) E., C., & V. 1967
36. *Heliodinium* cf. *voigti* Alberti 1961
37. *Dinogymnium* sp. 1
38. *Dinogymnium* sp. 3

39, 44. *Dinogymnium digitus* (Deflandre) E., C., & V. 1967

40, 41. *Dinogymnium westralium* (Cookson & Eisenack) E., C., & V. 1967

42. *Dinogymnium* sp. 4
43. *Diconodinium* sp. 1
45. *Gonyaulax* sp. 1
46. *Odontochitina striatoperforata* Cookson & Eisenack 1962
47. *Deflandrea acuminata* Cookson & Eisenack 1958

48, 51. Forma C. sp. 1

49. *Deflandrea microgranulata* Stanley 1965
50. *Deflandrea cooksoni* Alberti 1959
52. *Svalbardella lidiae* Gorka 1963
53. *Deflandrea* sp. 5
54. *Deflandrea* sp. 6

55, 56. *Deflandrea pannucea* Stanley 1965

57, 58. *?Deflandrea* sp. 1

59, 59a. *Deflandrea* sp. 4

60, 63. *Palaeohystrichophora* sp. 1

61. *Deflandrea magnifica* Stanley 1965
62. *Palaeohystrichophora infusorioides* Deflandre 1934

*Numbers correspond to those on Figure 3 and in Plates 1-6.

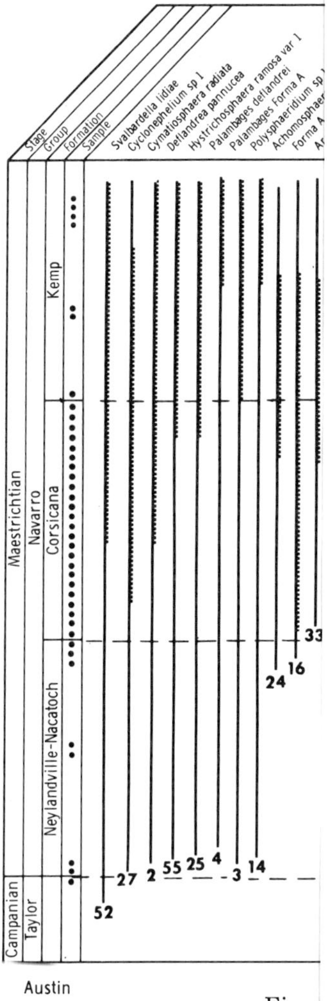

Figur
diagramm
and well
of Austin,
Formatio
samples a
A, B, and
in Table
the lower
differenti

occurrence of 18 species of the 33 indicated on the range chart (Fig. 3) in the Neylandville Marl-Nacatoch Sand. Seven of these eighteen species whose ranges terminate at the top of Zone A (*Odontochitina striatoperforata* (46)[2], *Heliodinium* cf. *voigti* (36), *Callaiosphaeridium* cf. *asymmetricum* (20), *Deflandrea*? *micracantha, Cannosphaeropsis fenestrata* (28), *Dinogymnium digitus* (39), *Deflandrea acuminata* (47)) were described in earlier literature from older rocks in other areas. None of these seven or *Palaeohystrichophora infusorioides* (62), which occurs in the lower part of Zone A, has been previously recorded from rocks as young as the Maestrichtian.

Although Zone A is rather rich in species, the number of individuals found in all except a few species is low, *Dinogymnium digitus* (39), *D. nelsonense* (35), and *Heliodinium* cf. *voigti* (36), when present in a sample, usually occur in great abundance. The zone is characterized in part by the abundance of these three species. The following species are considered good indicators of Zone A, whether they are found singly or in combination: *Heliodinium* cf. *voigti* (36), Forma C sp. 1 (48), *Odontochitina striatoperforata* (46), *Deflandrea* sp. 6 (54), *Cannosphaeropsis fenestrata* (28), *Deflandrea acuminata* (47). The division between Zone A and Zone B should prove to be of time-stratigraphic significance on a regional as well as a local basis. The sharp contrast in the dinoflagellate assemblage between the zones is probably indicative of an unconformity at the top of the Neylandville Marl, and the difference is not interpreted as facies-controlled.

Zone B. Zone B is characterized by a great diversity and abundance of species. The bottom of the zone is marked by the occurrence of 15 new forms that are present in the lowermost two samples of the Corsicana, also indicated in Figure 3. The top of the zone is less precisely defined, but it is marked by the youngest occurrence of seven species (Fig. 3). The species Forma B sp. 1 (30), Forma A sp. 1 (16), Forma A sp. 3 (18), *Diconodinium* sp. 1 (43), *Cordosphaeridium fibrospinosum* (22), and *Tanyosphaeridium* sp. 1 (11) usually occur in abundance in samples throughout the zone.

Zone C. The bottom of Zone C is marked by the greatest abundance of several species. The top of the zone is not indicated by the palynomorphs in the Austin composite section as the section

---

[2]Numbers in parentheses correspond to those on Figure 3 and Plates 1-6.

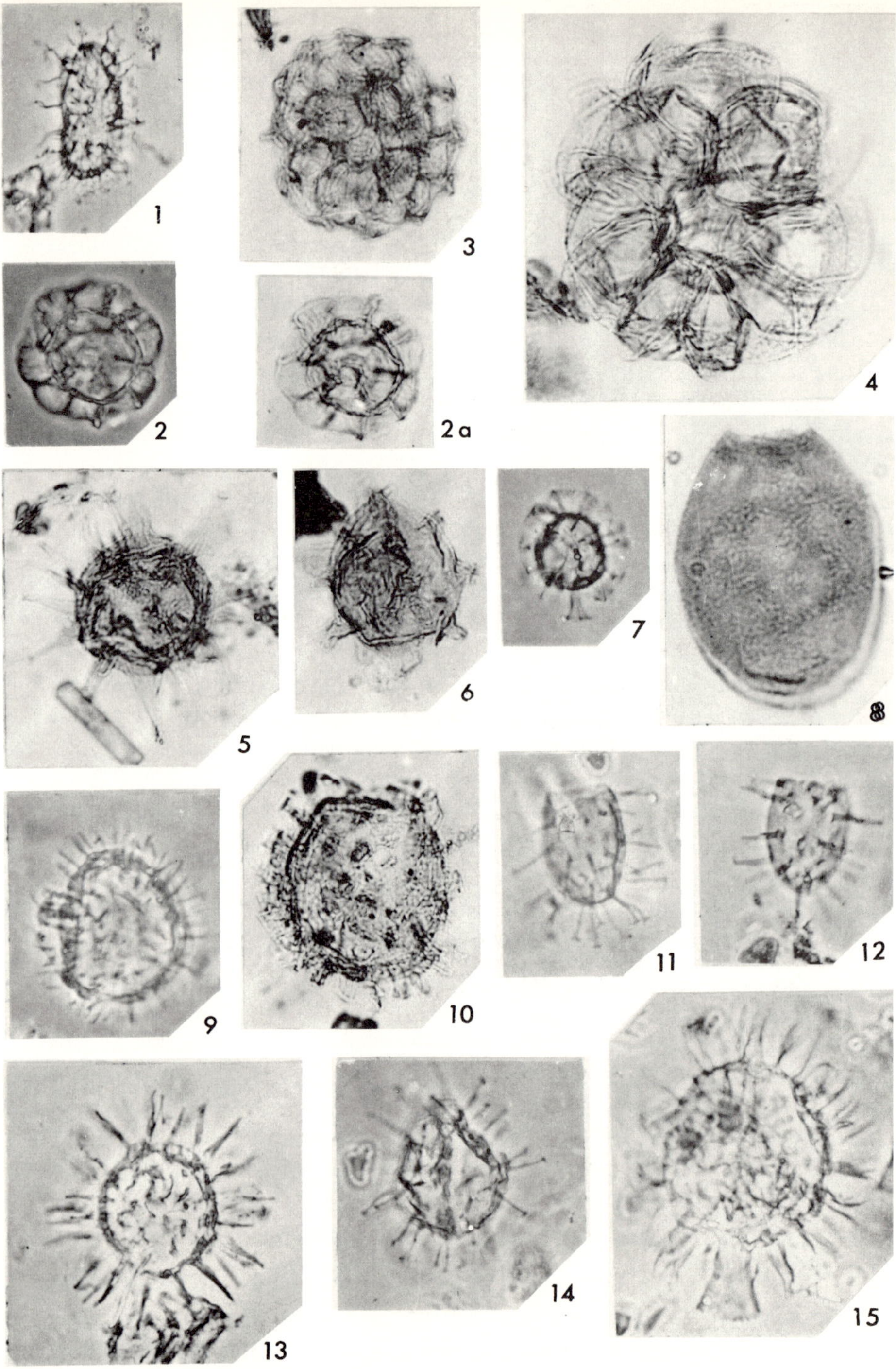

NAVARRO GROUP PALEOMICROPLANKTON

ZAITZEFF AND CROSS, PLATE 1
Geological Society of America Special Paper 127

PLATE 1. NAVARRO GROUP PALEOMICROPLANKTON

Figures ×500 unless otherwise indicated.

Subgroup Acanthomorphitae

Figure
1. *Baltisphaeridium* sp. 2 (phase contrast).

Subgroup Herkomorphitae

2, 2a. *Cymatiosphaera radiata* (2, phase contrast).

Class Chlorophyceae

3. *Palambages* Forma A.
4. *P. deflandrei.*

Subgroup uncertain

8. *Palaeostomocystis* sp. 1 ×1250.

Class Dinophyceae, Family Hystrichosphaeridiaceae

5. *Hystrichosphaeridium* sp. 1.
6. *H.* sp. 2.
7. *H. patulum* (phase contrast).
9. *Polysphaeridium* sp. 4 (phase contrast).
10. *P.* sp. 2.
14. *P.* sp. 1 (phase contrast).
11. *Tanyosphaeridium* sp. 1 (phase contrast).
12. *T.* sp. 2 (phase contrast).

Subgroup Acanthomorphitae

13. *Cleistosphaeridium* sp. 1 (phase contrast).
15. Forma F sp. 1 (phase contrast).

PLATE 2. NAVARRO GROUP PALEOMICROPLANKTON

All figures ×500.

Class Dinophyceae, Family Hystrichosphaeridiaceae

Figure

16, 17. Forma A sp. 1.
18. Forma A sp. 3.
19. Forma A sp. 2.
20. *Callaiosphaeridium* cf. *asymmetricum* (phase contrast).
21. *Systematophora* sp. 1.
22. *Cordosphaeridium fibrospinosum.*
23. *C. ?inodes.*

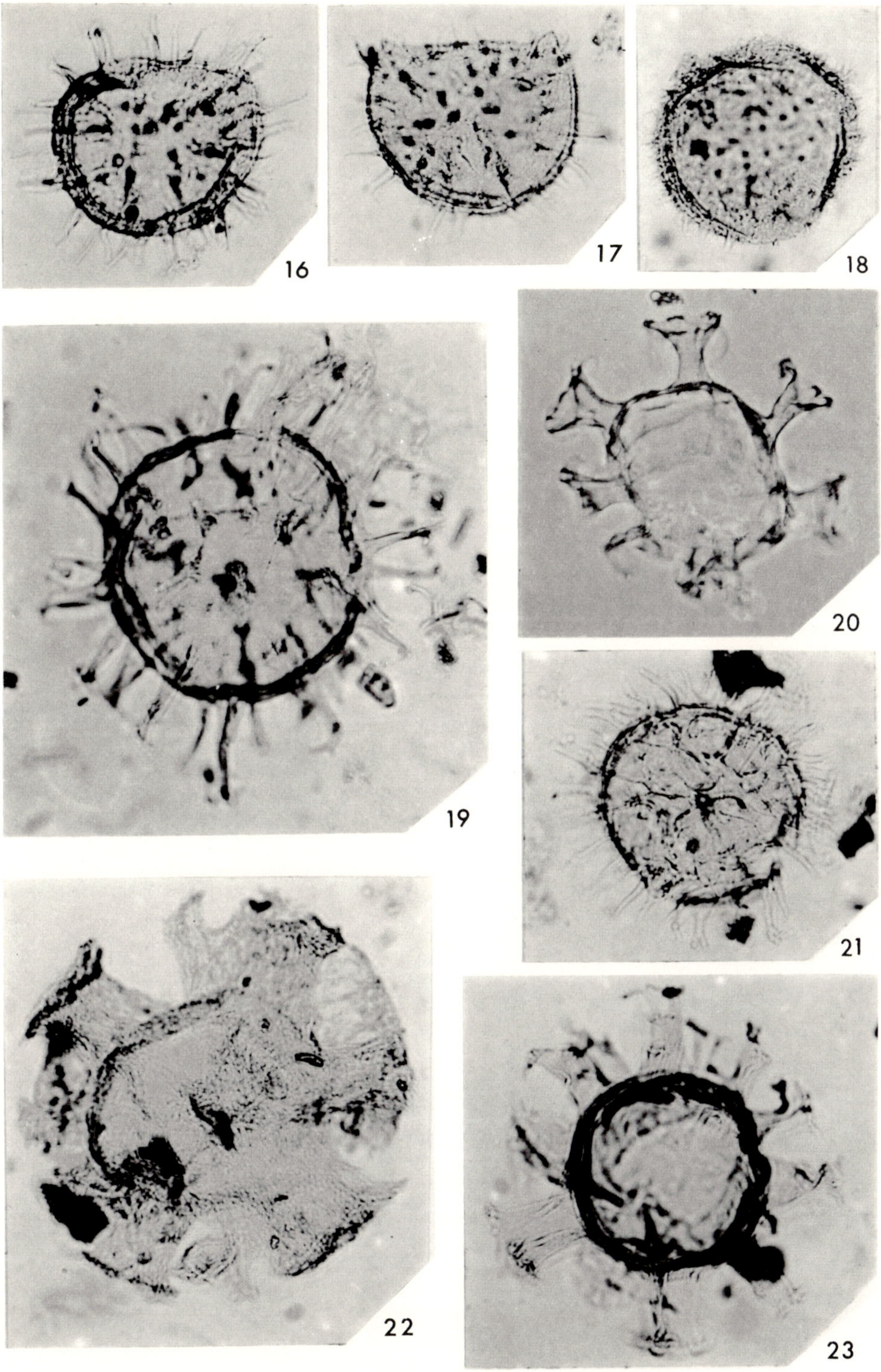

NAVARRO GROUP PALEOMICROPLANKTON

ZAITZEFF AND CROSS, PLATE 2
Geological Society of America Special Paper 127

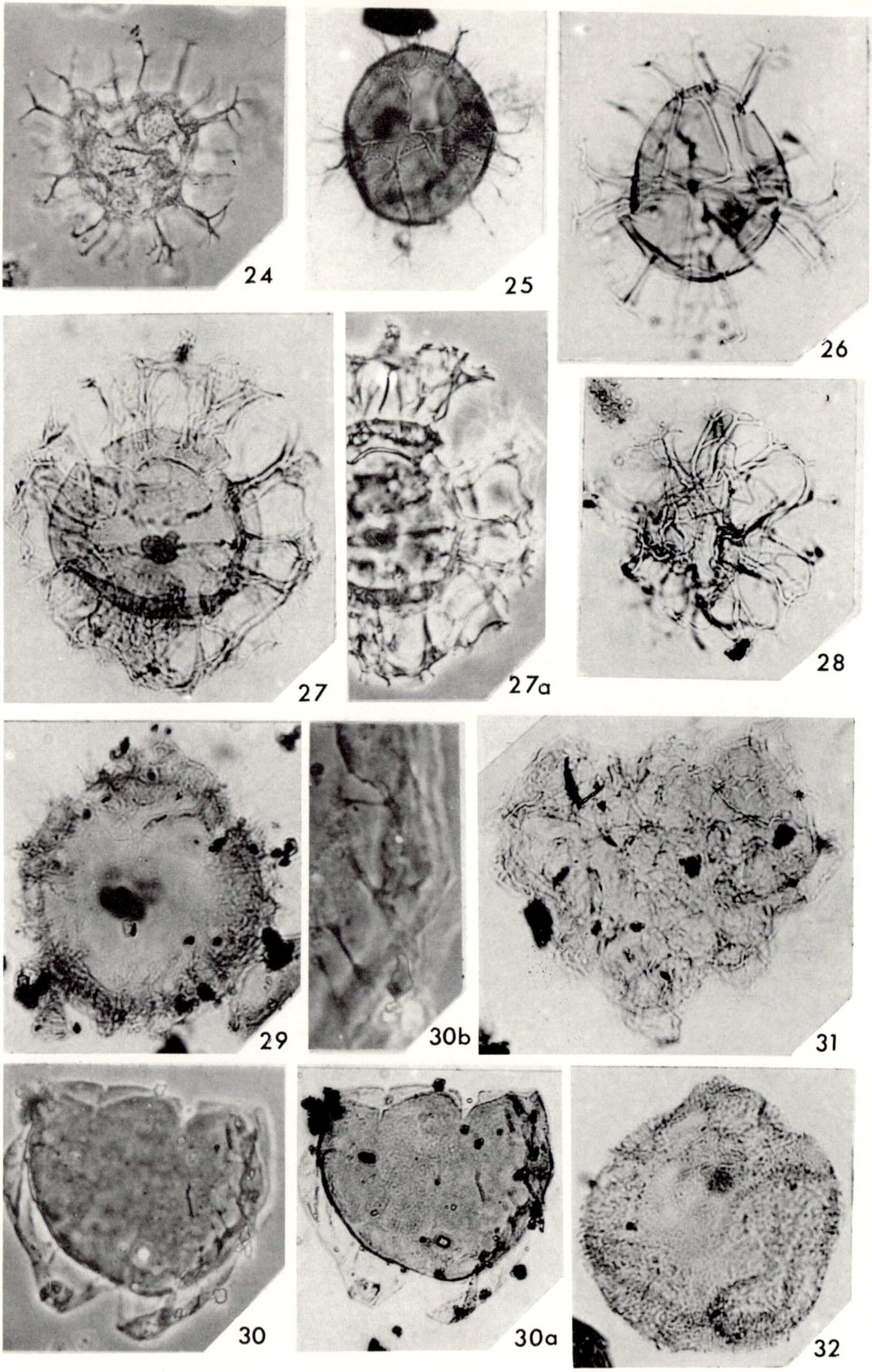

NAVARRO GROUP PALEOMICROPLANKTON

ZAITZEFF AND CROSS, PLATE 3

PLATE 3. NAVARRO GROUP PALEOMICROPLANKTON

Figures ×500 unless otherwise indicated.

Class Dinophyceae, Family Hystrichosphaeridiaceae

Figure

24. *Achomosphaera ramulifera.*
25. *Hystrichosphaera ramosa* var. 1.
26. *H. ramosa.*
28. *Cannosphaeropsis fenestrata.*

Family Areoligeraceae

27, 27a. *Cyclonephelium* sp. 1 (27a, phase contrast).

29. *C. divaricatum.*

30-30b. Forma B sp. 1 (30, 30b, phase contrast; 30b ×1250 showing fine hairlike processes attached to membranelike structure on ventral surface of central body).

31. *Tenua* sp. 1.

32. *T.* sp. 2.

PLATE 4. NAVARRO GROUP PALEOMICROPLANKTON

Figures ×500 unless otherwise indicated.

Class Dinophyceae, Family Areoligeraceae

Figure

33. *Areoligera senonensis.*
34. *A.* sp. 2.

Family Hystrichodiniaceae

36. *Heliodinium* cf. *voigti* (phase contrast).

Family Gymnodiniaceae.

35. *Dinogymnium nelsonense.*
37. *D.* sp. 1, ×1250.
38. *D.* sp. 3, ×1250.
39, 44. *D. digitus* (fig. 39, ×1250).
40, 41. *D. westralium* (40, phase contrast).
42. *D.* sp. 4.
43. *Diconodinium* sp. 1, ×1250.

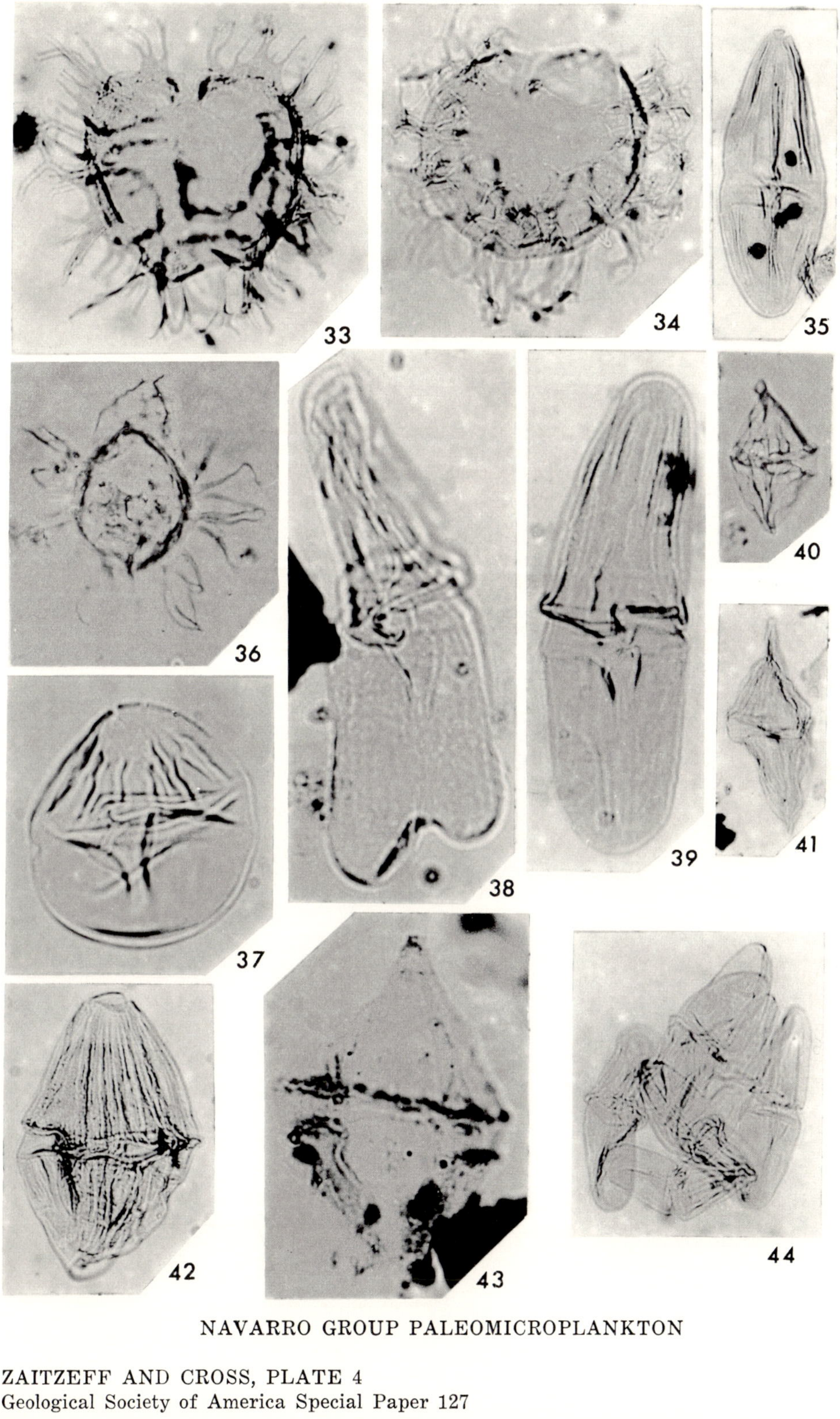

NAVARRO GROUP PALEOMICROPLANKTON

ZAITZEFF AND CROSS, PLATE 4

Geological Society of America Special Paper 127

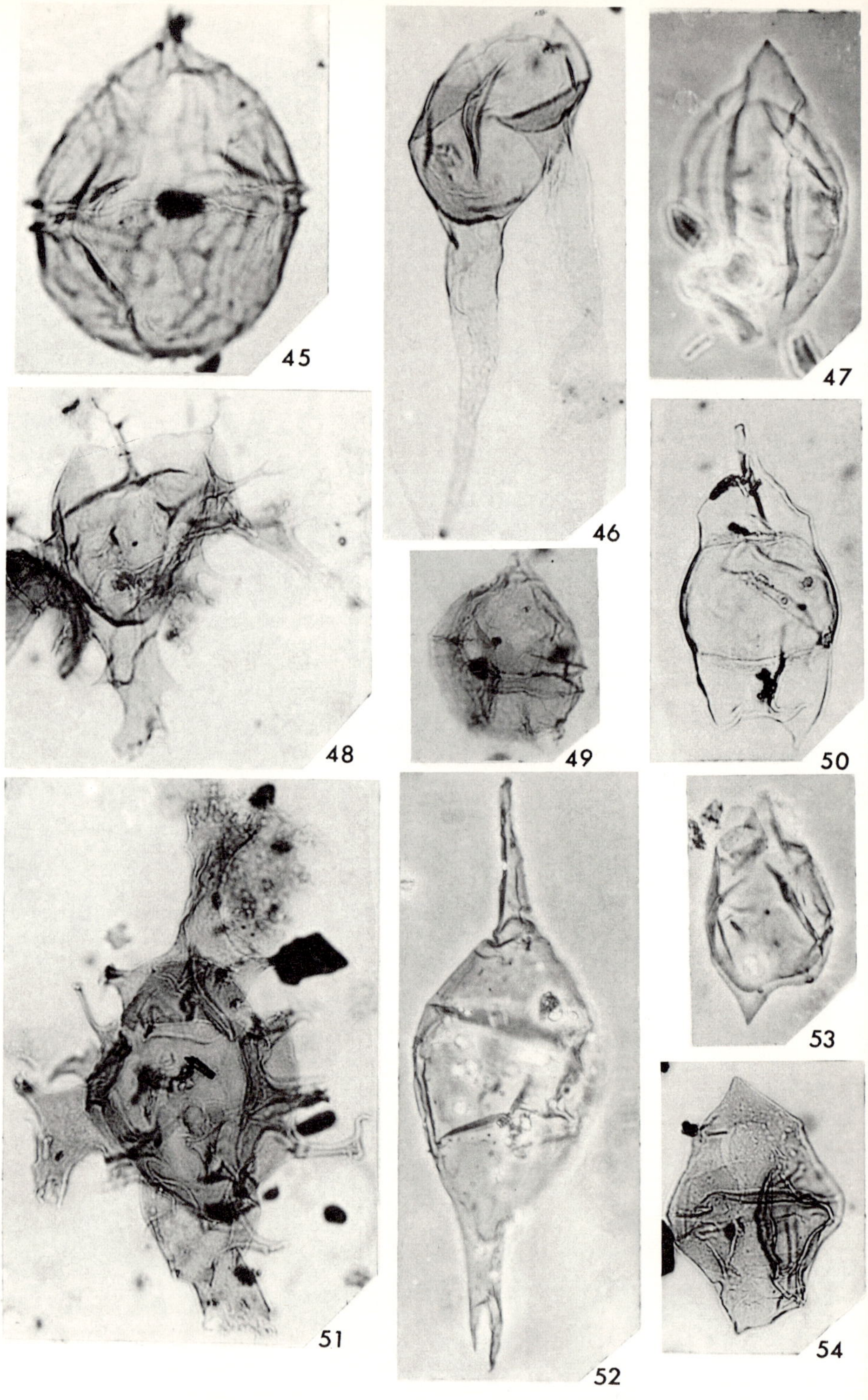

NAVARRO GROUP PALEOMICROPLANKTON

ZAITZEFF AND CROSS, PLATE 5
Geological Society of America Special Paper 127

PLATE 5. NAVARRO GROUP PALEOMICROPLANKTON

All figures ×500.

Class Dinophyceae, Family Gonyaulacaceae

Figure

45. *Gonyaulax* sp. 1.

Family Pseudoceratiaceae

46. *Odontochitina striatoperforata.*
48. 51. Forma C sp. 1.

Family Deflandreaceae

47. *Deflandrea acuminata* (phase contrast).
49. *D. microgranulata.*
50. *D. cooksoni.*
52. *Svalbardella lidiae* (phase contrast).
53. *D.* sp. 5 (phase contrast).
54. *D.* sp. 6.

PLATE 6. NAVARRO GROUP PALEOMICROPLANKTON

All figures ×500.

Class Dinophyceae, Family Deflandreaceae

Figure
55, 56. *Deflandrea pannucea* (phase contrast).
57, 58. ?*Deflandrea* sp. 1.
59, 59a. *D.* sp. 4 (59 phase contrast).
61. *D. magnifica.*

Family uncertain

60, 63. *Palaeohystrichophora* sp. 1 (phase contrast).
62. *P. infusorioides* (phase contrast).

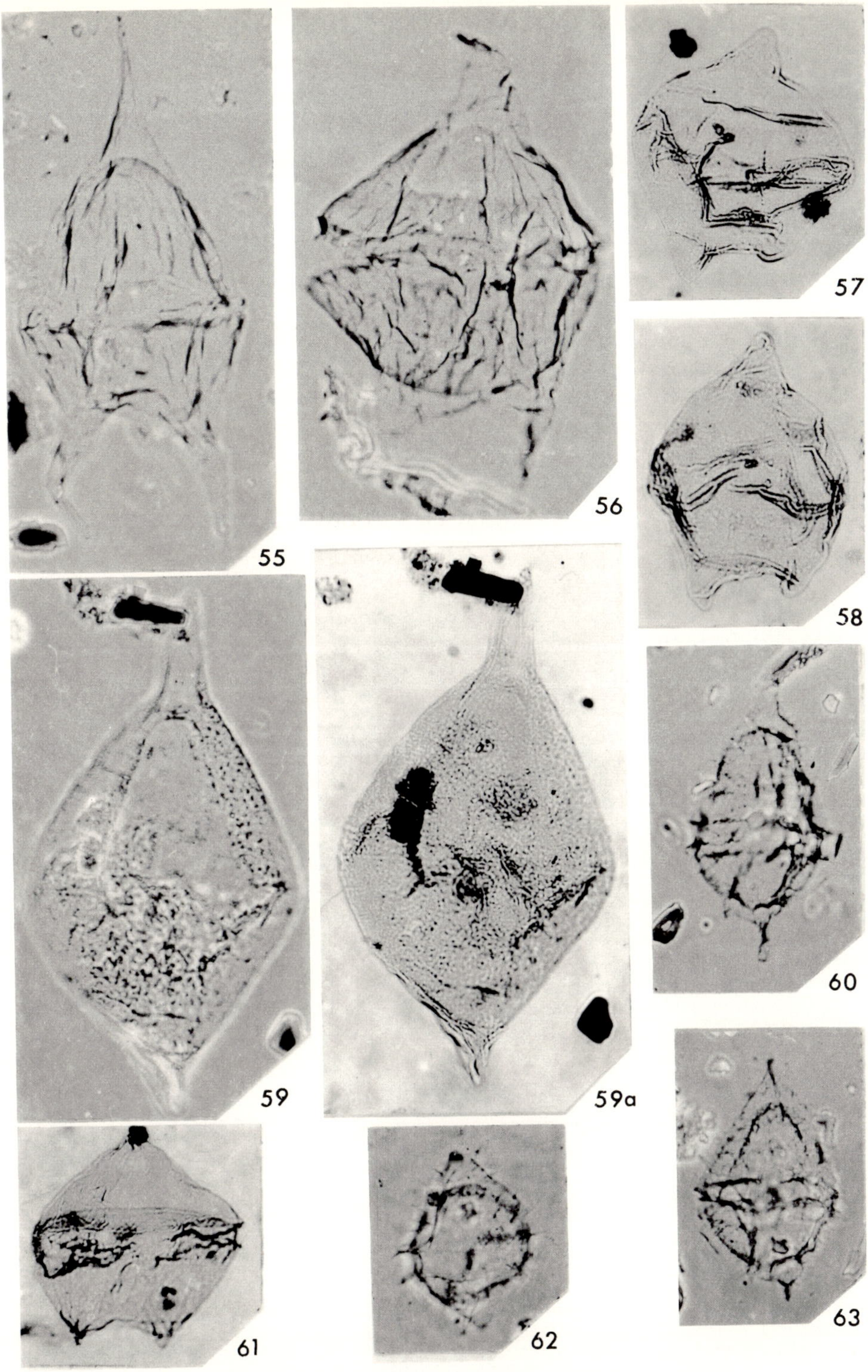

NAVARRO GROUP PALEOMICROPLANKTON

ZAITZEFF AND CROSS, PLATE 6
Geological Society of America Special Paper 127

was not sampled higher than the top of the Kemp Clay in this area and all 22 species present above the lowermost sample continue entirely through the portion of the Kemp sampled (Fig. 3). The youngest ranges of species first appearing in Zone C are not known, but many of the forms found, which have been previously described in the literature, are known to occur in younger sediments. There are two species illustrated, *Cyclonephelium divaricatum* (29) and *Dinogymnium westralium* (40), and a couple of others of lesser importance, which appear to be markers for the upper part of the zone.

Zonation of the Frio County Section

In the original palynologic study of the Frio County sections (Zaitzeff, 1967), 41 acritarchs and dinoflagellates were differentiated in the Olmos and Escondido Formations of the Navarro Group, 30 of which are illustrated here (Pls. 1-6). Their ranges are indicated in Figure 3.

Of the 20 species in the Austin sections whose ranges are limited to the Neylandville-Nacatoch rocks (*see* right side of Fig. 3), none is present in the samples studied from the Frio County sections. Only one species, *Palaeohystrichophora* sp. 1 (60), found only in the lower part of Zone B of the Frio section, is not represented in the Austin section. Of the 20 species in the Austin section that first appear at or just above the base of Zone B, none appears near the base of the Navarro strata in the Frio County area, but 19 of these appear above the middle, and 6 others, whose ranges begin earlier in Zone A in the Austin section, also appear for the first time above the middle of the Olmos. The ranges of 22 species appear to terminate below the top of the Escondido; 10 of these are species that do not extend above, or much higher than, the top of Zone B.

It is obvious that some discrepancies of range should occur because of facies represented by the sediments of the Austin and Frio County areas, but certain correlations are evident, as will be discussed.

Correlation of the Navarro Group, Austin and Frio Sections

Figures 4 to 9 show relative abundance percentages of selected dinoflagellate species used to correlate the two sections studied. Figure 10, a composite of this information, shows six horizons that can be identified in both sections; five horizons are based on positions of maximum occurrence of different species and one represents the uppermost occurrence of *Palaeohystrichophora* (62).

The termination of the range of *Palaeohystrichophora* is taken as the horizon at the top of a zone that occurs near the base of the Navarro sections at the two localities. *Palaeohystrichophora* sp. 1 (60) is very abundant in the lowermost samples of the Frio section and is not found above the 4254-foot depth in that well (Fig. 3). *Palaeohystrichophora infusorioides* (62) was recorded on the basis of a few specimens from the Neylandville Marl in the Austin section and was not seen in younger strata there.

Three horizons are differentiated that divide the Corsicana Marl into two zones in the Austin composite section and the same horizons are present in equivalent positions in the Frio County section. The base of the Corsicana is marked by a high occurrence of *Cyclonephelium* sp. 1 (27) where it represents 60 percent of the dinoflagellate assemblage (Fig. 4). This high abundance is reflected in samples from the Olmos Formation, Frio County. The next horizon identified in the Corsicana is the level where Forma A sp. 1 (16) (Fig. 6) attains its maximum abundance. This level may also be differentiated in samples from the Olmos Formation in the Frio section. A third horizon is differentiated near the top of the Corsicana Marl and in samples from the Olmos Formation. This horizon is identified as the time of maximum occurrence of Forma B sp. 1 (30) (Fig. 6).

The Kemp Clay in the Austin section is marked by two horizons that can also be identified in the Escondido Formation in Frio County. The lower is represented by a maximum occurrence of *Deflandrea magnifica* (61) (Fig. 7) and is associated with high occurrence of *Svalbardella lidiae* (52) (Fig. 8). Samples near the top of the Kemp Clay and the Escondido Formation contain a high abundance of *Hystrichosphaera ramosa* var. 1 (25) (Fig. 9), which is here taken to represent a correlative horizon.

These proposed correlations based on positions of maximum occurrence of single species are considered to be reliable horizon indicators for subdividing the Navarro Group.

In establishing these horizons, floods of occurrence of a single species can be used as a criterion for their recognition, and, particularly where occurrences are repetitious in a vertical sequence, trends of the microplankton spectra should be matched and the associated species considered.

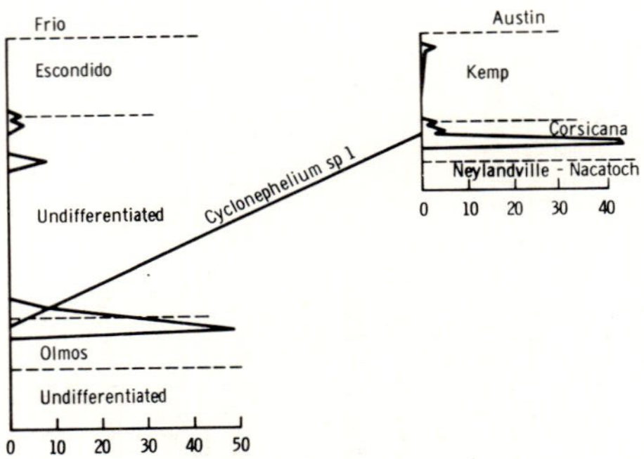

Figure 4. Relative abundance of *Cyclonephelium* sp. 1 shown as percentage of microplankton sum in the Austin and Frio sections. Line indicates correlations.

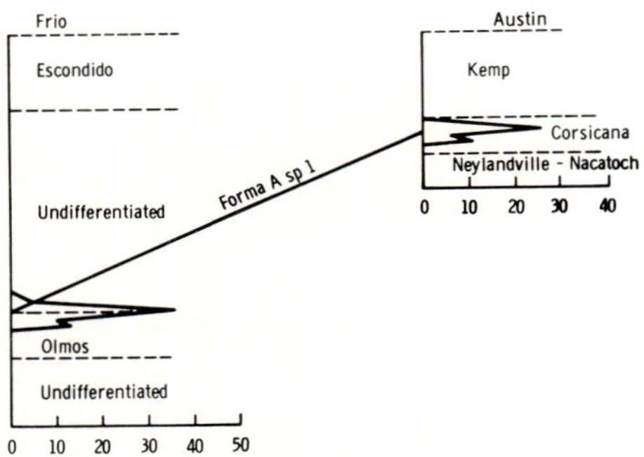

Figure 5. Relative abundance of Forma A sp. 1 shown as percentage of microplankton sum in the Austin and Frio sections. Line indicates correlations.

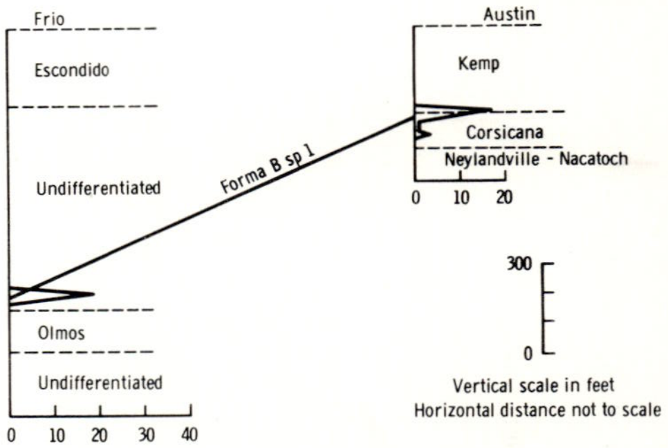

Figure 6. Relative abundance of Forma B sp. 1 shown as percentage of microplankton sum in the Austin and Frio sections. Line indicates correlations.

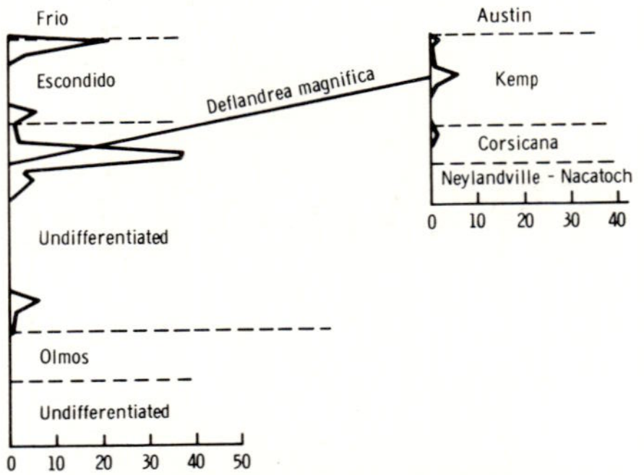

Figure 7. Relative abundance of *Deflandrea magnifica* shown as percentage of microplankton sum in the Austin and Frio sections. Line indicates correlations.

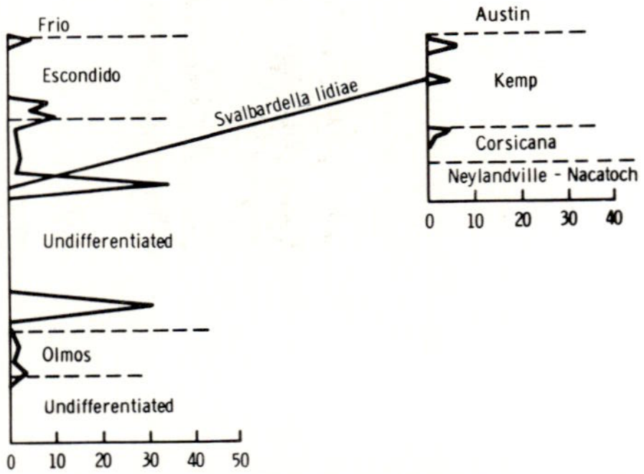

Figure 8. Relative abundance of *Svalbardella lidiae* shown as percentage of microplankton sum in the Austin and Frio sections. Line indicates correlations.

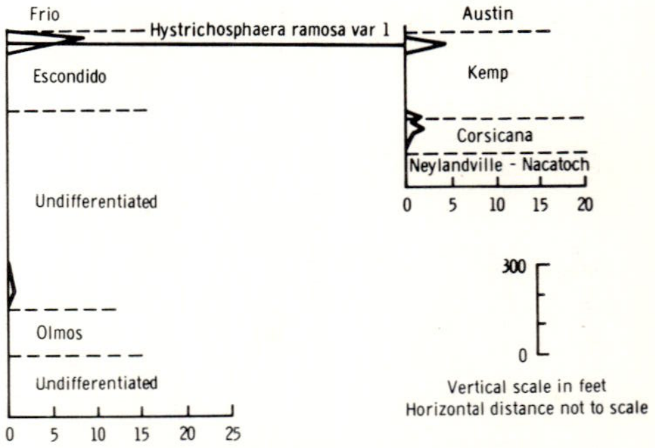

Figure 9. Relative abundance of *Hystrichosphaera ramosa* var. 1 shown as percentage of microplankton in the Austin and Frio sections. Line indicates correlations.

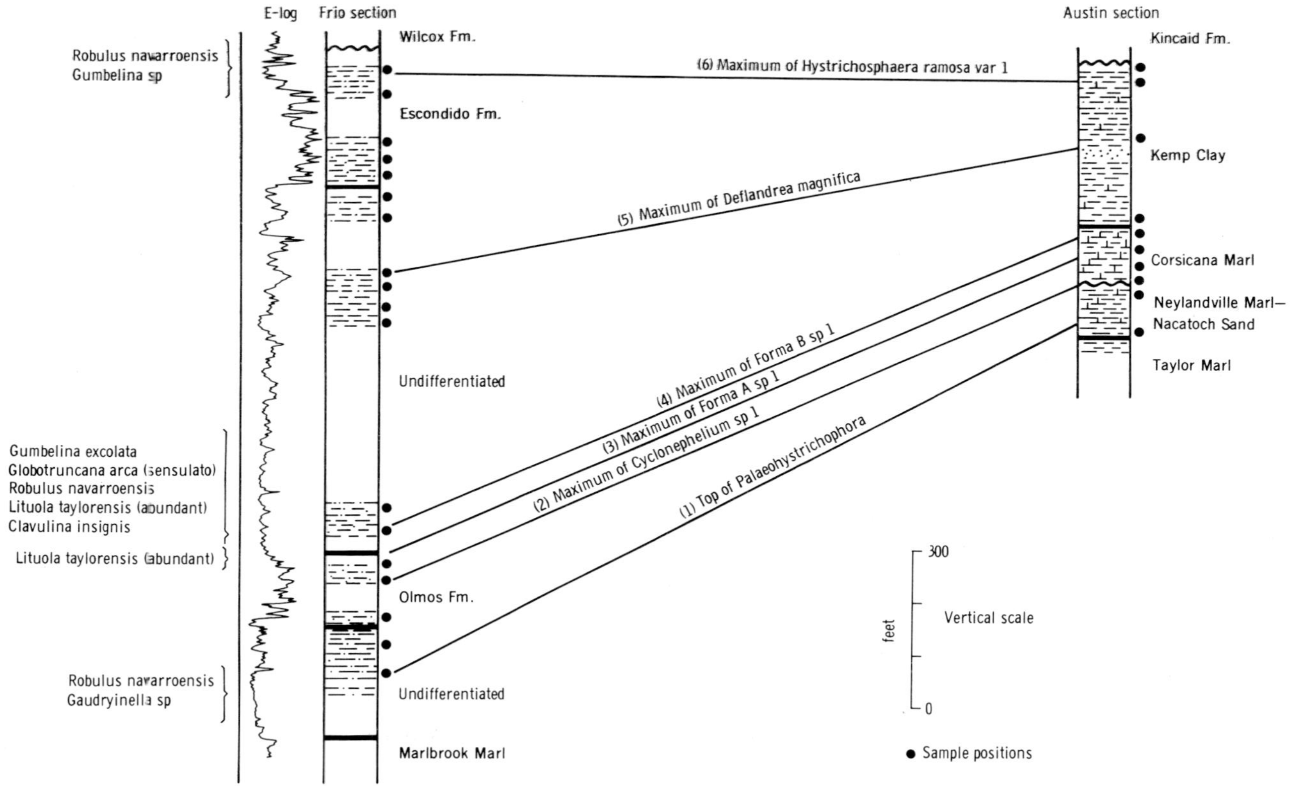

Figure 10. Correlative horizons in Frio (left) and Austin (right) sections. Horizontal distance between sections is approximately 135 miles. Because of close spacing, the position of only one of every three samples is indicated for the Nacatoch Sand-Neylandville Marl and Kemp Clay in the Austin section, and one of every 6 samples in the Corsicana Marl.

COMPARISON OF DINOFLAGELLATE-ACRITARCH ASSEMBLAGES WITH OTHER AREAS

The lithologic dissimilarities of the Austin and Frio sections seem to imply dissimilar environments of deposition, although the comparison of over-all dinoflagellate and acritarch assemblage characteristics from both sections shows extensive similarity. There are 81 species recorded from the Austin section. Of the 43 planktonic palynomorphs in the Frio section, in the original study (Zaitzeff, 1967), 41 are present among the 81 differentiated in the Austin section. Only two species, *Spinidinium* cf. *densispinatum* and *Paleohystrichophora* sp. 1, previously discussed, are recorded only from the Frio section. The Austin assemblage is more diverse in species, and the species are represented by larger numbers of individuals. The similarity of the assemblage of the Frio section to the assemblage in the Austin section, that is, 41 of the 43 dinoflagellates and acritarchs of the Frio section present in the Austin section, is striking when it is recognized that the two sections are approximately 135 miles apart.

In the Austin section, the limited vertical range and the assemblage composition of the palynomorphs allow differentiation of at least three zones. Zone A appears to support previous paleontological interpretation, based on Foraminifera, that the foraminiferal assemblage in the Neylandville Marl is dissimilar to the foraminiferal assemblages present in the other formations of the Navarro Group. Frizzell (1954) stated that the Neylandville foraminiferal assemblage shows 41 species in common with the Taylor Marl and the Navarro Group, 38 additional species in common with the Taylor Marl alone, 9 species in common with the other formations of the Navarro, and 2 species restricted to the Neylandville Marl.

Although analysis of the microplankton of the Taylor Marl was beyond the scope of this study, there are several species previously described from other areas that occur in the Neylandville in the Austin section whose ranges do not extend above the Neylandville Marl and are recorded from rocks of the same and older ages from other areas. One of these species is *Odontochitina striatoperforata,* which is reported from the upper Albian to Cenomanian of Australia. This species is also reported from the Cretaceous deposits of Graham Island, arctic Canada. A form similar to the Navarro specimens referable to *O. striatoperforata* was figured by Evitt

(1961, Pl. 6, figs. 15, 16) from a West Pakistan Senonian deposit. *Deflandrea acuminata* has been previously reported from Cenomanian to lower Turonian deposits of Australia. *Cannosphaeropsis fenestrata* has been described from the Cenomanian, Senonian, Aptian, and Albian rocks of Australia. *Palaeohystrichophora infusorioides* has been reported from Cenomanian to lower Turonian of Australia and the Cretaceous of France. *Callaiosphaeridium* cf. *asymmetricum* was described from the Senonian of France, Cenomanian of Britain, and Hauterivian and lower and middle Barremian of Yorkshire.

A study of the Neylandville microplankton shows there is possibly a closer relationship of the Zone A assemblage to that of the Taylor than to that of the Navarro. Twenty-two species have ranges that appear not to extend above the Neylandville rocks and, as indicated above, occur in rocks of equivalent or older ages elsewhere.

In comparing the Navarro dinoflagellate assemblage with other North American material, the Navarro assemblage, primarily of the Corsicana Marl, Kemp Clay, and Escondido and Olmos Formations, appears to most closely compare with that of the Red Bank Formation (Maestrichtian) of New Jersey. Although there is no published taxonomic work on the Red Bank Formation, except for a few species illustrated by Evitt (1961), one of us (Zaitzeff) had the opportunity to examine some of the material from that formation in Dr. Evitt's laboratory and make a few comparisons. Preliminary analysis indicates that there are several species that are probably conspecific with those in the Red Bank Formation. Forms in the Navarro that are also found in the Red Bank are: Forma A sp. 1, *Deflandrea cooksoni, Areoligera senonensis, Deflandrea* sp. 3, *Gonyaulax* sp. 1, *Tanyosphaeridium* sp. 1, *Polysphaeridium* sp. 1, *Baltisphaeridium* sp. 2, and *Cyclonephelium* sp. 1.

Stanley (1965) described a late Late Cretaceous and Paleocene pollen, spore, and dinoflagellate assemblage from South Dakota and illustrated several species that are closely related to the Navarro forms. Navarro taxa common to those described and illustrated in his work are: *Deflandrea microgranulata, Deflandrea magnifica, Deflandrea pannucea, Diphyes* cf. *colligerum, Hystrichosphaeridium tubiferum, Cordosphaeridium ?inodes,* and *Spinidinium* cf. *densispinatum.* These species all occur in the Cannonball Member (Paleocene), according to Stanley. Drugg (1967), in his report of the study of palynomorphs in the Upper Moreno Formation (Maestrichtian-Danian) of California, described 65 species in 28 genera of dinoflagellates and 17 species in 6 genera of acritarchs. Of those, 12 genera of dinoflagellates and 2 genera of acritarchs are also represented in the Navarro. Drugg distinguished 52 species in these 14 genera, including 12 in *Deflandrea* and 8 in *Hystrichosphaeridium,* whereas Zaitzeff (1967) differentiated 46 species in the same genera, including 11 of *Deflandrea* and 7 of *Hystrichosphaeridium.* In spite of these obvious similarities, however, only three species are identified in common with the two areas: *Dinogymnium nelsonense,* which seems to be common in and essentially restricted to the Maestrichtian beds of the Upper Moreno Formation, and *Deflandrea cooksoni* and *D. magnifica,* both listed as rare by Drugg but ranging upward from the Maestrichtian into the Danian. One species, *D. cretacea,* which Drugg found to be abundant in and a marker for strata of Maestrichtian age in the Upper Moreno, was not recognized in the Navarro paleomicroplankton. Another species, *Cordosphaeridium inodes* Klumpp, which appears to be correctly identified by Drugg, seems slightly different than the form listed here as *Cordosphaeridium ?inodes* from the Navarro. Drugg found *C. inodes* to be restricted to the Danian and abundant in the upper part of the Upper Moreno. Since the world-wide range of this species is Paleocene or younger (Paleocene of South Dakota; Eocene of Victoria, northwestern Germany, and Great Britain; and Oligocene of Germany), it is probable that the Navarro material is a new species with some similarity to the later *C. inodes.* An unnamed *Dinogymnium* species, species A of Drugg, which was rare but restricted to the Maestrichtian part of the Upper Moreno, is very close to if not identical to the Navarroan *Dinogymnium* sp. 4. An unnamed species of *Tenua,* abundant in the Danian Upper Moreno, appears to be very similar to *Tenua* sp. 1 from the much older Neylandville. Direct comparison should be made of these two before considering them identical.

PALEOECOLOGIC INTERPRETATIONS

The lithologic dissimilarities of the two sections studied indicate different environments of deposition. The Frio County section consists dominantly of dark-gray, slightly calcareous, arenaceous shales interbedded with sand sequences. Most of the samples made available for the study also contain abundant plant debris visible to the naked eye. Lithologically, the Austin section consists primarily of a medium-gray, fine-grained calcareous clay shale.

In the Austin composite section of the Navarro, the proportion of pollen and spores varies vertically in the strata comprising the group. Several workers have postulated that there is a correlation between number (absolute frequency) of pollen and spores in a sample and the distance from shore. Upshaw (1964) suggests that the determination of abundance ratios between microplanktonic organisms and entities of terrestrial origin would provide a qualitative means of estimating the relative distance from shore of sedimentary samples. Using this ratio, he was able to define marine and nonmarine zones in the Frontier Formation of Wyoming.

A study of this type of ratio in the Austin section (Fig. 11) indicates that there are almost continuous trends in the successive variations of this ratio upward through the Navarro. Such trends, or curves, are interpreted as showing changes in the relative distance from shore, or possibly water depth. These changes indicate transgressive and regressive phases of the Navarro sea in this area (Fig. 11).

The lower portion of the Corsicana Marl is considered to be deeper marine or farther offshore than any sections of the Neylandville Marl-Nacatoch Sand or Kemp Clay. The highest pollen-spore/microplankton ratios occur in the middle parts of the Neylandville Marl and Kemp Clay, suggesting two regressions of the sea. The top of the Kemp Clay and the bottom of the Neylandville Marl-Nacatoch Sand are marked by minor transgressions. There are several factors that may tend to invalidate such an interpretation, notably changes in prevailing winds and currents, proximity to deltas with their varying amounts of water runoff contributing variable quantities of land-derived plant material and sediments at variable rates to the depositional basin, and the proximity and density of vegetation to the coast (Muller, 1959; Cross and others, 1967). Analysis of benthonic and planktonic foraminiferal assemblages and planktonic/benthonic ratios here in respect to determining relative water depth closely correlates with the relative depth of water indicated by the pollen-spore/microplankton ratio (William N. Orr, oral commun., 1968).

Sections of the Austin and Frio sections are marked by floods or high occurrences of individual species that tend to dominate

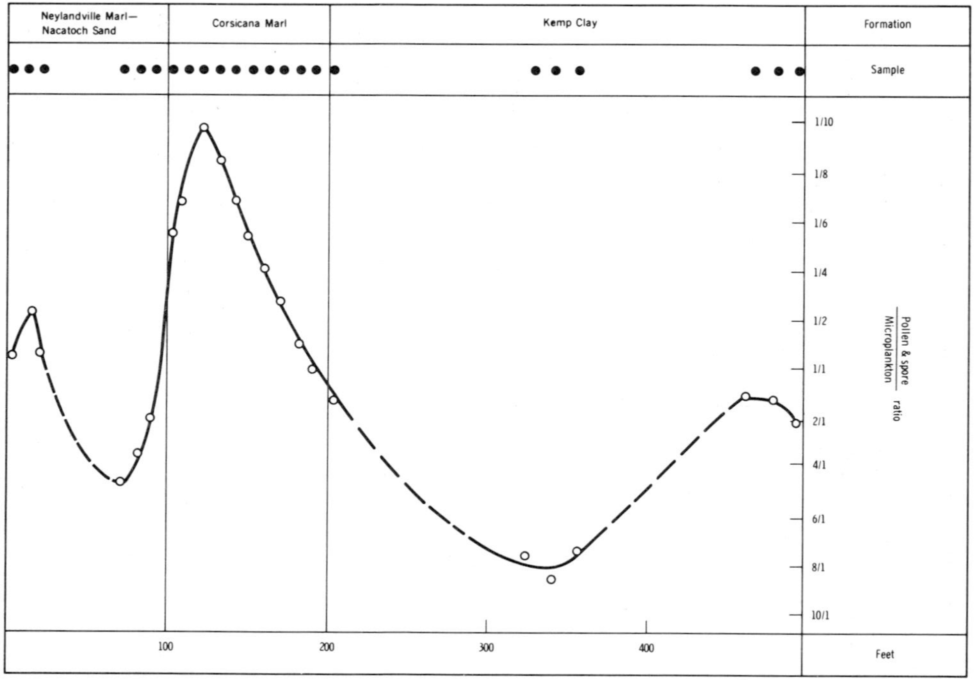

Figure 11. Ratios of pollen and spores to microplankton in the Austin, Texas, section indicate a major marine transgression during the Corsicana. The low relative frequency of land-derived palynomorphs to microplankton in the early Corsicana indicates that the samples studied contained sediments deposited either in deeper water or at a greater distance from shore, or both, in the Corsicana than in the samples from sediments laid down during marine regressions indicated by the higher pollen frequencies in Neylandville-Nacatoch and Kemp time.

some samples. Whether or not these represent fossil dinoflagellate "blooms" would be speculative, but the possibility exists and the idea is tempting. The amount of time represented by the sediments of a sample makes this quite unlikely as a logical explanation. Samples from the Neylandville Marl in the Austin area are dominated by the species *Dinogymnium nelsonense* and *Dinogymnium digitus*. The base of the Corsicana Marl and samples from the Olmos Formation in Frio County are marked by high occurrences of *Cyclonephelium* sp. 1 and Forma A sp. 1. The significance of these for correlative purposes has been discussed.

The occurrence of 41 species common to both sections suggests that these forms were widespread geographically and existed in diverse environments of deposition.

PALEONTOLOGIC AND STRATIGRAPHIC CONCLUSIONS

Palynologic examination of the Navarro Group sediments from east-central and southwestern Texas has revealed an abundant and varied dinoflagellate-acritarch assemblage. Analysis of the fossil assemblages indicates that some species have a restricted vertical distribution and that the relative abundance of species varies vertically within the group at both localities. These characters aid in the zonation and correlation and the interpretation of environment of deposition of the rocks.

Establishment of assemblages or types showing sequential variations in relative abundance is useful. The occurrence of unusually large numbers of particular species of dinoflagellates or acritarchs is considered to have time-stratigraphic significance and such occurrences can be used as horizon markers or point correlations. Five such horizons have been recognized here and are used as the basis for establishing five correlative horizons in the two sections.

Precise taxonomic interpretation of individual species is essential to this type of study, and consideration of the effect of environment on the morphology of individual species as well as genetic factors is requisite to good stratigraphic application or interpretation.

Comparisons of the Navarro assemblage with assemblages reported in other works indicate that the Corsicana Marl, Kemp Clay, Olmos Formation and Escondido Formation have a close affinity with the Red Bank Formation (Maestrichtian) of New

Jersey, whereas the Neylandville Marl-Nacatoch Sand (undifferentiated) from the Austin composite section contains a quite dissimilar assemblage and appears to represent an older age than the other formations. Many of the genera and species of the Neylandville Marl-Nacatoch Sand are referable to Senonian and older rocks.

ACKNOWLEDGMENTS

Much of the background information and part of the conclusions of this paper were originally prepared as part of a doctoral dissertation by Zaitzeff (1967) at Michigan State University. Some of the interpretations and conclusions have been made and the manuscript rewritten by A. T. Cross. We share responsibility for conclusions and errors of omission or commission that may be found.

Sincere appreciation is expressed to Dr. W. R. Evitt, Stanford University, and Dr. Graham Williams, Pan-American Petroleum Corporation Research Center, Tulsa, Oklahoma, for their examination of many of the slides and illustrations used in this study and suggestions concerning the morphology and taxonomy of a number of the dinoflagellate species.

Thanks are also given to Shell Oil Company for providing sample locality data for the Austin, Texas, composite section and to the Humble Oil and Refining Company for providing the subsurface core section of the Navarro Group from Frio County, Texas. Recognition and appreciation are due the Society of the Sigma Xi for a grant-in-aid that supported part of this work, and to the National Science Foundation for assistance under Project GA-429.

REFERENCES CITED

Adkins, W. S., 1932, The Mesozoic systems in Texas, p. 480-518 *in* Sellards, E. H., Adkins, W. S., and Plummer, F. B., The geology of Texas, Vol. I, Stratigraphy: Texas Univ. Bull. no. 3232, 1007 p.

Cross, A. T., Thompson, G. G., and Zaitzeff, J. B., 1967, Source and distribution of palynomorphs in bottom sediments, southern part of Gulf of California: Marine Geology, v. 4, p. 467-524.

Drugg, W. S., 1967, Palynology of the Upper Moreno Formation (Late Cretaceous-Paleocene) Escarpado Canyon, California: Palaeontographica, v. 120, abt. B, p. 1-71.

Evitt, W. R., 1961, Observations on the morphology of fossil dinoflagellates: Micropaleontology, v. 7, p. 385-420.

Evitt, W. R., Clarke, R. F. A., and Verdier, Jean-Pierre, 1967, Dinoflagellate studies III. *Dinogymnium acuminatum* n. gen., n. sp. (Maastrichtian) and other fossils formerly referable to *Gymnodinium* Stein: Stanford Univ. Pubs. Geol. Sci., v. 10, no. 4, 35 p.

Frizzell, D. L., 1954, Handbook of Cretaceous Foraminifera of Texas: Texas Univ., Bur. Econ. Geology Rept. Inv. 22, 232 p.

Leopold, E. B., and Pakiser, H. M., 1964, A preliminary report on the pollen and spores of the pre-Selma Upper Cretaceous strata of western Alabama: U.S. Geol. Survey Bull. 1160-E, p. 71-95.

Muller, J., 1959, Palynology of Recent Orinoco delta and shelf sediments: Reports of the Orinoco Shelf Expedition, 5: Micropaleontology, v. 5, p. 1-32.

Shumard, B. F., 1861, Descriptions of new Cretaceous fossils from Texas: Boston Soc. Nat. History Proc., v. 8, p. 188-205.

——1863, On the Cretaceous formation of Texas: St. Louis Acad. Sci. Trans., v. 2, p. 152.

Singh, C., 1964, Microflora of the Lower Cretaceous Manville Group, east-central Alberta: Geol. Div. Research Council Alberta Bull., v. 15, p. 1-238.

Stanley, E. A., 1965, Upper Cretaceous and Paleocene plant microfossils and Paleocene dinoflagellates and hystrichosphaerids from northwestern South Dakota: Am. Paleontology Bull., v. 49, no. 222, p. 177-384.

Stephenson, L. W., King, K. B., Monroe, W. H., and Imlay, R. W., 1942, Correlation of the outcropping Cretaceous formations of the Atlantic and Gulf Coastal Plain and Trans-Pecos Texas: Geol. Soc. America Bull., v. 53, p. 435-448.

Tasch, P., McClure, K., and Oftedahl, O., 1964, Biostratigraphy and taxonomy of a hystrichosphere-dinoflagellate assemblage from the Cretaceous of Kansas: Micropaleontology, v. 10, p. 189-216.

Upshaw, C. F., 1964, Palynological zonation of the Upper Cretaceous Frontier Formation near Dubois, Wyoming, p. 153-168 *in* Cross, A. T., *Editor*, Palynology in oil exploration, a symposium: Soc. Econ. Paleontologists and Mineralogists Spec. Pub. no. 11, 200 p.

Zaitzeff, James B., 1967, Taxonomic and stratigraphic significance of dinoflagellates and acritarchs of the Navarro Group (Maestrichtian) from eastcentral and southwest Texas: East Lansing, Michigan State Univ., Ph.D. dissert., 172 p., 31 pls.

Author Index

Subject Index